ACCLAIM FOR

HENRY PETROSKI

"Petroski is an amiable and lucid writer. . . . [He] belongs with the poets."
—John Updike, *The New Yorker*

"A triumph. . . . Reading *Engineers of Dreams* is akin to sitting at the knee of a favorite uncle who spins golden yarns of far-off places and events. . . . There truly is something here for everyone." —*Morning Star-Telegram* (Fort Worth, Texas)

"Henry Petroski is like a bright light sent from heaven."
—*Durham Morning-Herald*

"An engaging, entertaining history."
—*News and Observer* (Raleigh, North Carolina)

"Just as a good bridge weds sweeping visual grace with detailed mechanical calculations, *Engineers of Dreams* exhibits a rare mixture of eloquence and precision. That combination has made classics of Petroski's previous books, and his latest deserves no less of a reception." —*Invention and Technology*

"*Engineers of Dreams* makes [bridges] ever more marvelous."
—*Rocky Mountain News*

BOOKS BY **HENRY PETROSKI**

Engineers of Dreams

Design Paradigms

The Evolution of Useful Things

The Pencil

Beyond Engineering

To Engineer Is Human

HENRY PETROSKI

ENGINEERS of DREAMS

Henry Petroski's previous books include *To Engineer Is Human, The Pencil,* and *The Evolution of Useful Things*. He is the Aleksandar S. Vesić Professor of Civil Engineering and chairman of the Department of Civil and Environmental Engineering at Duke University.

ENGINEERS of DREAMS

ENGINEERS of DREAMS

GREAT BRIDGE BUILDERS AND THE SPANNING OF AMERICA

HENRY PETROSKI

VINTAGE BOOKS
A DIVISION OF RANDOM HOUSE, INC.
NEW YORK

to Catherine

FIRST VINTAGE BOOKS EDITION, OCTOBER 1996

Copyright © 1995 by Henry Petroski

All rights reserved under International and Pan-American Copyright Conventions. Published in the United States by Vintage Books, a division of Random House, Inc., New York, and simultaneously in Canada by Random House of Canada Limited, Toronto. Originally published in hardcover by Alfred A. Knopf, Inc., New York, in 1995.

The Library of Congress has cataloged
the Knopf edition as follows:
Petroski, Henry.
Engineers of dreams: great bridge builders
and the spanning of America. / Henry Petroski. — 1st ed.
p. cm.
Includes bibliographical references (p.) and index.
ISBN 0-679-43939-0 (hardcover)
1. Bridges—United States—History—19th century.
2. Bridges—United States—History—20th century.
3. Civil engineers—United States—Biography. I. Title.
TG23.P47 1995
624'.2'0973—DC20 94-48893
CIP
Vintage ISBN: 0-679-76021-0

Author photograph © Catherine Petroski
Book design by Brooke Zimmer

Random House Web address: http://www.randomhouse.com/

Printed in the United States of America
10 9 8 7 6 5 4 3 2 1

CONTENTS

PREFACE

This book tells the stories of engineers who have dreamed and engineers who have toiled, of bridges of celebrity and bridges of burden, and it is about the nature of technology in a human context. Some renowned engineers and some famous bridges have tended to overshadow their contemporaries and neighbors, but the full range of stories reveals that the lesser-known engineers have been of no less importance in shaping our built environment. Indeed, the personalities of all kinds of engineers, with their faults and foibles coexisting with their dreams and designs, have played as much of a role as has their technical know-how in bringing familiar bridges to fruition.

As is to be expected, only some of the bridges of which any engineer dreams get realized, but that is not to say that even the wildest schemes have not influenced others, and hence our roadscapes. A full understanding of how and why a great bridge came to be what it is where it is requires appreciating the often decades-long struggles that engineers have experienced with themselves, their colleagues, and their communities. In telling the stories of some engineers and some bridges, this book must necessarily tell the stories of many bridges and many engineers engaged in the professional, economic, political, and personal conflicts that occur in the technical, social, and cultural activities in which we all participate. When we see in the stories of bridges the full human dimensions of engineers and engineering, we also see more clearly the inextricable interrelationships between technology and humanity. As no person is an island, so no thing is an island. Certainly no bridge is an island.

And no book is an island. Many bridges were provided by many people on the way to this book's being realized, and I wish to acknowledge and thank at least some of them. Arthur Singer turned my rough sketch of an idea into a grant from the Alfred P. Sloan Foundation, which enabled me to travel to bridge sites, to gather illustrations, and to write. Ashbel Green, my editor at Knopf, has once again given me my head and his support. Anne T. Zaroff-Evans did a marvelous job of copy-editing, and Knopf's Jennifer Bernstein and Melvin Rosenthal also made the process from manuscript to book a smooth one, at least from my point of view.

There was also, of course, much help long before there was a manuscript, and libraries and librarians were, as always, remarkably tolerant of my inquiries. The wonderful collection of the Aleksandar S. Vesić Engineering Library at Duke University continues to provide resources and convenience of immeasurable value. Eric Smith, its former librarian, who was forever patient with my endless requests, located and obtained for me important materials so diverse that no one institution could ever be expected to contain them all. Rich Hines and Dianne Himler have continued to get to me the many odd library materials that are so essential in the final stages of preparing a manuscript. The resources and facilities of Duke's main library, the William R. Perkins Library, have once again been indispensable to me, as has the institution of Interlibrary Loan. I have also had much help from archives, historical societies, bridge authorities, and departments of transportation in locating information and photographs; the sources of these pictures are credited in the list of illustrations in the back of the book. Indeed, I am indebted to so many librarians, archivists, secretaries, assistants, and volunteers, at Duke and elsewhere, both known to me and anonymous, that I dare not begin to acknowledge them by name, lest I forget one.

I must, however, thank some other individuals by name. My brother, William Petroski, helped me early on to get a closer look at many New York bridges, and my sister, Marianne Petroski, gave me some helpful books. Stephen Petroski, my son and a student engineer, also helped me very early on by collecting essential material from newspaper indexes, and Ian Threlfall, a graduate student in civil and environmental engineering at Duke, later retrieved countless remarkably clear copies of articles from microfilm files. Margot Ammann Durrer kindly provided me with much material relating to her father, including letters and photographs. A host of engineers and friends of engineers have helped me with very useful material and leads, and I would like to thank especially Norman Ball, David Billington, Milton Brumer, Stephen Burges, Jameson Doig, Eugene Fasullo,

Steven Fenves, Henry Fischer, Jay Fredrich, Myint Lwin, Louis Miller, W. S. Persons, Allan Ryan, Thomas Sullivan, and Neil Wotherspoon. I also wish to thank my daughter Karen Petroski for her insights into scholarship. Finally, I am as always indebted to Catherine Petroski, my wife, for being my first reader and most constructive critic, and for understanding, at times perhaps even better than I, my writing habits and needs.

H.P.
Durham, North Carolina
September 1994

ENGINEERS of DREAMS

1

IMAGINE

Imagine a world without bridges. Imagine London, Paris, and Rome without dry paths across the Thames, the Seine, and the Tiber. Imagine Manhattan as an island with no hard crossings of the Hudson and East rivers. Imagine San Francisco without road communication across the gate to the north and the bay to the east. Imagine Pittsburgh wedged bridgeless between the Allegheny and the Monongahela rivers. Imagine Chicago without its massive lift- and drawbridges, or Amsterdam without its more modest canal crossings. Imagine Seattle without its long, low floating bridges, or St. Petersburg without its soaring cable-stayed structure arcing out over Tampa Bay.

Bridges and cities go together, in large part because so many of our greatest cities were founded where they are precisely because of the proximity of water. It is no mystery why so many settlements have grown up by rivers and bays, and it comes as no surprise that some of the oldest of them developed at important river crossings. Cambridge is one of the many English cities that date back to Roman times; a settlement was established there in A.D. 43. The location was that of a bridge over the navigable River Cam, on the road between Colchester and Lincoln. Oxford, another venerable English city, takes its name from its location as a crossing of the Thames. How many of our cities and towns have water words, like "port," "bay," and "haven," as part of their names? How many of our states share the names of the rivers that bound or bisect them? Some towns, like Iron Bridge in

England and Suspension Bridge at the Canadian border in New York, have even been named after the structures upon which they depended.

Water travel and commerce were highly developed long before there was the widespread erection of large bridges across navigable waters. Although today we transport so many products of manufacture and agriculture by railroad, truck, and airplane, we still "ship" the goods out and await new "shipments" of supplies. The priority of shipping and naval interests shaped the character of many of our port cities well into the twentieth century, until autobahns, autostradas, motorways, and interstate road networks focused attention elsewhere. But the water crossings of even the greatest roads still remain shaped by consideration for what happens in the water below.

Imagine Boston and Cambridge, Massachusetts, without bridges over the Charles and the early-morning rowers beneath them. Imagine Detroit without access to Windsor, its Canadian neighbor—by the oddity of local geography, to the south. Imagine Washington, D.C., without roads to Virginia across the Potomac and over its yachts. Imagine St. Louis—now with its arch, which is a bridge of sorts, bearing tourists to the sky—inaccessible across the Mississippi from Illinois. Imagine New Orleans, dry behind levees, but without a crossing of Lake Pontchartrain, or without the Huey P. Long Bridge across the lower Mississippi. Imagine Charleston without its serpentine Old Cooper River Bridge, known affectionately as Old Roller

A view of Pittsburgh, circa 1969, showing many of its bridges

Coaster. Imagine Philadelphia isolated by the Delaware River because it had no Ben Franklin or Walt Whitman bridge. Imagine Portland, Oregon, with its beautiful hills but without its crossings of the Willamette River. Imagine Florence with its Uffizi and its Pitti Palace but without their connection across the Ponte Vecchio or Venice without its Ponte Rialto or its Bridge of Sighs, so called because the sounds of the prisoners who passed over it between the palace and prison could be heard on the canal below.

Bridges have become symbols and souls of cities, and each city's bridges have been shaped by, and in turn shape, the character of that city. It is virtually impossible to go into a souvenir shop in San Francisco without being overwhelmed by images of the Golden Gate Bridge, on everything from T-shirts to spoons. The Sydney Harbour Bridge is as much a landmark of that city as is its famous harborside opera house. New York's Brooklyn Bridge is legendary—as is London Bridge, even though its stones have been reassembled in Lake Havasu City, in western Arizona, and the now incongruous landmark stands as one of the strangest monuments to our sense of possession over purpose.

Imagine the Golden Gate spanned by anything but the Golden Gate Bridge. Is it possible? The bridge's location, shape, proportions, scale, and color all seem so right for the site, and now it seems so for them. Is it possible even to imagine any other bridge between San Francisco and Marin County? Could, say, a copy of the Brooklyn Bridge, with taller towers and a longer span, have been cast across the gate? Or could a smaller version of the Golden Gate Bridge, color and all, have been erected between New York and New Jersey, where the George Washington Bridge now seems so naturally established? Yet this kind of questioning and imagining is precisely what engineers must do before any bridge exists. Some of the earliest proposals for bridges in New York and San Francisco looked nothing like what have since come to be such familiar features of those cities. Indeed, one nineteenth-century proposal for a crossing between New York and Brooklyn was a soaring arch, and an early idea for the Golden Gate Bridge was so ugly that it is a wonder any bridge there ever gained anyone's support.

Bridges define the approaches to cities, and passing over or under some of the world's great spans is an unforgettable experience. Many travelers from the north have their first view of San Francisco framed in the tunnel approach to the Golden Gate Bridge. To sail into New York Harbor today is to watch the Verrazano-Narrows Bridge grow to mythic proportions even before the Statue of Liberty comes into view. The first glimpse of the tallness of New York when driving south down the Palisades Parkway is of one of the monumental steel towers of the George Washington Bridge looming over the trees. Once within cities, the structures of great bridges often serve as landmarks and beacons for the disoriented tourist. If you are walking or driving about the canyons of New York, it is often possible to catch sight of the tops of the Brooklyn, Manhattan, Williamsburg, and other great suspension bridges whose necessarily tall towers once totally dominated the city's skyline.

Imagine traveling into, out of, or around a modern port city without bridges. Having known the speed of road communication that bridges make possible, we would have little patience with the reintroduction of long-since-displaced ferryboats. Tunnels, generally having a much lower traffic capacity than bridges, would need to be much more numerous than above-ground spans, and would burrow underwater every which way. But travel into or out of a city by tunnel is a much less dramatic, relaxing, or satisfying experience for the average driver or automobile passenger. Tunnels have dark connotations, and for many people the prospect of water rushing in is much more dreadful than that of a bridge falling into the water. There are of course some exceptional tunnel approaches, such as that which spirals down from atop the New Jersey Palisades into the Lincoln Tunnel under the Hudson River to New York, giving one of the best imaginable views of Manhattan's skyline. But generally, tunnel approaches cannot rival bridge approaches for the panoramas of great cities that they make accessible.

Bridges not only provide a balcony from which to appreciate the architecture of a place; they may also inspire its subsequent architecture. Though now long eclipsed in height, the towers of the Brooklyn Bridge, with their twin Gothic arches, seem still to dictate an architectural mood to lower Manhattan, and it is not hard to imagine the bridge's two stone towers having had something to do with the design of the twin steel towers of the World Trade Center. The arched Eads Bridge, constructed contemporaneously with the Brooklyn, might similarly be said to have influenced Eero Saarinen's brilliant concept of the Gateway Arch as a monument to the westward expansion of America across the Mississippi River through St. Louis. And the increasingly large lift and bascule bridges that began to cross the Chicago River around the turn of the century may have inspired that city's drive to build higher and higher skyscrapers in steel.

Nor is it only cities that rely on bridges. Imagine farm roads without culverts over which cows can pass from barn to field and back. Imagine mountain roads without suspension bridges only one person wide, to carry hikers and campers high and dry across a gaping gorge. Imagine backwoods roads without the narrow bridges that provide milestones in directions back to the main road. Imagine rural roads without the covered bridges that concealed so many lovers' trysts over rushing streams. Imagine Madison County without its bridges.

The Mississippi River at St. Louis, on July 4, 1982, with the Eads Bridge visible behind the Gateway Arch, and with the fireworks recalling the opening of the bridge on July 4, 1874

Though most of America's more than half a million highway bridges are small and anonymous, they may not be any less important to the local traffic than the Golden Gate and Brooklyn bridges are to their hordes. The engineers of our greatest spans began by designing our smaller ones. The scale may be different, but the process is essentially the same, and so these bridges have proved to be the training grounds for dreams. Furthermore, every bridge, small or large, is also an aesthetic and environmental statement. Its lines are important beyond its span; every bridge must not only bear its burden, whether cows or coal trains, but must also be able to withstand the burden of proof that, in the final analysis, society is better served, tangibly and intangibly, by the bridge's being there at all.

Imagine how a bridge can ruin a setting of natural beauty, whether the tranquillity of the countryside or the skyline of a city. Imagine what the wrong bridge across the Golden Gate might have done to that unique site. This is why place so often influences bridge design—for, contrary to the popular misconception, engineers are not insensitive to setting and aesthetics. The Rainbow arch bridge across the river gorge north of Niagara Falls was an appropriate form to mirror the rainbows ever present in the mist about the falls. Arch bridges can actually open up great spaces, as Navajo Bridge did over the Colorado gorge upriver from the Grand Canyon, providing to crossers views of Marble Canyon uninterrupted by any significant human artifact for as far as the eye can see. A second crossing, its steel structure again below the bridge deck, will also intrude only minimally on the natural beauty of the site. In Switzerland, the bridges of Robert Maillart and Christian Menn harmonize with the Alps in a different, yet totally compatible and successful way. In Tampa Bay, the replacement bridge for one that was rammed by a tanker is a soaring design whose pattern of towers and cables evokes the masts and sails of pleasure boats crisscrossing the bay. Though not a natural setting, the Tower of London so dominated the section of the Thames where a crossing was to be erected in the late nineteenth century that Tower Bridge was designed in consonance with the historic site, even at the risk of offending some structural purists with its stone-encased steel. Earlier in that century, Thomas Telford similarly respected the prior claim of Conwy Castle to the location of the river mouth in Wales for which he designed his suspension bridge with crenellated towers.

That there were bridges long before there were engineers does not diminish the achievement or the value of either. The earliest bridges were modest, instinctive, and imitative of nature; the latest are models of what we can achieve with experience and tools of which no primitive bridge builder may ever have dreamed. We can get some idea of the nature of the earliest bridge building by thinking of what is embedded in our own tradition, lore, and store of commonplace experiences. As infants, we have the grasping instinct, clutching at the air for something to take us over the void of separateness. We reach from mother to father and back as they take turns holding and bouncing us in their arms, swing bridges transporting us between them. As we grow, we learn that our own arms are bridges to everything. And so are our legs, as we crawl over obstacles between here and there, and then walk and run and skip and jump over space and time more in the joy of doing than in the joy of getting anywhere. We learn to walk along the sidewalk, avoiding cracks to save our mothers' backs—bridges all—and taking joy in counting how many great canyons in the concrete we have conquered without a fall. We learn from legends and lore how the gallant gentleman, if he did not carry his fair maiden across, threw his cape over the puddle, that the maiden might step dry to her destination. Even after we stop reciting nursery rhymes and we forget gallantry, we and our companions make a bridge fleeting in time when we step or jump across the water in the gutter in our way.

Long before there were fairy tales, at least as we know them today, nature provided models for bridges in the form of stepping-stones, arching branches, hanging vines, and fallen logs across streams. These found bridges were used by animals as well as men and women and their children, and eventually people learned to make their own bridges deliberately, placing stones step by step in streams, bending branches to a purpose, stringing vines in patterns of determination, and felling logs that did not fall by themselves. This was the work of the first bridge builders, and as their bridges grew and multiplied, so did the dreams and ambitions of the more reflective among the builders. Dreams became necessary when natural gaps became deeper than stones could fill and wider than vines and trees could reach. To bridge such gaps took more than imitating nature, it took the imagination and ingenuity that are the hallmarks and roots of engineering.

Almost three millennia ago, Homer wrote of bridges as commonplace achievements, mentioning in particular how armies crossed water on pontoon bridges. The Persian kings Cyrus, Darius, and Xerxes employed such structures about twenty-five centuries ago, as did Alexander the Great a century or two later. Among the earliest recorded specific bridges is one over the Euphrates at Babylon described by Herodotus, writing almost twenty-five hundred years ago. It was made of timber beams resting on stone piers. Engineering and technology have always advanced whether or not their achievements were recorded in words, and Greek and Roman bridge building, not to mention that of non-Western civilizations, long ago reached well beyond the limitations of the log as girder. The origins of the cantilevered or corbeled arch, which children who play with blocks still construct instinctively today; of the true arch, which we still admire in nature and in art; and of the suspension bridge, which is believed to have its roots in such diverse locations as China, northern India, central Africa, and South America, are lost to history.

Though some Roman bridges still stand after two thousand years—most notably the wonderful aqueducts, such as the one that dwarfs the marketplace in Segovia, Spain, and the magnificent Pont du Gard near Nîmes in southern France—many other ancient bridges have been lost to use and the elements. All bridges have always suffered a degree of wear and tear, of course; by the Middle Ages, there was widespread deterioration of the infrastructure of bridges whose materials or initial construction were not so fortunately chosen or carefully crafted as the most hardy of the Roman arches. One reason the aqueducts were less threatened by time was that they generally carried the constant load and laminar flow of water, rather than an ever-increasing and sometimes turbulent burden of people, animals, and vehicles. In the Middle Ages, as the conventional history has it, there appeared brotherhoods of bridge builders, in the form of congregations of clergy who had established themselves in remote monasteries in the hills to escape the barbarians. As some of them remain to do today, such congregations came to toil manually in their fields and vineyards to sustain themselves physically so they could continue to pray in their chapels and sustain themselves spiritually.

Among the monastic groups was the Altopascio Order, located near Lucca, Italy, on the ancient road between Tuscany and Rome. Members of the Altopascio wore embroidered on their robes an insignia resembling the Greek letter τ (tau), whose arms "were nicked or pointed in such a way that the vertical shaft may have represented an auger and the crossbar a hammer or ax," thus indicating a proficiency in carpentry. Since the order's Hospice of St. James was not far off the busy road in wild and dangerous country, travelers and pilgrims frequently sought refuge there. To serve these travelers, the Holy Roman Emperor Frederick II decreed in 1244 that the Altopascio "build and maintain upon the public pilgrim's highway" a bridge, thus prompting the name Fratres Pontifices. After the Fall of Rome, the Pope himself was known, of course, as Pontifex Maximus, the supreme bridge builder.

The fame of the Italian Brotherhood of Bridgebuilders spread, and in France a group of Benedictine monks established the Frères Pontiffes. According to tradition, their first settlement was on the River Durance, in southeastern France, at a treacherous ford called Maupas. After the *frères* built their bridge at this location, it became such a safe crossing of the Durance that the place name was changed from Maupas to Bonpas. As the work of bridge brotherhoods spread, so did the evolution of bridge types and construction techniques; eventually, the endeavor became a secular and moneymaking activity, as lotteries were held to raise funds for construction or tolls were charged to repay and reward investors, as well as to maintain the capital investment itself. The arch bridge, first in stone but later in iron, became the most common form by far, but that was to change with the development of engineering as a subject of study in its own right, and thus as a profession.

The familiar triangular roof truss—which, like all roofs, is really a bridge between walls and over house and home, barn and manger—has long been painted matter-of-factly in scenes both social and domestic, both rustic and religious. The wooden truss came in for attention as a true bridge with its discussion by Palladio in the sixteenth century. It was taken to new lengths in the eighteenth century in the hybrid arch-truss forms of the Swiss brothers Grubenmann, and it began to flourish in the nineteenth century, especially in America, where it was patented and thereby named by scores of inventors making use of ubiquitous timber, abundant iron, and fertile imaginations. These inventors and their trusses were among the last of the mechanic-builders; as spans of increasing length and strength were required for the advancing heavy railroads of the mid-nineteenth century, it took a sense of and a capacity for calculation before construction to achieve success in an increasingly competitive environment, for bridge building and everything else.

Squire Whipple, who was born in 1804 to the farming and mill-owning family of James and Electa Johnson Whipple in Hardwick, Massachusetts, has been called the "father of American bridge building" and the "father of iron bridges." Young Squire (his name, not a title) attended Hardwick Academy and the Academy at Fairfield, Connecticut, before going to Union College, in Schenectady, New York, where he earned his bachelor-of-arts degree in 1830. Whipple's education at Union actually predated its formal creation of an engineering course, which was announced in 1845 by President Eliphalet Nott, who had been serving simultaneously as president of the Rensselaer Institute, across the Hudson River in Troy. Since Rensselaer had been offering a program in civil engineering for a decade, Nott found he had a conflict of interest and resigned from the other school to serve Union for what would be a sixty-year tenure.

Union was a natural choice for Whipple's higher education. When he was a young teenager, his family had moved to Otsego County, New York, in which Cooperstown is located, and where young Squire farmed in the summer and taught school in the winter. Even though he attended Union before it offered a formal program in engineering, Whipple would have been expected to take a course in the elements of the science of mechanics, just as his contemporaries at Harvard would on their way to an A.B., and so he was as prepared as any of his time to see a truss not only as a bridge to be constructed but also as the object of study and calculation. After a decade of experience working on railroads and canals, Whipple patented a combination arch-truss bridge, and in 1847 published the first edition of his seminal *Work on Bridge Building*, which evolved into his definitive *Elementary and Practical Treatise on Bridge Building*. It was this work—which explicated his method of determining the distribution of forces in the various members of a truss, thereby making it possible to determine the most economical sizes of the parts to manufacture and ship to the location where they would be assembled—that earned him his appellations. In the association of bridge building with drawing and calculation and written argument before any construction was started, a new era was begun. From then on, the grandest dreams could be articulated and tested on paper, and thereby communicated to those who would have to approve, support, finance, and assist in designing a project that could eventually take years, if not decades, of planning and construction.

The stories of modern bridges are stories of engineers at their best, dreaming grand dreams of tremendous potential benefit to mankind and then realizing those dreams in ways consonant with the environment, both natural and previously built. Though there also have been misdirected

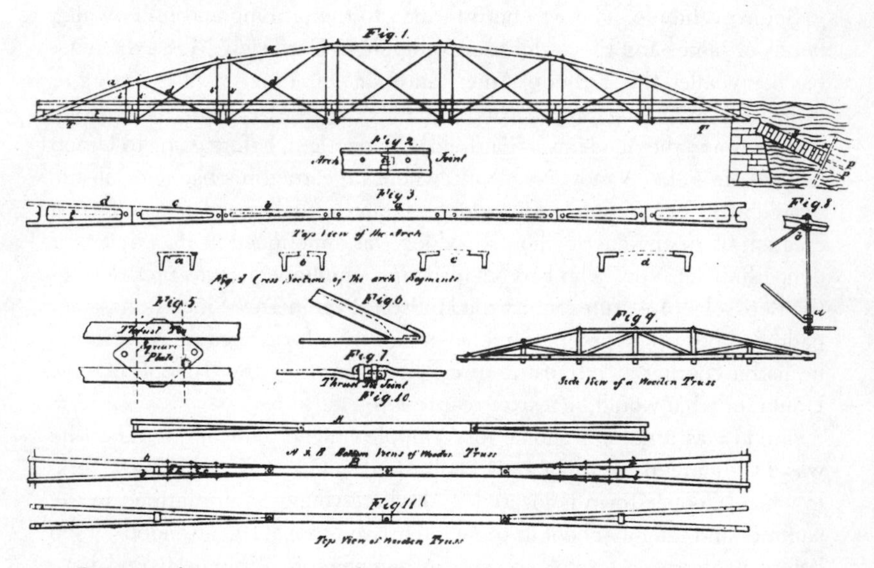

Drawings from a patent issued to Squire Whipple in 1841, one of many truss-bridge designs patented in the middle of the nineteenth century

schemes and pork-barrel projects and political corruption and disruption of neighborhoods associated with bridge building, the stories of the overwhelming majority of our grandest bridges are about technological daring and adventure and creative competition for the common good. Great bridges are conceived by great engineers; since there are often more than enough of these to go around at a given time in history, there are more often than not a plethora of proposals for bridges where there were not bridges before, frequently because the physical and intellectual challenges of the problem had been thought to be beyond the reach or means of the times.

Engineers are also people, of course, and so rivalries have developed among them for commissions to build the greatest bridges, but by and large the bridge engineers of a particular era have formed a kind of fraternity and an interlocking directorate of experts who work more in concert than in discord. Where one may have been the chief engineer, others will have served on a board of consultants. In another project, some of their roles will have been reversed. Thus the bridges of an era will often share certain characteristics, reflecting the collective wisdom and prejudices of the leading practitioners, while at the same time bearing the stamp of individuality of the leader of each particular project.

The generally acknowledged dean of American bridge engineers of the late nineteenth and early twentieth centuries was the Moravian-born Gustav Lindenthal. His masterpiece, Hell Gate Bridge in New York, built to carry a connecting railroad through New York City and thus between New England and the rest of the continent, was a training ground of sorts for the young engineers Othmar Ammann, born in Switzerland, and David Steinman, born on the Lower East Side of Manhattan in the shadow of the Brooklyn Bridge. Their stories, and those of American bridge engineers like Leffert Buck, Theodore Cooper, James Eads, Ralph Modjeski, Leon Moisseiff, the Roeblings, Joseph Strauss, John Waddell, and a host of others, reveal the way in which bridges are conceived and built and, in the process, tell the story of the flowering of engineering as a profession in America.

Telling the story of engineering through its engineers and their works was the method of Samuel Smiles, whose five-volume *Lives of the Engineers* was popular reading in Victorian times. He described his work as a history of inland communication, chronicling as it did the draining and reclamation of swampland, the development of harbors, the digging of canals, the pushing through of roads, and, finally, the building of the railroads and their concomitant bridge and tunnel structures. Mundane and pedestrian as the subject matter might otherwise have seemed, Smiles brought the adventure and altruism of British engineering alive and raised the status of the profession while at the same time inspiring new generations to creative lives of service to humankind. The stories of the American engineers have no less potential for bringing them alive as heroes of technology and culture, and no less potential for illuminating the process of engineering as an indispensable ingredient of civilization.

Try to imagine a world without engineers. In such a world, an absence of bridges would be among the least of inconveniences. Would there be a ready supply of food, for are farmers not soil and water engineers, and is agriculture not crop engineering? Would food be distributed very far beyond where it was grown, for how far could it go without roads or canals or ships or even containers in which to carry it—all such artifacts being the products of some kind of engineering, informal as it may be? Would food be refrigerated for shipment in summer or put away for the winter, for how long would it last without some form of preservation that involved engineering of a kind? And what of shelter? And what of human pride and pleasure and purpose in the construction of cathedrals and temples and monuments? Are any of these things imaginable without the ingredient of engineering, albeit rudimentary or informal?

To understand the works of engineers and engineering is to understand the material manifestations and progress of civilization. The monuments of ancient Egypt, Greece, and Rome, in turn, illuminate the nature of engineering in those cultures, which was in many fundamental ways the same as the nature of engineering today. To conceive and execute the pyramids, the Parthenon, or the Colosseum required the same kind of conceptual design and analytical mental projection that it takes to conceive and realize a grand stadium, skyscraper, or bridge today. Even if the scientific understanding and mathematical and computational tools of engineering have advanced beyond what must have been the wildest imaginings of the ancients, the basic ways in which engineers conceive of new designs and think about bringing them to fruition is essentially the same today as it has always been. And although science and mathematics and computers are likely to continue to develop beyond our most extreme prognostications, the conceptual and methodological aspects of engineering in the thirtieth century are likely to be little different from those we know today. This is why the history of engineering will always be relevant.

We can learn a great deal about ancient, modern, and future engineering by looking closely at virtually any artifact, from a safety pin to a jet airplane, but some made things are inherently more interesting than others, the stories about them more charged with human drama. Bridges are in this latter category, and there is no purer form of engineering than bridge building. Daring and distinctive suspension spans like the Verrazano-Narrows Bridge or the Golden Gate Bridge, which are so familiar to so many, have the shapes and proportions they do, not because of some architectural golden section or some abstract theory of space and mass. Rather, the greatest bridges look the way they do because physical constraints, engineering inspiration, and judgment have led to calculations concerning the relative strength and cost of foundations and towers and cables and anchorages and roadways and rights of way. That is not to say, however, that aesthetic and political questions do not also inform the calculations of the engineer, for they most certainly do, as we shall see.

Whereas some of the greatest skyscrapers, like Chicago's Sears Tower and John Hancock Center, are the result of close collaboration between architect and structural engineer, this is not generally the case. Large buildings and monumental structures are often sketched first by an architect, with an eye toward the visual, and engineers may be asked afterward to develop a structural skeleton to support the façade. This was the case with the Statue of Liberty. It was first suggested as a symbol of friendship between France and the United States at a dinner party in 1865 by the French historian and politician Edouard-René de Laboulaye, and another dinner guest, the sculptor Frédéric-Auguste Bartholdi, embraced the idea. On a trip to America in 1871, he identified the present site in New York Harbor, then, back in France, began to make models. In the meantime, money for the statue was raised in France through lotteries and dinner parties, while that for the stone pedestal upon which Liberty would stand was raised in America with the support of Joseph Pulitzer, the influential newspaper publisher.

Bartholdi, realizing that it would be impractical to ship a bronze or stone statue across the ocean, designed one to be made up of beaten sheets of copper that could be mounted on an iron framework. The design of this latter, hidden part of the statue was to be done by Eugène-Emmanuel Viollet-le-Duc, the French architectural critic whose practical bent had led him to

A Currier & Ives print, circa 1886, showing the Brooklyn Bridge across the East River and the Statue of Liberty in New York Harbor

write, among more theoretical works, a very basic book on how to build a house. But Viollet-le-Duc died in 1879 without completing the iron frame. Bartholdi then turned to Gustave Eiffel, whose engineering firm was, at the time, the designer and builder of some of France's most daring bridges. In the end, it was the bridge-building experience of Eiffel and his engineers that enabled the Statue of Liberty to be erected in New York Harbor, and to withstand the elements for over a century, as his tower has in Paris. The refurbishment of the statue for her centennial revealed that structural weaknesses that had plagued the monument and had closed Liberty's arm to tourists for so many years were due not to any structural miscalculation on Eiffel's part but, rather, to some alterations made during construction and to an electrochemical reaction between the dissimilar metals used for the statue's skin and skeleton. Much effort involved in restoring the one-hundred-year-old symbol went to addressing this problem.

Bridge designs cannot evolve the way the Statue of Liberty or glass-faced high-rise buildings do, from the artistic outside in. A great bridge is an engineering structure first, and only when its structural integrity has been established on the drawing board and through elaborate engineering calculations can architectural embellishments be considered. This is not to say that architects have no role in bridge design, for bridge engineers have a strong tradition of involving architects as consultants. Many of the distinctive visual features of the Golden Gate Bridge, including its sculpted towers and color, are owing to the involvement of the consulting architect, Irving F. Morrow.

The George Washington Bridge, when it was conceived in the 1920s, was to be twice as large as any existing suspension span, and so the towers had to be as tall as skyscrapers. Such massive structures demanded some special treatment, it was felt, and no less an architect than Cass Gilbert, designer of New York's Gothic-style Woolworth Building, was involved in the design of their façade. The full story of the George Washington Bridge will be told later in this book, but it is not giving away too much to say that the architectural stone façade was never applied to the towers, whose bare steel forms stand today as one of the masterpieces of modern bridge engineering. Imagine what the George Washington Bridge would look like with stone applied, and imagine what might have been its influence on later suspension bridges, almost all of which have been built with steel towers. Each great bridge influences each later one, and that is why it is necessary to understand the history of bridges and their engineers in order to understand present and future spans and perhaps something of their builders.

When the proportions of ancient bridges, having been arrived at by trial and error, were codified in stone according to rules that such architects as

Vitruvius and Palladio prescribed for buildings, then bridges could be designed as architectural edifices. Even the great Roman aqueducts, such as the Pont du Gard in southern France, could be built with little calculation of the kind required for designing a modern bridge, for each of the individual semicircular arches could be supported by the massive piers on either side of it, and construction was more or less a matter of piling arches like blocks one beside and one upon another until the valley was filled with bridge to the desired level. Though superficially analogous processes can be said to suffice for bridge building today, now each step in the construction must be weighed so that the incomplete structure is as able to support itself as the completed bridge. Because this simple fact was overlooked, the Quebec Bridge over the St. Lawrence River, planned to be the largest of its kind, spontaneously collapsed while under construction in 1907. Great suspension bridges can be constructed without falling only because elaborate engineering calculations determine the precise order in which the parts, which individually might weigh as much as a large locomotive, will be assembled.

The modern bridge-building era began in the late eighteenth century, with the daringly shallow stone arches built over the Seine by the French engineer Jean-Rodolphe Perronet, and with the revolutionary use of iron in British bridge building. What is generally considered the first iron bridge was built in 1779 across the River Severn at Coalbrookdale, where increasingly larger iron castings had been made by the Darby family of founders. The first iron bridge mimicked a stone arch, with connection details that suggested timber construction. When wrought iron became available in larger quantities and pieces, these were formed and assembled into chains to support a bridge that worked not on an arch but on a suspension principle. The increasing use of iron in bridges of ever-greater span led to increasingly innovative and daring designs, which more than once over the course of the nineteenth century culminated in a colossal failure. However, as the Victorian era was drawing to a close, advances in engineering, mathematics, and science had given bridge engineers a perspective and a collective set of tools that enabled them to tackle with confidence and success problems of bridging that had once been thought impossible.

This book is about how the late-nineteenth- and early-twentieth-century engineers did what they did to leave us a legacy of bridges that define our material environment, shape our cities, suburbs, and rural areas, and ordain our routes of communication over distance and time. That period of great bridge building, especially in America, coincided with the rise of the engineering profession, and so the story of bridges provides an excellent vehicle also for understanding the development of the engineer and engineering generally. How the engineer interacted with society in the process of conceiving, promoting, financing, designing, and building bridges serves as a paradigm for appreciating the nature of engineering endeavors, and thus provides a basis for understanding how technology and society interact today and can be expected to interact in the future. No bridge is an island, entire of itself, and the story of any bridge is the story of every bridge in that it involves a plethora of characters and circumstances. By considering the stories of a few of the most significant, though not necessarily the best-known, engineers and the bridges that they conceived and built over the last century or so, we can come to understand more fully the nature of the interaction of the engineer with the rest of society, of the relationship between technology and the rest of the stuff and ideas of the world.

From another viewpoint, fully understanding how bridges have been conceived, financed, and built requires a fully integrated view of technology, society, and culture. The financial link is often the crucial metaphorical span between the dream and reality of an actual bridge. Many a wonderful concept, beautifully drawn by an inspired structural artist, has never risen off the paper because its cost could not be justified. Most of the great bridges of the nineteenth century, which served to define bridge building and other technological achievements for the twentieth century, were financed by private enterprise, often led by the expanding railroads. Engineers acting as entrepreneurs frequently put together the prospectuses, and in some cases almost single-handedly promoted their dreams to the realists. In the early twentieth century, in larger cities like New York, there were needs for bridges to move citizens, increasingly in automobiles, from homes to workplaces and back, across rivers and bays that were becoming choked with ferryboats and other water traffic and sometimes ice, and so local and state governments began to get more and more involved in the building of great bridges. Debates over how to pay for them were common. When the Delaware River Bridge, now known as the Ben Franklin Bridge, was under construction in the mid-1920s, an argument between Philadelphia, which wanted a free bridge, and Camden, New Jersey, which wanted to collect tolls, brought progress on the structure to a standstill.

The stories of the building of great long-span bridges coincide with the rise of the steel industry. Beginning with the Eads Bridge, whose requirements for steel were almost too demanding for the fledgling industry and its up-and-coming barons, like Andrew Carnegie, the desire for stronger and stronger materials to make ever larger and relatively lighter structures drove research and development among competitive suppliers. Later, the introduction of concrete, first reinforced and subsequently prestressed, as an alternative to steel in some structures, provided a new element of competition that remains to this day. Whether a bridge should be steel or concrete in some cases can be a toss-up financially, and the decision becomes one of aesthetics, maintenance, or technological preference.

Though it is true that no individual engineer, no matter how great, can single-handedly do everything—from detailed calculations to supervision of construction—required to bring a major span to fruition, great bridges do appear to have had masterminds behind them, albeit masterminds with many helper minds. Indeed, the stories of the great bridges built in the half-century or so between the 1870s and the 1930s, the era when length records were set that remain unsurpassed or just barely surpassed today, are stories of recurring characters, both major and minor, who seem to have played a role in almost every bridge of any significance that was constructed during the period in which they flourished. There was also a necessarily large cast of supporting engineers, of course, and their roles in the realizations of dreams will be seen to be no less significant. However, the main action shows that a few handfuls of leading engineer-entrepreneurs, by the force of their personalities, talents, ambitions, and dreams, rose to or seized the leadership roles during the era of great bridge building. Yet these great engineers were also as much a product of the opportunities and circumstances of their times, which they often influenced themselves, as of their dreams and talents.

If the stories of bridges begin in dreams, they often reach a climax, at least formally, in celebration. The completion of a great bridge, especially one linking what theretofore had been so close to the eye and yet so far from the body, has traditionally been cause for celebration. The formal opening of the Eads Bridge on July 4, 1874, which began with a huge parade in the morning and closed with a grand display of fireworks in the evening, set the standard for subsequent American bridge openings. The opening of the Brooklyn Bridge in 1883 was the subject of many a lithograph, and its spectacular fireworks show was recalled by an equally spectacular one on the occasion of its centennial in 1983. Great suspension bridges and celebrations seem especially to go together, and the clearly distinct stages of construction provide various opportunities to acknowledge progress and achievement. Discrete ceremonies often mark the topping out of towers, the completion of footwalks for cable spinning, the finishing of the cables, and the placement of the final segment in the roadway.

A special rivet was put in place by the Prince of Wales when the Firth of Forth cantilever bridge was opened in 1890. Though the engineers, bankers, and politicians are often joined only by the press on such occasions, the opening ceremonies of a bridge can also be a veritable test of the bridge itself. Pedestrians have traditionally had the run of bridges on their first

The Golden Gate Bridge, on the occasion of Pedestrian Day in 1987, marking the structure's fiftieth anniversary

day, and re-created walks across them have marked their anniversaries. Throughout the course of its opening day, May 27, 1937, which was designated Pedestrian Day, about two hundred thousand people had the Golden Gate Bridge all to themselves, and they walked leisurely between San Francisco and Marin County. To celebrate the fiftieth anniversary of the bridge, another Pedestrian Day was held in 1987, and of the half-million or so people who showed up all at once, only a couple hundred thousand could get onto the bridge's main span at one time. It turned out to be the heaviest load the bridge had ever experienced, and the structure was visibly strained under the weight.

Unfortunately, our thoughts about bridges often end the day after such celebrations, and we tend to take these structures, once thought impossible to finance or build, for granted. Yet bridges are affected by their environment no less than people are, and the wear and tear of traffic, pollution, abuse, neglect, and just plain old age take their toll. It is implicit, and often made quite explicit, in the design of every product of engineering that there are limits to its health and strength, and therefore limits to what it can be subjected to. A recognition of those limits and regular checkups and inspections of the artifact are required, as is a certain amount of preventive maintenance and repair. To neglect this common sense is to find ourselves in the position in which we now are in America, with roughly one out of every five of our bridges said to be structurally deficient. A familiarity with the stories of our bridges not only can bring a fuller appreciation of their rich history and significance, along with an appreciation and understanding of the humanity of engineers and of engineering generally, but also can promote a greater enjoyment and pride in the contribution of bridges to our physical and cultural infrastructure, and a sense of obligation to maintain them. Imagine what our lives would be without bridges.

II

EADS

James Buchanan Eads was born on May 23, 1820, in Lawrenceburg, Indiana, which is in the southeastern part of the state, near the Ohio border, just a few miles west of Cincinnati, and, like that Queen City, on the Ohio River. The third child of Ann Buchanan and Thomas C. Eads, he was named after his mother's young cousin, soon to be a Pennsylvania congressman, who in 1857 would become the fifteenth president of the United States. Thomas Eads was a businessman looking for a business in which to succeed, and this led the family to move first just up the Ohio River to Cincinnati; then, when James was nine years old, down the Ohio to Louisville, Kentucky; and, finally, farther down the Ohio, along southern Illinois, to where the Ohio meets the Mississippi, and up that legendary river to St. Louis.

The accident of his birthplace and his forced travel on two of the most important waterways of the time seem to have greatly influenced young Eads, and he would spend most of his adult life engaged in pursuits that would keep him over, in, under, and around water. He would devise some of the grandest schemes of the nineteenth century to raise great masses of sunken riverboat cargo from the bottom, to flush the silt and sand from the entire middle of a continent out the mouth of the Mississippi, to build a bridge over a river that many said could not be crossed, and to carry fully laden oceangoing ships across the land between the seas. Of these dreams of James Buchanan Eads, only the last was not to be realized.

Young James, his two sisters, and their mother actually went ahead to St. Louis in September 1833, where she was to set up a household before the father arrived to open a general store in the bustling city. James is reported to have been fascinated by the riverboat voyage, and by the vastness and vitality of the Mississippi River. He is said to have told his mother then that he would build a steamboat in St. Louis, which seemed a reasonable ambition for a boy who had already built a model of a steamboat able to cross a pond, as well as models of steam engines, sawmills, and fire engines, in a small workshop that his father had fitted out for him in Louisville. The young lad is said also to have whispered something to himself on the Mississippi on the way to St. Louis, something that even his mother might have found more an idle dream than a possibility: "This is going to be my river."

First, however, the Eads family was to be the river's, for, while they lay asleep on the last night of their voyage, as the riverboat approached St. Louis, a fire broke out on board. To escape the smoke filling the confined spaces of the boat, the passengers rushed to the railing, from which they could see the city before them and feel the fire behind. The boat did remain afloat until it reached the dock, upon which the Eads family and other passengers then stood helplessly watching their possessions go up in flames. However, with the few resources with which she had escaped, Mrs. Eads was able to rent the upstairs of a house that faced the river. It was evidently large enough for her to take in boarders and so bring in some income.

Although he had attended school up until that time, thirteen-year-old James could not continue his education in St. Louis, because he had to work to help support his family. At first he sold apples to bring home some money, but soon he found a more substantial opportunity as a "boy-of-all-work" in the Williams & Duhring dry-goods store run by Barrett Williams, one of the men who took meals at Mrs. Eads's boarding house. James was evidently a bright, energetic, and well-mannered employee, to whom Williams took a liking. Thinking it a pity that someone like young James could not go to school, Williams gave his loyal worker the run of his library, which was located in a room above the store. James was told that in his spare time he could read at will among the books, which included works on physical science, mechanics, machinery, and civil engineering.

Had young Eads wanted to study engineering formally at this time in America, he would have had virtually no opportunities close to home, and thus it is hard to imagine a better opportunity, especially in the St. Louis area, than the one given him by Barrett Williams. Though an engineering school had been called for by General George Washington as early as 1778, the Military Academy did not become firmly established at West Point until 1802. There were only the beginnings of a few other formally established courses in engineering in the early 1830s, let alone schools where a young man like James Eads could get a degree in civil engineering.

There were, of course, the likes of the Franklin Institute, established in Philadelphia in 1825, where lectures on applied science and mechanics were offered, often in the evening and aimed toward individual betterment rather than toward a degree. Also, up and down the East Coast, there were the beginnings of what might have developed into engineering schools. In Vermont, in 1819, Alden Partridge, who in 1813 at West Point had become the first individual to hold the title of professor of engineering in the United States, established the American Literary, Scientific and Military Academy, which later was renamed Norwich University. Courses in civil engineering were offered there as early as 1821. In Maine, in 1822, Robert H. Gardiner started a lyceum that bore his name and offered courses of study preparatory to engineering, but it survived for barely a decade.

The most sustained effort was begun in 1824 at Troy, New York, by Stephen Van Rensselaer, who, as lord of over three thousand farms totaling almost half a million acres in New York's Rensselaer and Albany counties, was "the last patroon in full authority." Van Rensselaer founded the Rensselaer School "to qualify teachers for instructing the sons and daughters of farmers and mechanics, by lectures or otherwise, in the application of experimental chemistry, philosophy, and natural history, to agriculture, domestic economy, the arts and manufactures." By 1835, the Rensselaer Institute was authorized by the New York State Legislature to give instruction in "Engineering and Technology," and the new degree of "civil engineer," the first such in Britain or America, was granted that same year, to a class of four. By 1849, Rensselaer was the leading civilian engineering school in the country.

In the South, engineering was taught at the University of Virginia, which Thomas Jefferson had established in 1814 to teach natural philosophy, military and naval architecture, and technical philosophy. The first course in civil engineering at Virginia was offered by one of the school's original faculty members, Charles Bonnycastle, in 1833. He was joined in the newly established School of Engineering in 1835 by Barton Rogers, who in 1865 would become the first president of the Massachusetts Institute of Technology. Instruction in civil engineering was begun at the University of Alabama in 1837 by order of the board of trustees, who saw it as beneficial to the growth and maintenance of an increasingly important railroad network throughout the South. Among its first professors was Frederick Augustus Porter Barnard, who eventually became president of Columbia University. A School of Civil Engineering was begun at the College of William and Mary in 1836; Virginia Military Institute, which was modeled after the famous French Ecole Polytechnique, was started in 1839; and the Citadel was established in 1842 to teach both military and civil engineering.

But young James Eads was in no position even to dream of attending such schools. The more common route to becoming an engineer throughout the first half of the nineteenth century was to work on a project like the Erie Canal, begun in 1817, "completed" in 1825, and widened, deepened, and extended in the 1830s, when it had become jammed with traffic. Accordingly, "many of the fledgling surveyors and assistants who planned and completed the canal 'graduated' from the project as highly skilled engineers." Young men of more substantial means did go to Europe to study, rather than build, the great works of engineering, or specifically to France to learn engineering from a still more theoretical point of view. Many an engineer of the early nineteenth century also absorbed a great deal from his self-taught father.

One such engineer was Loammi Baldwin, who became widely known for his hydraulic works, and who was responsible for the navy drydocks at Charlestown, Massachusetts, and Norfolk, Virginia. Even though it would be said that he had "learned engineering through self-study and by working under his father, Loammi Baldwin I, on the construction of the Middlesex Canal," the younger Baldwin also studied mechanical subjects at Westford Academy, was a member of the class of 1800 at Harvard, and studied law. He practiced engineering before being admitted to the Massachusetts bar, then operated a law office in Cambridge from 1804 to 1807, and finally abandoned law to return to civil engineering. Following a visit to Europe to inspect public works, he opened an engineering office in Charlestown and became involved with the extension of Beacon Street beyond the Boston Common, the Union Canal, and other significant works. Though steeped in practical experience, he was also among the earliest of American engineers to call for state-supported schools to teach engineering theory.

Baldwin spent the period 1824–25 in a concerted effort to enlarge his father's civil-engineering library by augmenting it with British and French books, and he strongly advised that anyone "who would become an engineer must collect books." Although the young James Eads was in no position to buy, let alone collect, books, he did have access to Barrett Williams's library, where he could read well into the night. In this way, Eads, like many of his contemporaries, was able to lay the theoretical foundation for his own engineering education, which in his case would be completed on the river.

In time, the elder Eads joined the family in St. Louis, and his general store prospered modestly. But the restless Thomas Eads, evidently not content with that business venture, went into partnership with another man to buy some property up the river, near Davenport, Iowa, where they planned to open a hotel. In the meantime, eighteen-year-old James had become a salesman at the dry-goods store and elected to stay in St. Louis, where he had some cousins, and where he knew he had a steady income and the run of a fine (if necessarily limited) library. Before too long, perhaps when he began to exhaust the resources of Barrett Williams's books, James was drawn again to the river, and he signed on as a second clerk on the steamboat *Knickerbocker*. On a voyage, while rounding the bend from the Mississippi into the Ohio, the boat hit one of the countless snags in the water and went down.

I

The Mississippi River was notorious for claiming boats laden with personal and mercantile treasures, and Eads had had plenty of time while clerking on the *Knickerbocker* to reflect upon what was beneath the muddy waters. Many a person realized that whoever could salvage even a small fraction of the treasure sunk there could make a fortune, for shippers and insurance underwriters would pay anywhere between 20 and 75 percent of the net value of cargo salvaged, and anything sunk more than five years became the property of whoever could raise it. However, the treasure was elusive, for the constantly shifting sandy bottom of the river quickly covered up wrecks and their cargo and made them very difficult to locate—never mind to raise.

When he was twenty-two years old, Eads conceived of a scheme that would enable a diver to work underwater for extended periods of time, thereby allowing him not only to walk about the river bottom and locate wrecks but also to free valuable cargo. Eads effectively worked as an engineer by developing, in his head and on paper, the early ideas for his "sub marine" and diving bell. What he had read in Barrett Williams's books may have given him full assurance that enough air pressure could be pumped to a submerged diver to make the concept work, but since Eads did not possess the capital to realize the scheme by himself (a position quite familiar to engineers with dreams grander than their material resources), he took his design to potential investors and other entrepreneurs. In 1842, Eads approached Calvin Chase and William Nelson, St. Louis boatbuilders, and offered them a partnership. His investment would be the idea and the operation of the salvage craft, theirs the capital and experience to build the boat. They agreed, and Eads soon began the first of over five hundred explorations on the river bottom.

Eads's scheme involved the use of a modified snag boat, a double-hulled craft familiar on the Mississippi and so named because it was used to remove the many obstacles, or snags, that developed in the water. A diver was to descend to the river bottom in a diving bell supplied with air from the boat that served as the base of operations on the water. Divers had used diving bells successfully in calm lakes, and Eads engaged an experienced man to help him try out the scheme over a sunken barge loaded with about a hundred tons of pig lead. However, the swift currents of the Mississippi proved too much for the light equipment, and the diver found it difficult to maintain control underwater. Seeking a means of improving the operation, Eads went to the nearby town of Keokuk and obtained a forty-gallon whiskey barrel; he weighted its top down with a few hundred pounds of lead, and across its open bottom he attached a strap upon which the diver could sit. When the designated diver declined to use the contraption, Eads himself descended in what must have looked like so many mad Victorian inventions that would be illustrated years hence in the pages of *Scientific American*. He successfully gathered a quantity of lead into the barrel before signaling to be hoisted up, but by then he had ranged so far from the snag boat that the line to the derrick on the boat was overextended, and it capsized in the process. There were a few anxious moments before Eads was hauled to safety by hand, but once out of the water he commenced to modify the procedure and make improvements in the salvaging system. Future snag boats would carry not only an air pump but also a sand pump to expose wrecks and their cargo, in addition to heavy hoisting machinery to bring up safely the diver, the loot, and, in later modifications, whole riverboats.

Not only did Eads and his partners make a fortune in the salvage business, but he grew to know the nature of the river bottom between St. Louis and New Orleans perhaps better than any of his contemporaries. He was intimately familiar with the stretch of river below Cairo, Illinois, where he once spent four hours a day for two months, Sundays excluded, walking back and forth over a three-mile stretch of the river, until he found the wreck of the *Neptune*. Years later, in his 1868 report as engineer-in-chief of the Illinois and St. Louis Bridge Company, he would write from experience of the action of undercurrents and other phenomena along the river:

> I had occasion to examine the bottom of the Mississippi, below Cairo, during the flood of 1851, and at 65 feet below the surface I found the bed of the river, for at least three feet in depth, a moving mass, and so unstable that, in endeavoring to find footing on it beneath the bell, my feet penetrated through it until I could feel, although standing erect, the sand rushing past my hands, driven by a current apparently as rapid as that at the surface. . . .

> It is a fact well known to those who were engaged in navigating the Mississippi twelve years ago, that the cargo and engine of the steamboat America, sunk 100 miles below the mouth of the Ohio, was recovered, after being submerged twenty years, during which time an island was formed over it and a farm established upon it. Cottonwood trees that grew upon the island attained such size that they were cut into cord-wood and supplied as fuel to the passing steamers. Two floods sufficed to remove every vestige of the island, leaving the wreck of the America uncovered by sand and 40 feet below low-water mark. . . .

This kind of knowledge and experience would be invaluable later, when Eads had to determine how deep the piers would have to go to support a bridge over the Mississippi at St. Louis, and, still later, how to channel the waters at its mouth so that it would remain navigable past New Orleans and into the Gulf of Mexico. By observing the motion of the river bottom at many locations and under various conditions, he was able to formulate an unsurpassed theory of its behavior.

When he was not working on or in the Mississippi, Eads would sometimes return to St. Louis to visit his cousins Susan and Martha Dillon, especially Martha, whom he wished to marry. Although the salvage business was profitable, her father questioned James's financial and physical future in so risky an endeavor, and the marriage did not occur till 1845, after Eads had sold his part in the salvage business to invest in the land-based enterprise of running the first glassmaking factory west of the Mississippi. However, a poor financial climate and a scarcity of skilled workmen soon put Eads $25,000 in debt, and he returned to the salvage business in 1848.

James and Martha had two daughters and a son, but the boy lived only about a year, and Martha died of cholera shortly thereafter, in 1852, leaving Eads heartbroken. He immersed himself in work and became very rich and famous, but soon his own health began to deteriorate, and he was ordered by doctors to take a complete rest. He married his cousin's widow, Eunice Eads, traveled to Europe, and came back to work on the river again. After three more years, however, at the age of thirty-seven, with the Eads & Nelson Sub Marine No. 7 raising wrecks of all kinds, and with the salvage business one of the most prominent in the country, Eads became exhausted and was forced to retire. He did so in St. Louis, where he entertained some of the most famous visitors to the city and talked of politics, secession, and slavery, which Eads opposed. He did not agree with his second cousin, James Buchanan, who was then in the White House, on the Dred Scott decision, and when the Civil War came, Eads was happy that Missouri voted not to leave the Union.

Soon after the surrender of Fort Sumter in 1861, Eads, the expert on Mississippi River craft, was summoned to Washington by his friend Attorney General Edward Bates, for a conference regarding the use of gunboats on the river. Eads recommended that a base be established at Cairo, Illinois, that Confederate commerce be blockaded, and that a snag boat be converted into an armed steamer protected by cotton bales. The proposal was referred to the War Department, but instead of a snag boat three wooden steamers were employed as the nucleus of the Mississippi fleet. Eads became the successful bidder to build seven five-hundred-ton, 175-foot-long armored wooden gunboats, whose hulls were to be divided into fifteen watertight compartments, and whose boiler and engines were to be protected with iron plates two and a half inches thick. Though the boats were supposed to be completed in two months, the last of them took over twice that time to finish.

The first gunboat completed, the *St. Louis*, was launched on October 12, 1861, and fought in the battle against Fort Henry on February 6, 1862, thus predating the more famous battle between the ironclads *Monitor* and *Merrimac* by over a month. In the meantime, independent of the War Department, General John Charles Frémont ordered the conversion of two steamboats to ironclads. Thus Eads was able to implement his own plans to convert a snag boat, which resulted in the "most powerful of the western ironclads," the *Benton*, with sixteen guns protected by as much as three and a half inches of iron. He later wrote to President Lincoln that "the *St. Louis* was the first ironclad built in America. She was the first armored vessel against which the fire of a hostile battery was directed on this continent, and so far as I can ascertain, she was the first ironclad that ever engaged a naval force in the world."

2

Before the war, the Baltimore & Ohio Railroad had reached Illinois Town, later known as East St. Louis, Illinois, thus establishing a continuous rail line from the East to the Mississippi River. Proposals followed to build a bridge across the Mississippi into St. Louis, Missouri, thus opening a rail route to the West that would compete with the one through Chicago. It would be hard to say when exactly the first idea for a bridge might have crossed anybody's mind, but as early as 1839 at least one engineer had not only thought about it but done enough preliminary calculations to write to William Carr Lane, the mayor of St. Louis, outlining a proposal for a bridge that would cost no more than $600,000 to erect.

Charles Ellet, Jr., was born in Penn's Manor, Pennsylvania, in 1810 and studied in Paris at the Ecole Polytechnique before commencing engineering work on railroads and canals in America. Around 1836, he turned his attention to the study of suspension bridges, perhaps inspired by the completion in 1834 of the 870-foot wrought-iron wire suspension bridge across the Sarine Valley at Fribourg, Switzerland, then the longest bridge span in the world. In 1842, Ellet would complete the Fairmount Bridge across the Schuylkill River in Philadelphia, the first suspension bridge in America to employ strands of wire rather than iron chains or eyebars to hold up the roadway, and in 1849 he would build the record 1,010-foot-span wire suspension bridge across the Ohio River at Wheeling, West Virginia. The deck of this latter bridge was to be destroyed by the wind in 1854, but in his 1839 proposal for St. Louis, Ellet had the utmost confidence in such designs.

Upon receipt of Ellet's letter, the mayor submitted it to the members of the St. Louis City Council with the request that a committee report on the proposal. Since the mayor noted that "Mr. Ellet promises leaving the city in a few days," a speedy report was clearly his wish, and the joint committee of three delegates and two aldermen reported within six days. According to Calvin Woodward, dean of the Polytechnic School of Washington University, in his definitive history of the St. Louis Bridge published in 1881, the committee's recommendation was to accept Ellet's "proposition to make surveys and soundings, and to furnish full drawings and estimates, and present three hundred printed copies of the same to the city for the sum of $1,000."

Ellet evidently stayed on in St. Louis to investigate three possible locations for his bridge, all of which had rock on the St. Louis side of the river, thus ensuring firm foundations there. In midstream and on the Illinois shore, he found that the sounding auger could not be driven more than twenty feet below the water, and thus Ellet reported that the riverbed was "superior to the soil which sustains some of the most celebrated stone bridges in Europe" and firm enough to drive piles into for the foundations of piers. The proposed bridge was to have three towers, with a central suspended span of twelve hundred feet and two side spans of nine hundred feet each. The length of cables required would thus be within the limits of a suspension bridge, which Ellet calculated to be one and one-fifth miles, and for the Mississippi spans he specified ten cables, each comprising twelve hundred one-eighth-inch-diameter wires gathered into a cylinder of about five inches in diameter. Though the final estimate of $737,566 was less than 25 percent higher than the original one, which they seem not to have balked at, the mayor and City Council used cost as an excuse to reject what they must have feared was an overly ambitious technical scheme:

"The time is inauspicious for the commencement of an enterprise involving such an *enormous* expenditure of money." Their instincts were correct, of course, for, as Eads would soon discover on the turbulent bottom of the Mississippi River, the foundations of Ellet's bridge would have been scoured away, possibly even before the cables were in place.

The state of bridge building at midcentury was changing as rapidly as the bed of the river itself. As the railroads spread their routes throughout Britain, America, and elsewhere, they came to use ever heavier and more powerful locomotives to carry ever-increased loads, and thus the suspension bridge was generally thought to be too flexible and too susceptible to wind damage to be considered a viable and reliable railroad structure. This is what led Robert Stephenson, in the mid-1840s, to design and build in northwestern Wales a revolutionary bridge type of such massive proportions and strength that it carried trains not over but through its great tubular girders, which spanned almost five hundred feet between piers and about fifteen hundred feet total over the Menai Strait. The Britannia Bridge was a marvel of engineering, but it was an extremely expensive undertaking, costing a total of 600,000 pounds sterling by the time it was completed in 1850, and so improvements, by way of spanning similar distances with lighter structures, became imperative. Yet, though British engineers like Isambard Kingdom Brunel and Thomas Bouch designed lighter and lighter girder bridges that carried heavier and heavier railroad trains, the British generally

The Britannia Bridge, in northwestern Wales, with the Menai Strait Suspension Bridge visible about a mile to the north

shied away from the suspension bridge for railway applications. Some Americans, however, did not.

John Roebling was educated as an engineer in Germany, having received a degree in civil engineering from the Royal Polytechnic School in Berlin in 1826, but for philosophical reasons he emigrated to America in 1831 with the intention of starting an agrarian community. He settled first in western Pennsylvania; when the utopian experiment did not work out, he turned to manufacturing wire rope for towing barges on canals and some small suspension bridges that carried the canals over rivers. In 1841, he published a paper discussing the "comparative merits of cable and chain bridges," in which he described many of the then widely known failures of suspension bridges, arguing that the incidents showed an engineer what he had to design against. By 1854, Roebling had completed a bridge with an 810-foot span over the Niagara Gorge, which demonstrated incontrovertibly that an efficient and economical cable suspension bridge could indeed be built to carry heavy railroad trains. Shortly after the Niagara Gorge Suspension Bridge was completed, Roebling proposed a very long-span suspension bridge over the Mississippi at St. Louis, but there was not sufficient financial support for any kind of bridge at that time. More than a decade later, Roebling would propose several other designs, including combination suspension-and-arch types, but there was little support for these proposals either.

After the Civil War, however, the City Council declared that it had "become indispensably necessary to erect a bridge across the Mississippi River at St. Louis, for the accommodation of the citizens of Illinois and Missouri, and the great railroad traffic now centering there," and the city engineer, Truman J. Homer, was instructed to draw up possible plans and estimate costs. He did report, only four days later, and was able to make a recommendation based on an idea he had actually conceived several years earlier and had since spent "much thought and extended inquiry upon." Homer condemned suspension bridges and proposed the building of a tubular bridge, slightly larger than Stephenson's Britannia, with carriageways on either side of the main tube and footpaths above it. For forty-nine weeks of the year, the clearance above high water would have been at least forty-four feet; Homer argued that steamboats could pass under such an obstruction, and that, in any case, the chimneys of steamboats could be made so they could be raised and lowered at will. The cost of Homer's tubular bridge was given at over $3 million. An earlier scheme, proposed by the Mississippi Submerged Tubular Bridge Company, for a tunnel under the river, might have cost even more. Neither was built.

The urgency for a bridge at St. Louis had been driven by earlier developments elsewhere. In 1856, the first rail bridge across the Mississippi was completed at Rock Island, Illinois, which was just about due west of Chicago, on the Chicago & Rock Island Railroad, thus promising an uninterrupted route westward. St. Louis boatmen reacted by filing lawsuits, "charging that bridges across navigable waterways were public nuisances, navigation hazards, and unconstitutional restraints on interstate commerce." In the meantime, the river had also been bridged by railroads at Dubuque and Burlington, Iowa, and at Quincy, Illinois, only a hundred or so miles upriver from St. Louis. The Missouri River was also bridged, at Kansas City, thus allowing St. Louis to be bypassed entirely by railroads on the way to its historic trade territories. Although complaints were rising that "it cost nearly half as much to ship a barrel of flour fifteen hundred feet across the river as it did to ship it upstream twelve hundred miles from New Orleans," ferryboat interests at the city without a bridge insisted they could continue to serve St. Louis commerce by floating entire railroad cars across the Mississippi on barges. However, the expenses and interruption of continuous rail service created bottlenecks in Illinois Town, and business was lost to the northern routes. One newspaper editor is reported to have said that geography had been undone by technology.

Even though the population of St. Louis was on a par with Chicago's two hundred thousand in the mid-1860s, in the commercial race the Missouri city was trailing and falling further and further behind. Illinois ranked second in railroad-track mileage in the early 1860s; Missouri was fifteenth among the thirty-seven states, with only 983 miles laid by 1867. Still, there were five railroads from the east and three from the west converging on St. Louis, and no continuous river crossing to serve them. Local newspapers and civic leaders began frantically to call for a bridge, which they argued not only would help St. Louis replace Washington, D.C., as the nation's capital but also would enable it to become "the future Great City of the World."

Neither in the heat of community boosterism nor in calmer times can bridges be erected wherever one pleases. In order to throw a bridge over a navigable waterway between two states, one has first to secure the appropriate enabling legislation. Thus, as an initial step, bridge promoters had obtained a charter for the St. Louis and Illinois Bridge Company, having secured the authorization of the two states in 1865 and that of the federal government in 1866. Like many a bridge charter, this one made certain specifications about the structure, which "might be a pivot or other form of drawbridge or else one of continuous spans." If the bridge did not pivot or open, it had to have spans of no less than 250 feet and it had to provide no less than forty feet of headway above the city directrix, which was a curbstone at the foot of Market Street indicating the level that record flood wa-

James Buchanan Eads, as he was pictured in A History of the St. Louis Bridge

ters had reached in 1828, and which defined "the datum plane for all city engineering in St. Louis."

It was not uncommon, once one private company was formed to build a bridge and obtained a charter, that a rival firm soon also was established and sought a charter of its own. In the case of St. Louis, the competition was embodied in a Chicagoan, Lucius Boomer, and his Windy City backers. They exerted pressure on the Illinois Legislature to rescind the charter of the St. Louis and Illinois Bridge Company and give Boomer's deliberately named Illinois and St. Louis Bridge Company the exclusive right for twenty-five years to build a bridge from the Illinois shore. If such a bridge were actually built, toll revenues would effectively flow from St. Louis business interests to Chicago investors. Even if Boomer's group did not complete a bridge, or if it sold its charter to St. Louis steam- or ferryboat operators, the effect would be to cause St. Louis to fall further behind Chicago in mercantile activity.

Among the contemporary movers and shakers in St. Louis was James Buchanan Eads, whose interest in bridges to this time was mainly in how they might obstruct the waterway. However, he took the potential commercial threat from Chicago as a call to action, and since a bridge was believed to be inevitable, he encouraged support of the original, and local, bridge company. A committee went to the Illinois state capitol in Springfield to lobby against what came to be known as the Boomer bridge bill. Southern Illinois legislators, who understood the importance of St. Louis to their own economic future, helped to get compromise legislation passed specifying that the exclusive building rights of Boomer's firm would lapse if a bridge was not begun in two years or finished in five.

Early in 1868, Boomer began to make noises about what kind of bridge his company would build. An earlier bridge of his had collapsed in 1855, killing Calvin Chase, one of Eads's original salvage partners, along with many other prominent St. Louis businessmen and politicians en route to a convention in Jefferson City. This time Boomer involved a consulting engineer, Simeon S. Post, of Jersey City, New Jersey, whose reputation was sound; his proposed bridge was to consist of six spans of an iron-truss design he had patented in 1863. The term "truss" designates any arrangement of beams, rods, cables, or struts that are connected together to form a rigid framework, thus enabling relatively long and stiff bridges to be built with a minimum of material. Wooden-roof trusses are of such construction, but, perhaps because they are concealed, they were an often overlooked kind of bridge. The idea of a truss as a bridge in its own right had been made explicit in the Renaissance.

In his sixteenth-century book on architecture, the Italian architect Andrea Palladio illustrated the wooden truss as a "most beautiful contrivance."

A truss bridge and some terminology used to describe its various parts

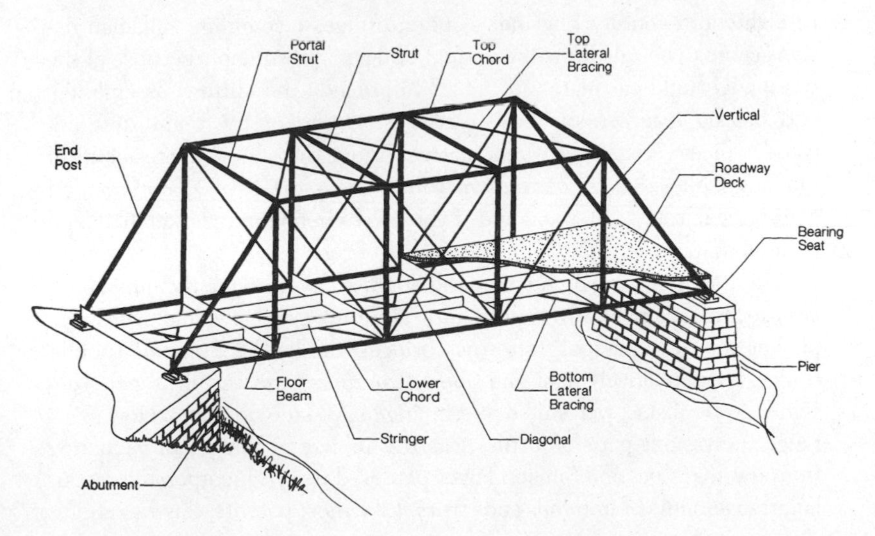

A variety of truss types employed in bridges

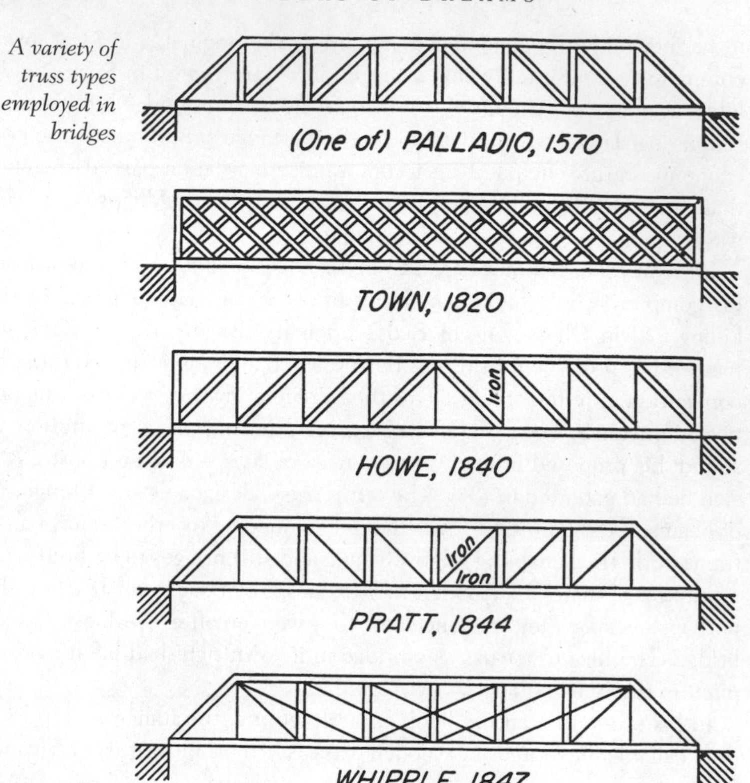

In eighteenth-century England, wooden bridges resembling Palladian designs came to be called "mathematical bridges," presumably because of the forethought and calculation that had to precede the cutting, assembling, and bolting together into an effective structure of the many different wooden pieces. Today, the Mathematical Bridge that allows the residents of Queens' College a very convenient route across the River Cam is one of Cambridge's tourist sites and one of the most photographed, sketched, and painted of its structures.

With the increasing production and application of iron in the nineteenth century, trusses naturally evolved into a plethora of types and styles employing the new material. Iron-truss bridges, unlike the Britannia tubular bridge, were relatively light and open structures, and yet, if properly designed, were just as well suited to carrying heavy railroad trains. How to arrange the various parts of a truss was the subject of many patents dating from the 1840s on, and Simeon Post's patented design incorporated modular arrangements of iron rods and struts. Like most patents, this was an improvement on the prior art. Post's arrangement of the components allowed for the expansion and contraction of the iron so that traffic and temperature changes would "not produce injurious effects upon the structure, and in this manner obviating one of the most serious objections to the universal use of such bridges."

3

James Eads had never built a wooden bridge, let alone one of iron, and perhaps he had never even dreamed of doing so. His interests were more in and on the water than over it. Indeed, to Eads, putting the piers of a bridge in the water meant providing obstacles to river traffic, as did the superstructure they supported, but his concern for the commercial future of St. Louis overcame his preference for an unobstructed river. He never did lose sight of the importance of river traffic to the city, however, and so he could not imagine or endorse any bridge design that would have obstructed the waterway more than it had to, even temporarily during construction. Thus Eads would certainly have balked at the traditional means of building stone-arch bridges and even some of the newer iron-truss bridges, whereby a timber scaffolding known as "centering" or "falsework" needed to be erected first, and might have to remain in the main waterway for an unconscionable length of time until the great bridge being assembled atop it reached the point where the stone or ironwork was self-supporting and the centering could be struck or the falsework disassembled.

As early as 1866, Eads's recommendations for bridge legislation included a minimum width of six hundred feet between piers and a minimum headroom of fifty feet above high water, "measured in the center of the span," to minimize interference with river traffic. Such a specification, which effectively ruled out a truss bridge of any kind, may well have reflected Eads's reading of the unrealized proposal made early in the century by the British engineer Thomas Telford for an iron arch to provide a clearance of sixty-five feet above the water at the crown while spanning six hundred feet over the Thames River at London. Legislation as strict as Eads's suggestion did not pass, but a minimum span length for a crossing of the Mississippi below its confluence with the Missouri, just above St. Louis, was fixed at five hundred feet. This was the size arch Eads himself eventually proposed to build.

Even though no arch greater than about four hundred feet had ever actually been constructed, the reputation of Telford, who was the first president of the Institution of Civil Engineers and who is buried in Westminster Abbey, made his dream almost as good as reality, at least to Eads; besides,

Thomas Telford's 1800 proposal for an arch bridge across the Thames

the iron bridges of modest span that Telford did complete in Wales and Scotland were masterpieces. However, aesthetic models were ideals to be challenged by competition and economics. A convention of civil engineers, including Post, which was lavishly hosted by Boomer in St. Louis in August 1867, praised his bridge proposal while cautioning investors that Eads's had "no engineering precedent." Eads countered with an appeal to Telford's generally acknowledged sound judgment and authority, which he felt did indeed provide "some 'engineering precedent' to justify a span of 100 feet less," six decades later. He went so far as to state that it was "safe to assert that the project of throwing a single arch of cast steel, *two thousand feet* in length, over the Mississippi, is less bold in design, and fully as practicable, as his cast iron arch of 600 feet span." Eads's own reputation for sound engineering judgment, albeit with projects other than bridges, and his confidence, coupled with Boomer's increasingly transparent attempts to manipulate public opinion, led to negotiations between the two rival bridge companies, which consolidated in 1868. Their stock was combined, a new board of directors was formed, with equal representation from each side of the river, and the name Illinois and St. Louis Bridge Company was adopted, with Eads as engineer-in-chief.

Some insight into what must have helped sway support to Eads can be gained by reading his report of May 1868 to the president and directors of the company, in which he consistently writes of "your Company" and "your Bridge." As are virtually all reports by successful entrepreneurial engineers like Eads, the twenty-five-thousand-word document is technically concise and sound, a model of clarity and persuasion, and totally accessible to the general reader. Eads said as much in his opening statement to the directors:

> In view of the great importance of your enterprise, the deep interest manifested in it by our citizens and the public generally, and because the plans

> adopted by you have been frequently misrepresented and unfairly criticised, I have deemed it proper that everything of interest connected with my department should be placed in such form as to be clearly understood, not alone by your stockholders, but also by every person of ordinary intelligence in the community. I have, therefore, endeavored to explain the plan of the structure, the principles involved in its construction, and the reasons for its preference, in the simplest language I can command, and with an avoidance, as far as possible, of the use of all technicalities not understood by every one.

Under the topic of location, Eads explained why his bridge was to be sited at Washington Avenue, rather than a few blocks north, where "Mr. Boomer's bridge" was to be located. The convincing arguments had to do with access to the center of population in St. Louis, the pre-existence of streets that were able to absorb all the traffic that would concentrate at the bridge approach, the cost of the connecting tunnel needed to carry the railroad trains through the center of the city without interfering with carriage and foot traffic, and the location along the wharf that would minimize interference with riverboats. Eads also led the reader through an elementary discussion of the principle of the lever in order to demonstrate "the economy of the arch, over the truss, for long span bridges." In fact, Eads saw the arch as a kind of limiting case of the truss, with stone abutments serving to take the thrust that in a truss would otherwise have to be resisted by increasing amounts of iron, which naturally added weight and thereby cost to the structure.

To show that his bridge design was not "needlessly extravagant," Eads proceeded to illustrate "enough of the general principles involved" in bridge building to allow "anyone with ordinary intelligence" to judge for himself the value of the arch over the truss for the St. Louis crossing. He began his remarkably concise and well-illustrated exposition with the "simplest of all the mechanical powers, the *lever*," which is familiar to everyone in the form of a simple balance scale. The action of a large weight on a short arm is balanced by a small weight on a longer arm, with the ratios of weights and arm lengths simply related to each other. If the short arm of the lever was bent, or canted, relative to the long arm, Eads argued that the same ratio of pulls or pushes was necessary to maintain equilibrium, and the supporting wall or abutment into which the canted lever was anchored could supply whatever force was needed. This canted-lever principle was the same as that articulated by Galileo in his seventeenth-century investigations into the strength of materials, and it would play a central role in bridge building later in the nineteenth century. But for Eads in his 1868 report to the president

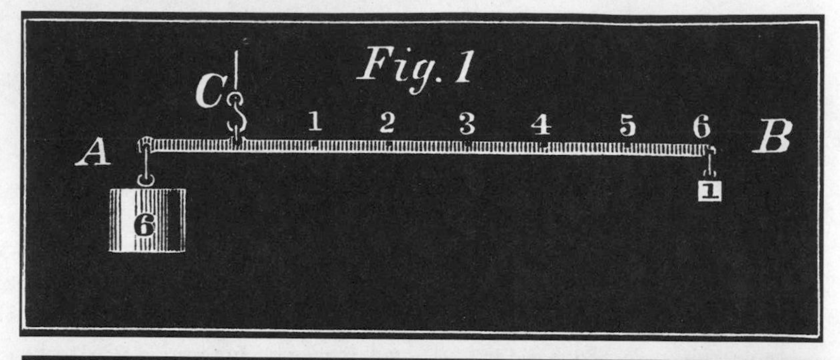

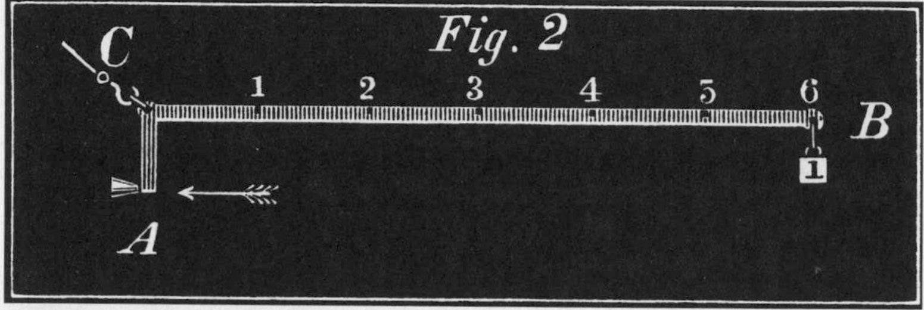

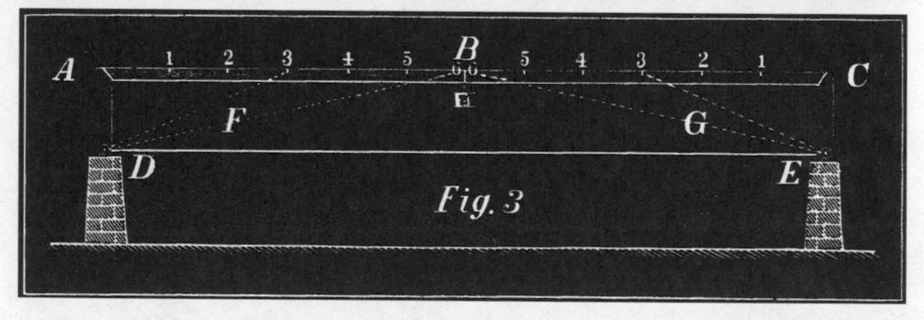

Three figures from the report of James B. Eads, showing the principle of the lever, the "canted lever," and back to back canted levers tied together by a lower chord to make a truss bridge

and directors of the Illinois and St. Louis Bridge Company, it was only a means to an end.

Eads next argued that, if two canted levers were placed tip to tip, with their short arms resting on piers and tied together with a structural element known as a chord, there would result an elementary truss bridge, in which the pull in the chord would be opposed by a compression of the longer lever arms. Like all successful engineers, Eads understood the important and nontrivial practical implications of the truism that the structure worked if it did not fail: "If the upper member fails to resist the crushing force, or the lower one is rent asunder, the truss must fail." Such simple reasoning provided the basis for calculation of how much material, and hence how much money, was required to build a bridge.

With his explanatory bridge constructed on paper, Eads led his readers to observe that the pair of canted levers could be replaced by slanted straight members connecting one end with the other, thus saving material and money. At this point, the bridge looked like the roof structure of a house, which every reader of Eads's report must have recognized as a kind of bridge in which they had long had confidence even if they lacked a complete understanding of its principles. As with larger roofs, the simple triangular arrangement has to be supplemented with cross bracing in order to keep the lines of the timbers from bending and breaking under their burden, and hence the familiar roof truss. In the 1860s, however, when railroad bridges had to span much greater distances than roofs, and without the help of excessively high and costly peaks, the flatter bridge truss made of iron evolved. The forces that had to be resisted by the various members of the truss depended upon its proportions, and these in turn affected the cost of the bridge. The proportions of height to length of bridge that would "insure the greatest economy" depended upon the type of truss, and they had been found to vary from about one-to-eight to one-to-twelve, thus making all sorts of bridges share a certain sameness of outline. Indeed, to Eads at least, the term "truss" included "every known method of bridging except the arch," which provided the ultimate economy.

To use less material than a truss of triangles, the top of the bridge could be curved, thus leading to the "bow-string girder," and Eads led his readers to observe that its bottom chord became unnecessary, thus reducing the cost, if the bridge piers or abutments could provide the forces to keep the bow from flattening out. However, for such a long bridge as Eads proposed at St. Louis, such an unbraced arch would be too flexible, especially under the action of heavy railroad traffic, and so he argued that it could be stiffened by means of a form of trussing between it and the roadway. Finally, in the ultimate refinement of his argument, Eads replaced the single heavy arch with a pair of lighter arches, themselves trussed together to provide enough stiffness so that the roadway could be supported on lighter members, thus reaching a further economy of materials.

According to Eads, there were two general kinds of arch bridge—the "upright" one, which he reasoned to from the principle of the lever, and the "catenary or suspended arch," which was otherwise known as a suspension bridge. Even though Roebling's Niagara Gorge Bridge had carried railroad traffic for almost fifteen years, some disagreement still remained among engineers about the safety of suspended spans. However, there was little doubt that the suspension bridge, or suspended-arch principle, was the most economical means of spanning long distances, such as that over the Mississippi between Illinois and Missouri. Eads addressed the question of upright versus suspended arch later in his report, and he weighed the pros and cons of using iron in tension and compression.

Though the repeated loading of iron in tension, such as occurs in a suspension bridge, was then known to be capable of leading to failure by fatigue, Eads and other engineers understood that they could avoid this by keeping the loading levels sufficiently low; but this meant using more material, and hence resulted in a greater cost. Since the ultimate strength of conventional iron was greater in tension than in compression, in the final analysis the suspension bridge, which relied on the former, was the most economical type for long spans. However, the upright arch, which Eads clearly favored on philosophical if not aesthetic grounds, would become economically competitive with the suspended arch if it could be made of a material whose capacity to carry a load without yielding in compression was not so inferior as was that of iron to its capacity to carry a load in tension. There was in fact a new material that fell into this category; Eads specified that his upright arch be made of cast steel.

As everyone knew, temperatures in St. Louis ranged from well below zero in the winter to well over a hundred degrees Fahrenheit in the summer sun, and this also had to be considered in bridge building. Since iron and steel expand when heated and contract when cooled, a five-hundred-foot arch would rise and fall eight inches with the passing seasons. Similar movements could result from heavy loads crossing the bridge. Uneven heating of the bridge by the sun, or asymmetrical loading of the bridge, could also push and pull it every which way over time. If this kind of potentially destructive movement could not be accommodated, it could destroy abutments, buckle rails and roadways, and eventually tear the structure apart. Eads was clearly concerned about this aspect of his bridge, as indicated in several patents for improvements in bridges that he took out while the bridge was under construction, in the late 1860s and early 1870s. The drawings for the earlier patents, though clearly showing an archlike bridge, are titled "truss bridge," which further emphasizes Eads's view of the arch, especially when stiffened, as a supremely efficient limiting case of a truss.

No engineer, in Eads's day or now, can conceive, promote, design, defend, and build a major bridge without help. A good deal of the attention of an engineer-in-chief like Eads is necessarily taken up in matters of a political, financial, and public-relations nature, and there is not enough time, let alone experience or talent, in one person's lifetime of days to carry out all the minute calculations and prepare all the detailed drawings that enable the proper parts to be designed, ordered, and assembled into a finished bridge, whether mathematical or not. Early in his report, Eads acknowledged his debt to some of the most important contributors to the engineering endeavor.

Henry Flad was born in Rennhoff, Germany, in 1824. After graduating from the University of Munich in 1846, he worked for the Bavarian government. He served as captain of engineers in the Parliamentary Army during the 1848 revolution, then fled to America. Flad landed in New York City in 1849 and worked there as a draftsman until getting involved with railroad engineering and moving west with the railroads. He enlisted in the Union Army and served in an engineering regiment, rising from private to colonel. After the war, he worked on plans to improve the water supply of St. Louis,

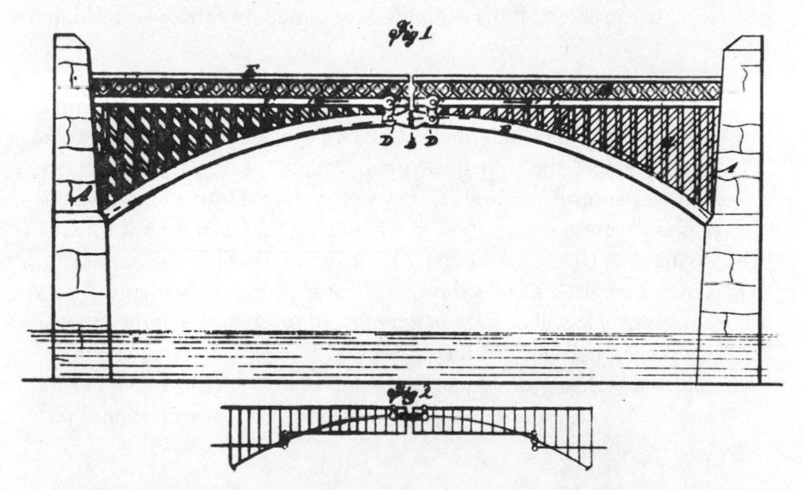

Drawings from one of Eads's several patents, this one employing the cantilever principle and showing how broadly he applied the term "truss bridge"

and then was engaged as chief assistant to Eads for carrying out "the mathematical investigations and calculations for the Bridge." Flad was in turn aided in the design by Charles Pfeiffer, who had recently emigrated from Stuttgart, where his thesis on the theory of arch-bridge design had won a prize. Indeed, in addition to his German degree of "civil engineer," Pfeiffer may also have brought with him the experience of the Koblenz Bridge, completed in 1864, for Pfeiffer "based his first series of calculations on the equations developed for Koblenz and modeled his first sketches directly on the German bridge." In 1869, Eads and Flad jointly were issued a patent for an "improvement in arch bridges" that relieved the thrust of the arch on its piers, "thus permitting a light construction of such piers." Like Eads's earlier patents, the drawing was headed "truss bridge." A few years later, Flad was issued a patent in his own name for an invention that enabled the stay cables essential to the construction process to be maintained at a uniform tension even as temperature changes affected a bridge structure under construction.

As Eads, Flad, and Pfeiffer knew, the essence of sound engineering lay in clearly stating the assumptions upon which calculations are based so that they may be checked at all times for lapses in logic and other errors. It is thus imperative that engineering premises be set down clearly, and that the calculations that follow be systematically and unambiguously presented, so that they may be checked by another engineer with perhaps a different perspective on the problem. Eads explained the procedure that was used in his office:

> After careful revisions by Col. Flad, the results obtained from time to time were submitted to me; and, finally, to guard against any possible error in the application of the principles upon which the investigations were made, or in the results arrived at, they were referred by me to the patient analysis and careful examination of Chancellor W. Chauvenet, LL.D., of the Washington University, formerly Professor of Mathematics in the U.S. Naval Academy at Annapolis. His certificate, affirming their correctness in every particular, will be found appended to this report. For the interest this gentleman has taken in the enterprise, for the care bestowed in examining and verifying the scientific data required for the work, and for many valuable suggestions and simplifications in the investigation, I feel under many obligations.

The title, degree, and affiliations of William Chauvenet, who was serving as consulting engineer to the project, were given, of course, to establish his authority and integrity, just as his certificate was to attest further to the correctness of Eads's report. In virtually all large engineering projects there is one or more consulting engineers of vast experience and irreproachable authority, who provide either the basis for carrying out the details of the design or the imprimatur that some less prestigious engineer's design is sound. The profession of engineering, like all professions, rests upon the exercise of sound judgment, and the interaction of engineers of varying degrees of experience must involve a delicate balance between recognizing and accepting something as correct, on the one hand, and recognizing that it might not be done in exactly the same way by another engineer—including the one doing the checking.

Engineers are also human, of course; sometimes they cannot entertain the idea that a competing design may be an equally correct, if different, solution to a problem. An earlier consultant to Eads, Jacob H. Linville, was an undisputed authority on railroad bridges. Having become the engineer for bridges and buildings for the Pennsylvannia Railroad in 1863, he had by 1864 completed the first long-span (320-foot) truss across the Ohio River, at Steubenville, on the West Virginia border. When Eads sent him the preliminary drawings for his own long-span arch bridge, Linville wrote back disapprovingly of the design: "I cannot consent to imperil my reputation by appearing to encourage or approve of its adoption." He further wrote, "I deem it entirely unsafe and impracticable, as well as in fault in the qualities of durability." Linville was perhaps more concerned with his own reputation as a proponent of more conventional trusses than with the safety of Eads's design, and he was perhaps a bit disingenuous in condemning what he may not have fully understood. Not surprisingly, Linville suggested for St. Louis a truss bridge, one that could be assembled on pontoons, floated into place, and then raised into position by hydraulic jacks, the way Stephenson's Britannia tubes had been, thereby obviating the objection to scaffolding obstructing the river during construction.

Eventually, however, with the opposition of Linville, Boomer, and the Convention of Engineers assembled to discredit Eads a thing of the past, he was able to proceed with the plans endorsed by Chauvenet and the board of directors. First the bridge's piers had to be built up from the rock below the river bottom, for Eads believed that no other foundation would survive the scour of the Mississippi. Indeed, the need to sink the piers so deep gave Eads additional incentive to construct the smallest possible number of them in the river, and thus he had had to specify the longest practicable spans. This kind of trade-off is common in large bridge design and often has an enormous effect on the total cost of the bridge. However, later in 1868, before any final decisions had been made on how to sink the piers, Eads became ill with a terrible cough and, under a doctor's orders to seek complete rest, tendered his resignation to the bridge company. But it was not accepted, and so work on the bridge, which had actually begun as

Eads's proposal for a bridge across the Mississippi River between St. Louis, on the left, and Illinois Town

early as August 1867 with the construction of the western abutment, was halted when Eads left for a restful trip to Europe.

4

After six months, Eads came back briefly to New York in order to conduct some financial negotiations, then returned to Europe, where he discussed plans for his bridge with British and French engineers and visited construction sites. It was on this trip that Eads became acquainted with the relatively new technique of using a *plenum pneumatic*, or pneumatic caisson, for sinking foundations underwater, and he inspected at Vichy a bridge-construction site employing the procedure. The customary method, which Eads had had in mind initially, was to erect a cofferdam enclosing an area from which the water could be pumped out and the riverbed excavated to bedrock. Among the drawbacks to the cofferdam, however, was that it was open to the air, and thus exposed to the elements and subject to flooding. The new caisson method employed what was essentially a gigantic inverted box into which compressed air was pumped to keep the water out while the work went on inside. As the river bottom was excavated, Eads planned to pile heavy stones atop the caisson, thus erecting the pier at the same time that its weight pushed the caisson into the riverbed. The scheme appealed to Eads's experience with submarine salvaging, being not unlike a gigantic diving bell or inverted and weighted barrel, and the sand pump he had devised for salvage work would also provide an efficient means of getting the excavated debris out of the caisson.

Thus, though the pneumatic-caisson method had been used for almost fifteen years to sink over forty piers in Europe, in America Eads was going to improve upon the concept and extend it to greater depths than ever before. The first caisson was launched from its own construction site, floated into position, and sunk in October 1869. Work inside the chamber was to continue smoothly day and night for five months, through the winter. As the exotic construction progressed, "a visit to one of the air-chambers under the piers was one of the principal attractions that St. Louis had to show to visitors." One retrospective description captured the experience of descending into the caisson:

> For a while one felt perfectly comfortable in this underworld—a world such as no mythology and no superstition ever dreamed of. The transition, indeed, became apparent by pain in the ears, bleeding at the nose, or a feeling of suffocation; but these inconveniences and seeming dangers, in-

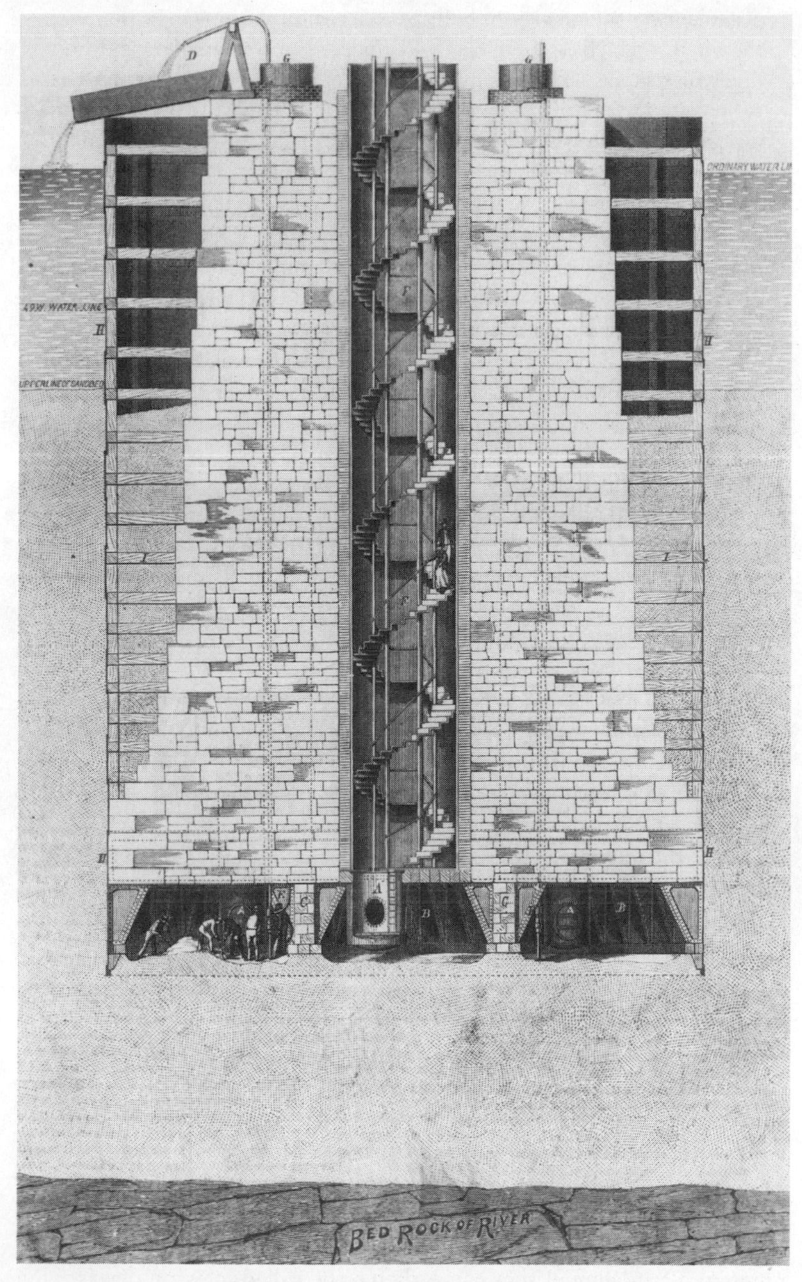

A caisson being sunk for the St. Louis Bridge

> evitable upon such a visit to hell, were insignificant in comparison with the interest which it offered. It was undertaken by hundreds and hundreds of visitors, including many ladies, and none returned from that depth without carrying along with them one of the most remarkable reminiscences of their whole life. Shrouded in a mantle of vapor, labored the workmen there loosening the sand; dim flickered the flames of the lamps, and the air had such a strange density and moisture that one wandered about almost as if he were in a dream. For a short time all this was extremely interesting and delightful, but it was not long before the wish to escape again from this strange situation gained the upper hand over the charm which it exercised. Gladly did the visitor, after a quarter of an hour, re-enter the air-lock, with an unfeigned feeling of relief, to watch the air beginning to escape from this chamber. At once the door behind him leading from the caisson closed by the denser air, and fastened as firmly as if there was a mountain behind it. The compressed element escaped whistling from the air-lock; the air within was more and more equalized with the air without; a few minutes, and they were of equal density; then the door, no longer pressed against its frame by the dense atmosphere, opened to the winding stairs, and the visitor came forth taking a long breath, and, to use one of Schiller's words, once more "greets the heavenly light," which shone from far above down the shaft.

When the caisson had penetrated sixty-six feet, a telegraph terminal was installed inside it, "where all things hideous are," enabling workmen there to communicate at all times with "those that breathe in the rosy light." Eads used this telegraph as a promotional device to send greetings to bridge directors in New York, and visitors in the caisson sent messages to friends in the "outer air." Such antics and poetic rantings demonstrate how casually the caisson was treated, and no doubt many of the visitors had a good laugh when a flask of brandy one of them had taken down into the depths, and there drunk from, exploded violently when they returned to the outer air. This, however, was but an amusing precursor of more serious events to come.

When the caisson reached a depth of seventy feet, the workmen began to experience some difficulty climbing the stairs to the surface. As the caisson was sunk deeper, men suffered increasing attacks of cramps and paralysis, which were thought to be due to insufficient clothing or poor nutrition. In March 1870, when the caisson had reached ninety-three feet, the air pressure inside it was about four times what it was in the open air, and workmen began dying upon emerging from the caisson, or after being hospitalized for an ailment that came then to be called "caisson disease" but today is known as "the bends." Eads asked his family physician, Dr.

Alphonse Jaminet, to look after the workmen, but Jaminet himself became paralyzed one day, having spent time down in the caisson and come up after only a few minutes in the air lock.

Perhaps somewhat to his own surprise, Jaminet recovered, and began to conduct research into these mysterious attacks. He shortly concluded that the major cause was too-rapid decompression in the face of a drastic difference in air pressure between the submerged caisson and the outer air above. Thereupon he placed restrictions on the amount of time the men could work inside the caisson, and on the speed with which the pressure in the air lock could be reduced. The west pier of the bridge, having to go to a depth of only eighty-six feet, was sunk with fewer incidents. Fatalities would also occur in sinking the caissons used in the piers for the Brooklyn Bridge, however, and Andrew H. Smith, a former army doctor and specialist at the Manhattan Eye and Ear Hospital, would be engaged as surgeon to the New York Bridge Company. Smith, like Jaminet, would eventually recognize too-rapid decompression as a prime contributor to the problems being experienced in the deep caissons.

In time, the physiological condition of the bends began to be understood and guarded against, but other aspects of work on large construction also imperiled the health and safety of workers and engineers alike. A bridge from New York to Brooklyn was John Roebling's dream, but while he was involved in surveying to fix the location of the Brooklyn tower, in final preparation for beginning work on the pier, his toes were crushed when a ferryboat bumped into some piles of the slip on which he was standing. The symbolic significance and irony of the accident were probably not lost on the senior Roebling, who had had to fight for years the opposition of the ferryboat interests. He soon developed lockjaw, or tetanus, of which he died on July 22, 1869, a little more than three weeks after the accident.

John Roebling's bridge did not die with him, of course, and it was under construction contemporaneously with that of Eads. The chief engineer was Roebling's son Washington, who was born in 1837 and graduated from Rensselaer Institute in the Class of 1857. He gained practical experience in the family's wire-rope mill, and in assisting his father in building the suspension bridge across the Allegheny at Pittsburgh, just before the Civil War. It had been said that the elder Roebling was upset when Washington enlisted in the Union Army, first in the New Jersey militia and later in a more active New York artillery regiment, for which he would build bridges and ascend in balloons to observe enemy movements. He was to become a veteran of such battles as Second Bull Run, Antietam, Chancellorsville, and Gettysburg, and attained the rank of lieutenant colonel, by brevet, before resigning in 1865 to assist his father with the completion of the Cincinnati and Covington Bridge across the Ohio River, now known as the Roebling Bridge.

While Washington was still in the army, he had met Emily Warren, sister of his general, at an officers' ball. She captured his heart at first sight, and the two were married early in 1865. At his father's request, the young Roebling couple went to Europe to make a study of pneumatic caissons, in preparation for the construction of the Brooklyn Bridge. They also visited such then famous suspension bridges as Telford's across the Menai Strait and the Clifton Suspension Bridge erected across the Avon Gorge, near Bristol, as a memorial to Isambard Kingdom Brunel. Roebling was critical of the towers of each, mostly on aesthetic grounds, and found the deck of the Menai bridge to be very light and subject to vibrations. On the Continent, he received a grand tour of the Krupp ironworks in Essen, Germany, where he was shown an eyebar that had been made up especially for the occasion of the American engineer's visit. Such eyebars were essential links in tying the steel strands of wire that make up suspension-bridge cables into the anchorage.

Throughout his travels, Washington Roebling wrote of what he was learning to his father, who was back in Trenton thinking about the great East River bridge and how it would be designed and constructed. The clear necessity of deep foundations for the towers, in particular, was of prime concern, and Washington's information about caissons, which he spelled "cassoons," was invaluable. Colonel Roebling, as he came to be known, also visited St. Louis, early in 1870; he and Captain Eads, who had become so designated in recognition of his exploits on the river, discussed caisson design. Subsequently, the roughly contemporaneous design of caissons to establish foundations for the St. Louis and Brooklyn bridges resulted in some conflicting claims, and in 1871 Eads sued Roebling for patent infringement. The rivalry led to a series of letters in the English trade journal *Engineering*, to which Eads first wrote in April 1873 to "correct some statements" made by Roebling in a pamphlet on pneumatic-tower foundations that he had recently published.

The crux of the matter concerned the location of the air lock, which Eads had put within the caisson's chamber, thus making ingress and egress relatively convenient and obviating the need to pressurize the vertical access shaft, or even to make it airtight. What irritated Eads was Roebling's statement that the idea was not new with Eads but had been used earlier in Europe, although admitting that "the first practical application . . . on a really large scale in this country" was in the St. Louis Bridge. Eads contended that Roebling had neglected to distinguish between iron caissons and those upon which masonry was piled, and that he had thus not given sufficient credit to the American innovation. Eads's second letter to *Engineering* on the subject included drawings of Roebling's Brooklyn caisson, which had the air lock outside the air chamber, and his New York caisson, launched about a year later, in May 1871, which had its air lock inside the air chamber and was clearly similar to Eads's design. Eads explained his interest in carrying on about the matter, while incidentally getting in some digs against Colonel Roebling and providing some insight into the nature of what constituted engineering practice, then as now:

> I trust I shall not be understood as finding fault with Colonel Roebling for copying my plans in his New York caisson. On the contrary, I hold it to be the duty of an engineer to use the surest and most economic methods which are known in accomplishing his work, and, if possible, to improve upon those methods. Nor is the lack of inventive talent, whereby it is frequently possible to improve on the plans of others, or to devise new ones, at all necessary to constitute an able engineer, or to insure professional success. It is of much greater importance that the engineer, on whom rests the responsibility of a work, should be competent to select the best devices proposed by his assistants, or used by others, than to be able to invent novel ones himself. The obligation to adopt the best is, however, not incompatible with a generous regard for the rights or merits of others, and his professional reputation will never suffer by giving such credit to them as may be justly due.

The exchange between Eads and Roebling may have been especially emotional because of the bad experiences both men had had personally, and because of their workmen's suffering from the bends and other complications in the pressurized caissons. Late in 1870, a fire had broken out in the Brooklyn caisson, and Roebling worked himself to near-exhaustion directing operations to extinguish it. Spending so much time in the compressed and smoke-filled air caused him to suffer an attack of the bends and remain paralyzed for some hours. He recovered, only to suffer a much more serious attack in the New York caisson in 1872, when he appeared to be close to death. He recovered again, however, or appeared to, and went back to active work on the bridge. But by the end of the year, he was in extreme pain and frequently sick to his stomach, and he and Emily went to Europe for his health. He never did fully regain his strength, and was to spend three years in Trenton while the bridge towers were completed. Eventually, he became bedridden, in a room that overlooked the bridge under construction, with Emily effectively serving as assistant engineer and intermediary between the incapacitated chief engineer and his lieutenants on the construction site.

Though an engineer like Eads or Roebling dominates the story of a given bridge at a given time, that engineer is typically not only one among many working on similar problems contemporaneously, but also constantly dependent upon the assistant engineers responsible for the various parts of each larger project. Among the young men Washington Roebling had hired as assistants shortly after his father's death was Francis Collingwood, Jr., who in 1855 had graduated first in his class from Rensselaer Institute, where the younger Roebling had met him. Collingwood had worked as a city engineer in his hometown of Elmira, New York, and in the family jewelry business before being asked by Roebling to help the self-taught engineer William Paine on the Brooklyn caisson. Though Collingwood agreed only to a short engagement, he became so involved that he was among those engineers believed to have been capable of taking charge of the work in the event of Roebling's death, and remained associated with the work until its completion in 1883. Collingwood's distinguished career on the Brooklyn Bridge project, and later as a consulting engineer, is remembered today in the American Society of Civil Engineers' Collingwood Prize, which he instituted and endowed in 1894 to recognize young engineers who publish technical papers of exceptional practical merit. The subjects of prize-winning papers can range from the foundation of piers to the capacity of the superstructure they support. Whether engineers write such papers or not, the problems they deal with, as Eads and his colleagues knew, must be addressed in designing and building the artifacts.

5

As papers alone do not make a design, so piers alone, difficult as they may be to construct, do not make a bridge. Eads wanted the superstructure between and over the piers to be as solid as the masonry it would bear against, and in his report he wrote as follows about the strength of the bridge under the heaviest load that he could conceive its ever having to carry: "The arches have been designed with sufficient strength to sustain the greatest number of people that can stand together upon the carriage way and foot paths from end to end of the Bridge, and at the same time have each railway track below covered from end to end with locomotives."

In addition to all the people and locomotives it might have to carry, Eads's bridge would have to carry its own weight, of course. For many a bridge, in Eads's time as well as before and since, the weight of the structure itself can be so many times greater than what could physically be crowded onto it that a considerable proportion of its strength must go toward just holding itself up. The weight of a structure is also a principal determinant of its cost, and so there are clear advantages to using material that is strong relative to its weight. Before the 1860s, steel was not generally available in sufficiently large amounts or pieces to be incorporated in bridge structures, but Eads believed that the most economical arch would be made up of cast-steel tubular segments. He thus specified steel for the major structural elements, making his bridge the first to incorporate so much of the high-strength material.

The Keystone Bridge Company of Pittsburgh was given the contract to build the superstructure. Keystone grew out of what Andrew Carnegie called the first iron-bridge company, the firm of Piper & Schiffler, which he organized in 1863 with the help of the engineer Jacob Linville, the "hustling, active mechanic" John L. Piper, and "a sure and steady" Mr. Schiffler, whose first name Carnegie seemed to have forgotten in writing his autobiography. In 1865, the Piper & Schiffler firm was absorbed into Keystone, a name Carnegie was "proud of having thought of as being most appropriate for a bridge-building concern in the State of Pennsylvannia, the Keystone State." Among the first major contracts of what came to be Keystone was the enormous 320-foot-span truss bridge over the Ohio at Steubenville. By the time Eads approached them, the very successful company had an established reputation but a "bad credit" rating, according to Bradstreet's, because they never borrowed any money.

Andrew Carnegie and James Eads were perhaps more alike than either of them might have wanted to admit. Like Eads, Carnegie finished his formal education at age thirteen, when his family sailed from Scotland to America to seek better opportunities. They settled in Pittsburgh, where young Andrew began working as a messenger boy. He had to stay on in the office every other night, thus getting home as late as eleven o'clock, and that "did not leave much time for self-improvement, nor did the wants of the family leave any money to spend on books." It so happened, however, that a Colonel James Anderson "announced that he would open his library of four hundred volumes to boys, so that any young man could take out, each Saturday afternoon, a book which could be exchanged for another on the succeeding Saturday." Andrew found that Colonel Anderson's largess extended only to "working boys," however, "and the question arose whether messenger boys, clerks, and others, who did not work with their hands, were entitled to books." Not to be deterred, Andrew wrote a letter to the Pittsburgh *Dispatch*, arguing for a more inclusive policy, and his persistence paid off: he was allowed to check out a book a week. The memory of this experience no doubt prompted an older and more affluent Carnegie to give so many libraries to communities.

As youths, Carnegie and Eads read in different subjects, the former reveling in history and essays, the latter in science and mechanics. Yet great engineering projects tend to bring such disparate traditions and personalities together, and by the time the two men met over the plans for the St. Louis Bridge, they were equally headstrong, albeit each in his different sphere of endeavor. Carnegie remembered Eads as an "unusual character," whom he characterized as "an original genius *minus* scientific knowledge to guide his erratic ideas of things mechanical," but Eads's own reports and achievements belie this assessment. Carnegie's point of view was no doubt influenced by Linville, who advocated trusses, like the one at Steubenville, over arches for large bridges. Linville told Carnegie that if the St. Louis bridge were built on Eads's plans, it would "not stand up; it will not carry its own weight." Keystone could not sell a truss bridge to the customer Eads, however, for, according to Carnegie, he "was seemingly one of those who wished to have everything done upon his own original plans. That a thing had been done in one way before was sufficient to cause its rejection." Eads's response to such criticism was reflected in his report, where he wrote, "Must we admit that because a thing never has been done, it never can be, when our knowledge and judgment assure us that it is entirely practicable?"

Eads wanted a steel arch, not unlike the one in Koblenz, and Keystone did agree to construct it for him, but this was not to be easy. After approval of the plans, the next step lay in raising the money. This obstacle was overcome thanks to Carnegie's "first large financial transaction," the sale of some bridge mortgage bonds to Junius S. Morgan, the American financier in London and father of J. Pierpont Morgan. Getting the steel parts fabricated and assembled into the bridge was another matter. John Piper, who began addressing Eads not just as "Captain" but also as "Colonel," came to refer to him in succession as "Mr. Eads," "Jim Eads," and occasionally "Damn Jim." When Eads insisted on steel over iron parts, it took considerable time and money to get them made to meet the specifications. The joke that had been current when the bridge was still a proposal could thus continue to be told:

> *First gentleman:* "How much would the bridge cost?"
> *Second gentleman:* "Seven million dollars!"
> *First gentleman:* "How long will it take?"
> *Second gentleman:* "Seven million years!"

Once the steel parts were made and shipped to St. Louis, they were erected without interfering with traffic on the river. The method employed

The St. Louis Bridge under construction, showing the cantilever principle employed

had been suggested over a half-century earlier by Telford as a means of constructing a five-hundred-foot cast-iron arch across the Menai Strait without the use of any scaffolding in the water, and in St. Louis it was implemented by extending the halves of each arch—part by part—equally on either side of the central piers, supporting the heavy mass by guy wires that passed over temporary towers erected atop the piers—until the arch was completed and could support itself. Eads had at one point urged the use of catenary cables, slung over towers much like a suspension bridge, to support the partial arch, but the steadily changing weight of the arch as it progressed out from each tower would have called for constant readjustment of the curve. Early in 1871, Linville proposed to Walter Katté, engineer in charge of Keystone's western office, in St. Louis, that direct guys to the towers and backstays be used, along with Henry Flad's scheme of using hydraulic rams to adjust cable tensions, thus effectively employing the cantilever principle to support balanced back-to-back arch sections until the completed arches could support themselves. Original estimates were that the arches could be completed in a few months, but delays in receiving parts slowed the work considerably. Since financial assumptions had been based on projected revenue once the bridge was completed, it was imperative that steady progress be made toward that end. Incentives were offered to Keystone to close the arches by January 1, 1874, and to have the bridge ready for traffic by March.

By the end of the summer of 1873, the halves of the arches were approaching each other, but the critical operation of putting the last piece of steel in place was thwarted by the exceedingly high seasonal temperatures, which had so expanded the metal that the gap was too small for what might be called the steel keystone. Eads was in Europe at the time, recuperating from a condition characterized by chronic coughing and hemorrhaging lungs, and the task fell mainly to Henry Flad and a young engineer named Theodore Cooper, who earlier had been responsible for making sure that the steel mills were producing the proper material for the bridge parts. When the late-summer heat showed no signs of letting up, a wooden trough was constructed along the entire length of the arch, so that ice might be packed in and cool the steel enough to contract it—and thus open the gap wide enough for the last piece to be inserted. This scheme did not succeed, however, and it was necessary to resort to an alternative plan that had been devised by Eads.

Earlier in the year, he had applied for a patent involving the erection of arches that incorporated a screw mechanism capable of raising the completed arch a few inches, so that the supporting cables would be slackened and could be removed. By cutting the ends off the too-long arch ribs, threading them, and inserting the screw mechanism into the sprung arch, it was possible to close the arch on September 17. After that, work went relatively smoothly and quickly, and in January all the cables were removed and the arches became self-supporting. The Keystone Bridge Company announced that the bridge was to be available for pedestrian use in late April, but then Carnegie reconsidered allowing any traffic on the bridge until his company had been paid. It was another month before the upper roadway was opened to pedestrians and fifteen thousand people paid a nickel apiece for the privilege of being among the first to walk across the Mississippi at St. Louis. By July 2, the rails had been completed on the lower deck, and fourteen heavy locomotives, in two sets of seven, were driven back and forth for five hours, first on both sets of tracks and then in one long line. Such a procedure, during which engineers monitor the structure's behavior, constituted what is known as a proof test.

The formal opening of the bridge, scheduled for July 4, 1874, was to be a gala affair. Invitations announced that President Ulysses S. Grant would attend, but he canceled at the last moment—too late to have his picture and references to him deleted from memorabilia of the event. The citizenry might have been more disappointed if James Eads had not attended, however. His portrait alone graced the invitation; it dominated a broadside issued for the occasion; and a fifty-foot-high portrait of him hung from the bridge, with the notation that the Mississippi had been discovered by Mar-

A contemporary photograph taken after the arches of the St. Louis Bridge became self-supporting

quette in 1673 and spanned by Eads in 1874. A parade almost fifteen miles long, with a great line of carriages leading a procession of representatives of various trades and occupations, crossed and recrossed the bridge.

Eads began his speech at the opening ceremonies by admitting his belief that "the love of praise," whether "a frailty or a virtue," was "common to all men," that he was not exempt from its fascination, and that it served as a "laudable stimulus to effort." Love of praise was, furthermore, "the grand motor which actuates the mind of man to attempt the accomplishment of worthy deeds," and the building of what that day was the greatest bridge in America was certainly a worthy deed. Eads, however, like many a chief engineer at the dedication of a great work, recognized that he had not built the bridge himself, and he spoke explicitly of the help he had received:

> Yon graceful forms of stone and steel, which prompt this wonderful display, stand forth, not as the result of one man's talents, but as the crystallized thought of many, aye, very many minds, and as the enduring evidence of the toil of very many hands; therefore I would forfeit my self-respect and be unworthy of these pleasing evidences of your good will, if on this or any other occasion, I should appropriate to myself more than an

> humble share of the great compliment you are paying to those who created the bridge.

Eads spoke of himself that day as the representative of a "community of earnest men, whose combined labor, brains and wealth, have built up this monument of usefulness for their fellow-men." He mentioned few facts and figures about the bridge, about which already so much had been written. Rather, he spoke of the design and construction process that assured him and his fellow engineers that the bridge was safe:

> Everything which prudence, judgment and the present state of science could suggest to me and my assistants, has been carefully observed, in its design and construction. Every computation involving its safety has been made by different individuals thoroughly competent to make them, and they have been carefully revised time and again, and verified and re-examined, until the possibility of error nowhere exists. . . . When the first arch was closed, Mr. J. S. Morgan, of London, whose firm has supplied so many millions for this work, and whose confidence in it has contributed so much to its success, wrote me, hoping that the closing of the arch had made me as happy as it had him. I replied that the only happiness I felt was in the relief that it afforded my friends, for I *knew* it would be all right. . . .

Eads also confessed that he had felt no great relief when the piers reached bedrock, or when the first heavy locomotives were driven over the finished bridge, for he "had felt no anxiety on the subject." He felt "justified in declaring that the bridge will exist just as long as it continues to be useful to the people who come after us, even if its years should number those of the pyramids." He explained with some technical detail how the sun and temperature caused various parts of the bridge to be relieved of strain at different times of the year, so that any piece of steel could be "easily taken out and examined, and replaced or renewed, without interrupting the traffic of the bridge."

Among those individuals Eads singled out in connection with making the dream of the bridge a reality, the financier Morgan was the one most prominently mentioned. Several bridge-company directors were also acknowledged by name, complete with titles, for their "unswerving confidence and kindness." The longest list of names, however, comprised those of the assistant engineers, their assistants, and others who provided "indispensable services." Eads's listing of these by surname only was no doubt a sign of his familiarity with them: "Flad, Roberts, Pfeiffer, Dwelle, Cooper, Devon, Gayler, Schultz, Wieser, Smith, McComus, Wuerpel, Klemm, and a host of others, earnest, faithful and accomplished." He also acknowledged James

Andrews, master mason; Walter Katté and his foreman, McMahon, "the skillful engineer" who swung the bridge's steel arches into place; William S. Nelson, responsible for the caissons beneath the piers; and Charles Shaler Smith, for the Illinois approach to the bridge. There was no mention of Andrew Carnegie. Eads closed by observing what has been heard at many a bridge dedication ceremony, that "a great work is rarely erected without the sacrifice of human life," and he remembered those who had died during the construction project, especially those who had died in the caisson work.

That evening, there was to be a spectacular display of fireworks set off from the bridge, and hotels and steamboats had long advertised their verandas and decks as the best viewing locations. The pyrotechnic artist W. W. Judy was pictured along with President Grant and engineer Eads on a memorial broadside, but the show was a great disappointment. One of the "grand Temple pieces" was to be "of the bridge itself surmounted by allegorical figures representing Missouri and Illinois, clasping hands." Another was to be of "Eads in the Temple of Honor, flanked by a locomotive and a steamboat, being crowned by Genius." The illuminated "Phantom Train," however, which was to traverse the entire bridge as a climax to the day's events, was nowhere to be seen. The organizing committee had apparently economized in its contract with Judy, who came to be called "Judy Iscariot." Many of the actual trains expected to cross the new bridge were also in fact phantoms, for it would be almost a year before the first regular passenger train contributed to the revenue. This, of course, was contrary to estimates made when the bridge was being financed, and soon the company went bankrupt.

The official name of the bridge, as it appeared on invitations, was the "Illinois and St. Louis Bridge," but Chicago newspapers delighted in calling it the "Chicago and St. Louis Bridge." The definitive history of the engineering enterprise, published in 1881, called it the "St. Louis Bridge," but then as now it was known to all as the "Eads Bridge," making it one of the few major structures in the world named after their engineers. Ironically, in 1924, the Eads Bridge was identified as among the works of the U.S. Army Corps of Engineers, "conducted principally, or in most important executive or advisory capacity," by West Point graduates. This was, of course, a gross misattribution, and it did not escape the notice of Arthur E. Morgan, the fractious first chairman of the Tennessee Valley Authority. Morgan was a self-educated engineer whose work in drainage and flood protection gave him a special perspective on the work of the Corps. From 1920 to 1936, he was president of Antioch College, where he fostered the school's work-study plan, and he wrote books on topics ranging from religion and science to a critical history of the Army Corps of Engineers.

6

The Corps of Engineers had little control over the Eads Bridge, of course, for it was planned and built before there was a law requiring prior approval by the secretary of war and the chief of engineers for bridges over navigable waters. Not that the secretary and the Corps did not try to influence the bridge. As the arches of Eads's bridge were closing over the river, the operators of Mississippi steamboats had complained to William Belknap, the easily bribed secretary of war who would be impeached in 1876, that the bridge would interfere with their tall stacks, which, "flamboyant in their gaudy paint and gilded fretwork, were the boatmen's pride, cherished trademarks of the western steamers." The Corps of Engineers was ordered by Belknap to look into the matter of the bridge as an obstacle to navigation, and its report recommended that one of two actions be taken: (1) a deep-water canal be constructed on the Illinois side to bypass the bridge, or (2) the bridge be dismantled. Furthermore, all future bridges across the Mississippi at St. Louis were to be of the truss type. Eads appealed to President Grant, whose birthplace near Point Pleasant, Ohio, was less than fifty miles up the Ohio River from Eads's Lawrenceburg, and Grant suggested that Belknap "drop the case." Though the Corps persisted, Eads prevailed.

Eads's nemesis in the Corps of Engineers was General Andrew Atkinson Humphreys, who had concurred in the recommendations of the Corps's board. Humphreys and his associate, General Henry Larcom Abbot, had in 1861 first published their *Report on the Physics and Hydraulics of the Mississippi River;* it had since become the "bible" of the Corps, and its authors the Corps's authorities on questions of river improvements. To challenge Humphreys and Abbot was to invite trouble, as Eads was to learn. A year before his bridge was completed, a convention of congressmen, governors, and interested citizens was convened in St. Louis to discuss the Mississippi River, whose mouth was constantly silting up and hindering shipping. The Corps of Engineers, which had been working for forty years to keep the channel open, were now proposing a canal from New Orleans to the sea. Eads joined the delegates on an excursion to New Orleans and the delta, and he proposed a system of jetties as a quicker, more economical, and more effective solution. It was within two months of Eads's proposal that his bridge at St. Louis had become the subject of retrospective scrutiny by the Corps of Engineers.

Whereas the action backed by Humphreys and Abbot was to build a canal from New Orleans to the Gulf of Mexico, Eads proposed extending parallel jetties from one pass of the river into the Gulf. He recommended first constructing long parallel screens of willow branches, held in place by piles under the water, in order to slow the current through the obstacles enough so that the sand it carried would be deposited. Properly arranged, this barrier would cause the water to flow more quickly in the unobstructed channel, and the increased velocity would itself in time scour out a deeper channel.

A Board of Army Engineers was instructed to report on the canal scheme, which it endorsed, and on the idea of jetties, which it condemned. Only General John Gross Barnard, president of the board, dissented. He considered his fellow board members to be engineers in name only, engineers "in a narrow executive capacity," but lacking in "the wide induction of experience [and] the wide observation of travel to see and judge other engineering works—[and in] the *indispensable* familiarity with *what engineering is* in its practical developments all over the world which alone can give any insight into an opinion on a great engineering question." Furthermore, and perhaps most culpably, they did not even "*know their own deficiency*." There were certainly extra-engineering reasons behind the opposition to Eads, however, no doubt fueled by his lengthy review, occupying twenty-five pages of small type in his collected addresses and papers, of the Humphreys and Abbot "bible." In this review, which first appeared at the height of the canal-jetty debate—in 1878, in *Van Nostrand's Engineering Magazine*—Eads stated that Humphreys and Abbot's book, then recently reissued, "contains certain grave errors," which Eads proceeded to expose, "touching the navigation of the river and the reclamation of its alluvial basin."

The official Corps of Engineers estimate was that a canal would cost $13 million whereas Eads's scheme would run to twice that amount, and his jetties would constantly have to be extended into the Gulf of Mexico to maintain channel depth. Eads countered with a proposal to secure a 350-foot-wide channel at a depth of twenty-eight feet in half the time it would take to construct a canal, and with payment from the government commencing only after he had achieved a certain measure of success; the total bill was to have been about $10 million. This presented a classic choice between a government project and private enterprise. Not surprisingly, the Corps continued to oppose Eads's proposal, and a protracted legislative battle ensued. Eads revised his proposal, promising a deeper channel through the preferred Southwest Pass of the river through the delta, at a lower cost of $8 million. The final legislation gave approval to Eads's latest financial offer, but authorized jetties only at the smaller South Pass.

In an address at a banquet in honor of the passage of the 1875 Jetty Act to Improve the Mouth of the Mississippi, Eads spoke eloquently of his conviction that his system would work:

> If the profession of an engineer were not based upon exact science, I might tremble for the result in view of the immensity of the interests which are dependent upon my success. But every atom that moves onward in the river, from the moment it leaves its home amid crystal springs or mountain snows, throughout the 1,500 leagues of its devious pathway, until it is finally lost in the vast waters of the Gulf, is controlled by laws as fixed and certain as those which direct the majestic march of the heavenly spheres. Every phenomenon and apparent eccentricity of the river, its scouring and depositing action, its caving banks, the formation of the bars at its mouth, the effect of the waves and tides of the sea upon its currents and deposits, are controlled by laws as immutable as the Creator, and the engineer needs only to be assured that he does not ignore the existence of any of these laws, to feel positively certain of the result he aims at.
>
> I therefore undertake the work with a faith based upon the ever constant ordinances of God himself; and so certain as He will spare my life and faculties for two years more, I will give to the Mississippi river, through His grace, and by the application of His laws, a deep, open, safe, and permanent outlet to the sea.

Eads was expressing the same confidence in the engineering method with regard to opening a deep channel at the river's mouth as he did in his bridge, of course, and his success was virtually guaranteed, because he knew the laws governing the fluidity of the river even better than he had come to know those governing the solidity of steel. Constructing the jetties took Eads's own capital, however, and he had to fight continued battles with the Corps of Engineers to get paid for work done according to specifications. The jetty system, called "the most difficult piece of engineering in river hydraulics," was ultimately a tremendous success. After four years of work, in 1879, the South Pass channel reached a depth of thirty feet, a depth greater than required even in busy New York Harbor, and it was said that "the savings on transportation of one year's cotton crop alone was equivalent to the cost of the entire jetty project."

With the establishment of the South Pass channel, Eads had thus added another monumental engineering achievement to his life's work. But, although he was not yet sixty years old, his health had suffered from the many extended periods of time he had worked underwater, first in his diving bell but especially in the compressed air of the caissons at St. Louis, which would remain the deepest such atmosphere at which workers toiled for almost a century of bridge building. Nevertheless, Eads was not the sort of person to retire, and one of the major engineering issues of the day continued to interest, if not obsess him.

A common thread to all of Eads's work related to the efficient transportation of goods, upon which he believed national prosperity depended. In a speech at the dedication of the Grand Hall of the Merchants' Exchange in St. Louis, in 1875, he articulated his fervor for the subject:

> The key-note of our national prosperity is sounded in the simple words, "Cheap Transportation." They should be stamped upon the stripes of our national banner and thrown to the breeze from every farm-house, mill, and factory throughout the commonwealth. Schoolboys should be taught that the superior facilities for cheap transportation secured to Phoenicia, Athens, Venice, Genoa, the Florentine Republic and Holland, the commerce of the world. Each retained it until its rival became a cheaper carrier; and it is a notable fact that art, refinement, literature, history and eloquence attained in each State their highest development during its commercial sway.

Great civil-engineering projects to facilitate transportation and communication—including roads, harbors, canals, and bridges—were essential to the prosperity to which Eads was referring, of course. He was echoing the growing spirit of the nineteenth century, which recognized the importance of the engineer in "directing the great sources of power in nature for the use and convenience of man," as Thomas Tredgold had defined civil engineering a half-century earlier for the purposes of obtaining a Royal Charter for the Institution of Civil Engineers. This formalization of the profession was the natural culmination of the realization that the work of engineers had in fact "changed the aspect and state of affairs in the whole world." But engineers like Eads did not think the job was yet completed, for there remained many obstacles to cheap transportation, especially between the East and West Coasts.

Increasing international commerce in the latter part of the nineteenth century created worldwide interest in a canal across Central America, to reduce the time and risk that ships took in transporting people and cargo between the Atlantic and Pacific Oceans. By 1855, a fifty-mile railroad across the Isthmus of Panama—the first transcontinental railroad—presented an alternative to the thousands of sea miles (and added perils) it took to get around the southernmost part of South America. Of course, unloading ship cargo onto railroad trains and reloading it onto ships at the other terminus was as costly as ferrying rail freight across rivers without bridges. By the late 1870s, a private French company had been formed to explore options, and the prospect of an Isthmian canal, promoted by "*Le Grand Français*," Count Ferdinand de Lesseps, who had been responsible for the Suez Canal, offered some promise amid great engineering controversy.

Needless to say, the problem of interoceanic communication and de Lesseps's scheme attracted the attention of American interests generally and James Eads in particular. In an 1880 address before the House Select Committee on Inter-Oceanic Canals, Eads offered his opinion:

> The question of the practicability of opening a tide-level waterway through the American isthmus is simply a question of money and of time. If sufficient money were supplied, and time enough were given, I have no doubt that, instead of the narrow and tortuous stream which Count de Lesseps proposes to locate at the bottom of an artificial canon [*sic*] to be cut through the Cordilleras at Panama, engineers could give to commerce a magnificent strait through whose broad and deep channel the tides of the Pacific would be felt on the shore of the Caribbean Sea, and through which the commerce of the next century might pass unvexed, from ocean to ocean.
>
> The science of engineering teaches those who practice it how the forces of nature may be utilized for the benefit of mankind, and it is the duty of the engineer when charged with the responsibility of solving an important engineering problem, by which his fellow men are to be benefited, to consider carefully how the desired results can be most *cheaply* and most *quickly* secured. Therefore, it is his duty to consider every method for the accomplishment of the end in view which science and nature have placed within his power, and to select from the fullness of their stores such methods as the precise teachings of mathematics and a knowledge of the laws which control the forces of nature assure him will certainly accomplish the desired result in the least time and for the least money.

The method Eads had come up with was not a canal but a *ship railway*, in which fully laden vessels would be loaded onto great flatcars and pulled by teams of locomotives on multiple parallel tracks across the Isthmus of Tehuantepec. Such a trans-Mexican route would save ten thousand miles over the sea route via Cape Horn, and more than a thousand miles over the Panama Railroad route. Eads believed the ship railway could be completed long before de Lesseps's canal, and at a lower cost, and the American engineer spent the remainder of his life promoting his novel scheme. Other American interests included a more conventional canal over a Nicaraguan route. The political and technical debate outlasted Eads: he died on March 8, 1887, in Nassau, in the Bahamas, where he was seeking support for his final, unrealized dream. No other engineer, no matter how young, could

James B. Eads shortly before his death, in a photograph by Emil Boehl

take it up with the vigor of Eads. Had he chosen to devote his more youthful energies to a trans-Mexican ship railway, we might not have the bridge that to this day memorializes him on his river. There would no doubt have been a bridge across the Mississippi at St. Louis before long, and certainly by the end of the century, but it would not have been Eads's bridge. However, because his is the bridge that was built, it constituted a legacy from Eads and his assistant engineers, not only to the people who used and benefited from it, of course, but also, in its technical achievement, to the entire bridge-building fraternity, which everywhere in the last decades of the nineteenth century possessed the dreams, ambitions, and unbridled energies of youth.

COOPER

James Eads was an anomaly among bridge builders, in that his involvement with the genre began and ended with a single example, albeit one of historic proportions. His first and most ardent love was the Mississippi River itself, and he appears to have backed into bridge building more because of his civic involvement with the mercantile movers and shakers of St. Louis than because of any long-harbored dream to build a bridge greater than any other. Eads was, however, a consummate engineer, and once he got involved with the problem of bridging the Mississippi, it was as driving a challenge as was raising wrecks from the river's bottom or channeling the river itself to the Gulf.

Most engineers involved with bridge building in the late part of the nineteenth century and the early part of the twentieth could not have such a fleeting romance with iron and steel. With the railroads continuing to expand and to increase the power of their locomotives and the size of their rolling stock throughout this period, there was a constant need for stronger, larger, and more innovative bridges, and for engineers to do everything from creating their designs to putting up their superstructures. In America, these engineers were by and large of a different generation from Eads. Theodore Cooper was to be among those of the newer generation who rose to the top of the profession and became involved with building what would have been the greatest bridge in the world.

Theodore Cooper, as a member of Rensselaer's Class of 1858

Theodore Cooper was born in 1839 in Cooper's Plain, in the western New York State county of Steuben, which would give its name to the finest products of the Corning Glass Works, now located nearby. Unlike Eads, who came from an itinerant family with a precarious financial future, Cooper was born into one of permanence and purpose. The son of John Cooper, Jr., a practicing physician, and Elizabeth M. Evans, young Theodore was one of nine children who grew up on land their parents settled on shortly after their marriage in Pennsylvania. The Coopers, among the earliest settlers of Steuben County, developed an estate there on land inherited from the elder Cooper's maternal grandfather. When young Theodore decided that he wanted to study engineering, he did not have to rely upon a benefactor's library for books. Rather, he traveled about 150 miles east-northeast to attend Rensselaer Institute and study toward the degree of civil engineer in that still-young institution. He graduated in 1858, in the class after Washington Roebling, who was two years older.

Washington Roebling was, of course, the son of a famous engineer, who not only served as the young man's mentor but also provided him with opportunities to gain invaluable work experience as an assistant on such projects as the Allegheny suspension bridge in Pittsburgh and, after service in the Civil War, on the bridge between Covington, Kentucky, and Cincinnati, over the Ohio River. Such privileged apprenticeships prepared Washington Roebling well to take over construction of the Brooklyn Bridge after his father died. Beyond his Rensselaer degree, Theodore Cooper had no such personal entrée into engineering. Rather, he started his career as an assistant engineer on the Troy & Greenfield Railroad and on the Hoosac Tunnel project in northwestern Massachusetts.

A tunnel through Hoosac Mountain, which would be an important link on the route between Albany and Boston, had been proposed as early as 1819. In 1825, the younger Loammi Baldwin had identified a location near North Adams, Massachusetts, at which a five-mile tunnel could be driven almost due east through the mountain at a cost, he estimated, of no more than a million dollars. At the time, this was too dear for the Massachusetts Legislature, but the Boston & Albany Railroad Company began the task in 1848. It was expected to be a five-year project, but that was overly optimistic. In 1856, when little progress had been made, Herman Haupt, a seasoned engineer who had been trained at the U.S. Military Academy and gained much experience on railroad bridge and tunnel projects, was prevailed upon to help with the Hoosac Tunnel, by raising funds for it as well as overseeing its completion, and he resigned from the Pennsylvania Railroad to do so. It was in 1858, the year Haupt attacked the mountain with renewed vigor and with an improved pneumatic drill, that Theodore Cooper came on board to begin his engineering career. After three more years of work on the tunnel, however, it was only 20 percent complete, and, amid charges of corruption and mismanagement, the commonwealth of Massachusetts took over the project. With the outbreak of the Civil War, Cooper, who had risen to the position of assistant engineer, left the tunnel project to join the U.S. Navy. (The tunnel was not to be finally blasted through until 1874. After it opened to rail traffic the next year, its route could be advertised as the "shortest line between the east and west.")

As an assistant engineer in the navy, Cooper was ordered to the gunboat *Chocura,* then still under construction in Boston. He spent almost four years attached to the *Chocura,* which saw action at the siege of Yorktown and the battle of West Point, served as a guardship on the Potomac, and took part in the blockade off Fort Fisher and along the Texas coast, among other campaigns. Cooper was ordered to the Naval Academy, then at Newport, Rhode Island, in June 1865, and then to Annapolis, Maryland, where the academy reopened that fall, as an instructor in the recently formed Department of Steam Engineering. He was in charge of all new construction at Annapolis for the three years he spent there, then undertook a two-year tour of duty in the South Pacific on the *Nyack*. Finally, after returning to

Annapolis for two more years, he resigned his position as first assistant engineer to work for James B. Eads, with whom he may have become acquainted during the latter's many trips to nearby Washington.

Cooper was appointed by Eads in mid-1872, first as inspector of the steel being made at the Midvale Steel Works, and later as inspector of construction at the Keystone Bridge Company, where the parts were finished and tested before being shipped to St. Louis. These were important responsibilities; if the steel was not made to specified standards and did not have the same strength and flexibility as assumed in the design calculations, then all the engineering predictions of the finished bridge's behavior were invalid. Such assignments in Pittsburgh were common beginnings for promising young engineers, but Cooper was already approaching his mid-thirties, and he must have been anxious for even more responsible work. By the end of the year, he was sent to the construction site in St. Louis to supervise the erection of the parts whose quality he had assured, and it was in this position that his reputation among bridge builders became more visible.

When Cooper arrived in St. Louis, the superstructure of the bridge was well along, the trussed ribs arching almost one hundred feet over the river. Walking about on the incomplete superstructure must have been nerve-racking and was certainly dangerous, but Cooper gained a reputation for personally inspecting the erection of the superstructure daily. Indeed, he was absent only on one "wet, snowy day, when everything was covered with ice," because he was "too stiff in the joints" from a fall he had taken a few days earlier. According to Cooper's own account, he "tripped on an unbalanced plank" and fell ninety feet into the river, but "escaped uninjured, excepting a stiffness resulting from the shock." He afterward elaborated on the incident to Calvin Woodward, the principal historian of the Eads Bridge, who related the story as follows:

> He was conscious (he said) of its taking him a long time to fall 90 feet. He thought of the probable force with which he would strike the water, and rolled himself in to the shape of a ball as much as possible. He struck the water he hardly knew how, and went very deep into the river,—nearly to the bottom, he thought. After what seemed another long interval, he reached the surface and struck out vigorously for shore. He then found that he still held in his hand the lead-pencil which he was using when he stepped on the treacherous plank. A boat from the East Abutment soon picked him up. In an incredibly short time he changed his clothes and walked into the office of the company as though nothing had happened.

Several weeks after his fall, Cooper examined a tube that a workman had reported broken and found another tube broken also. With Henry Flad sick in bed at the time, Cooper ordered emergency measures taken to keep the unfinished bridge from collapsing, and he telegraphed Eads in New York of the alarming development. Eads first wondered if all their calculations had been in error, and if somehow extremes of temperature were straining the metal beyond its limits. After reflection, however, he recognized that relaxing somewhat the cables supporting the unfinished arches would relieve the tubes in the arch ribs of part of the strain that was causing them to break, and he telegraphed instructions back to Cooper. The cables were adjusted accordingly, and the danger passed. This incident was prophetic of a similar one thirty-five years later, when Cooper's career was at its zenith. However, in that case, the absentee chief engineer would not be so fortunate in getting his telegram through to his assistant.

After the Eads Bridge was completed, Cooper moved about for a while, in a manner not unfamiliar to engineers today. He served as superintendent of the shops for the Delaware Bridge Company, in Phillipsburg, New Jersey; worked as assistant general manager of the Keystone Bridge Company, in Pittsburgh; designed and built (i.e., oversaw the building of) the Laredo Shops of the Mexican National Railroad; remodeled and rebuilt the furnace plant of the Lackawanna Coal and Iron Company, in Scranton, Pennsylvania; and designed and built the Norton Cement Mills at Binnewater, New York. He was approved for hiring as an assistant to Wilhelm Hildenbrand, who had made the first drawings of the Brooklyn Bridge under the direction of John Roebling and who served as principal assistant engineer of the construction of the bridge under Washington Roebling, but it is not clear how extensively Cooper worked on that project. Nevertheless, with his broad and varied experience, he was able to establish himself as a consulting engineer in New York City in 1879.

I

The year 1879 is among the most infamous in the history of modern bridge building, for it was on the last Sunday of that year, December 28, that the Tay Bridge disaster occurred, an event that immediately affected the character of bridge design and construction endeavors throughout the world, was to affect Cooper in the twilight of his career, and to this day influences the way bridges are built and look. As with all bridges, the history of the Tay begins long before its name became familiar, even to engineers, and the nature of the bridge itself, or its influence on subsequent bridges, cannot be completely understood without understanding the circumstances surrounding its origins, design, and construction.

In the 1850s, travel to the east coast of Scotland above Edinburgh was a grueling ordeal. Although the town of Dundee was only forty-six miles north of Edinburgh as the crow flies, even in the best of weather a railroad journey took over three hours, because there were two wide and unbridged estuaries to cross. Passengers who left Edinburgh on the 6:25 a.m. to Grafton had to change there to a paddlewheel steam ferry, which carried them across the Firth of Forth to Burntisland; here they boarded a train that took them through Fifeshire to Tayport, then disembarked and boarded another ferry to carry them across the Firth of Tay to Broughty Ferry, where they finally caught a train that took them to Dundee. This was largely the route of the North British Railway—a successful line in the south, but one that was fast losing business in Scotland to the Caledonian Railway. The Caledonian route to Dundee was a much more comfortable journey, albeit a longer one, taking passengers well westward out of their way, through towns like Stirling and Perth, where the two firths had narrowed to rivers and were more readily bridged.

The word "firth" means "estuary" and is related to the old Norse word "fjord." Unlike the Scandinavian arms to the sea, however, the Scottish firths are located not so much between steep cliffs as between gentle hills. This made ferry landings rather easily established, and would have made bridges and their approaches relatively uncomplicated, if the firths were not so wide near Edinburgh and Dundee. It may hardly have crossed the minds of railway directors and shareholders in the middle of the nineteenth century that bridges could be thrown across such great stretches of water, no matter how shallow these might be, and they accepted the limitations of "floating railways," in which railroad cars fully loaded with coal or other commodities would be uncoupled at the water's edge and rolled onto ferries, to be carried across the firth and then coupled to new locomotives for the continuation of the journey.

Thomas Bouch, who devised the scheme of the railway ferries, was born to a sea captain and his wife in Cumberland in 1822. At the age of seventeen, young Bouch became associated with the famous Stockton & Darlington Railway, designed and constructed by the engineer George Stephenson, whose portrait now appears on the English five-pound note. Railroads had been used for some time to move loads of coal and other heavy materials between mines, forges, shipping points, and the like, often in small hand-pushed or horse-drawn cars. When steam locomotives came to be used, it was clear that railroads could also serve to transport goods and people between towns. The Stockton & Darlington was among the first

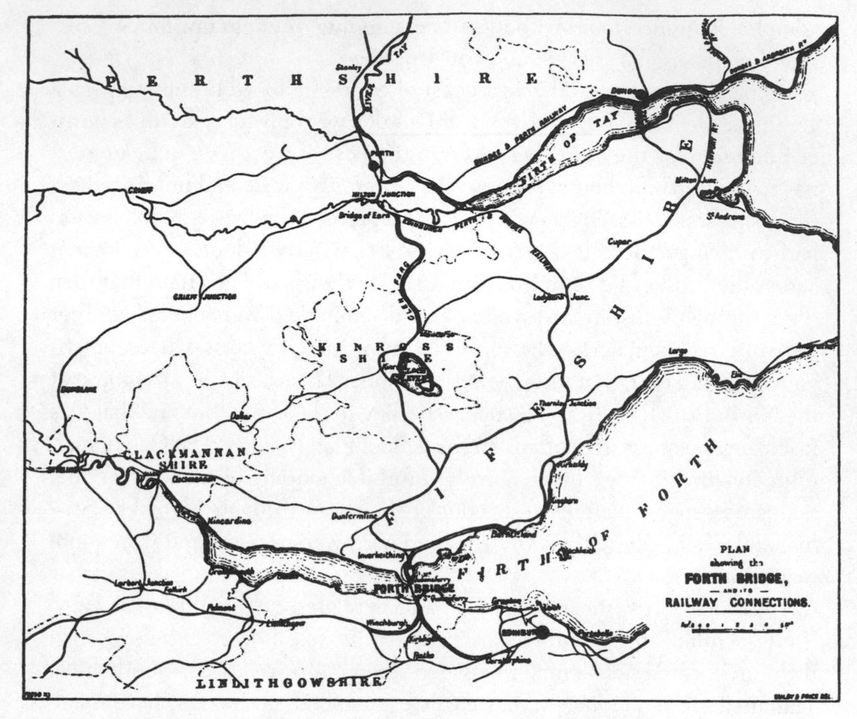

East coast of Scotland, showing railway connections around the Firth of Forth and the Firth of Tay, circa 1890

railways authorized by Parliament to be operated also for public transport, and September 27, 1825, when the first ticket was sold on the line, is often taken as the birthdate of the railways as we know them—true competitors of canals and carriage roads.

In 1849, when he was twenty-seven, Bouch went to Scotland to become traffic manager and engineer of the Edinburgh & Northern Railway. He went on to design railway viaducts in the highlands, a tramway system for Edinburgh, and the railway ferries across the firths of Forth and Tay. When the Edinburgh & Northern was taken over by the North British Railway in 1854, Bouch proposed to the directors a bridge across the Firth of Tay, but some considered a two-mile-long structure "the most insane idea that could ever be propounded." Opposition to the idea of a bridge at Dundee was not so much ridiculed as feared, however, in the town of Perth, about twenty miles up the Tay, where shipping interests worried about a bridge downstream blocking their access to the sea and their livelihood on the interior railroad. And one naysayer presciently articulated the fears of many along the wide, cold, and windy estuary: "The tremendous impetus of the icy blast must wrench off the girders as if they were a spider's web, or hurl the whole erection before it." In the final analysis, the long struggle between the Caledonian and North British railways for the Scottish traffic was the deciding factor.

It is said that the first time Bouch looked across the firths of Forth and Tay, he was convinced that they could be bridged, and subsequently spent his energies trying to convince others. The economic value in opening up a continuous rail link along the eastern Scottish coast was evident to all, and the final decision turned on the uncertain balance between projected cost and potential benefit. According to one account, at least, Bouch was driven by more than economics:

> [T]he simple reason for his enthusiasm was that he was a dreamer, and the most determined type of dreamer who must build what he dreams. Perhaps in the darkness at night he already believed it built, and could have put out his hand from the sheets and touched its cold iron and masonry. All creative work has its greatest reality while it is still in a man's mind, before he begins to execute it.

In July 1870, "after some twenty years of hawking his dream," Bouch learned that royal assent had been given to the bill authorizing what was effectively a North British undertaking to bridge the Tay, and one year later he watched the laying of a cornerstone. The bridge was to consist principally of latticed girder or truss spans not unlike those that engineers like Simeon Post had only a few years earlier proposed to cross the Mississippi River at St. Louis. At Dundee, however, the Tay was over a mile wide, and since the combined railroad and firth alignments necessitated an oblique crossing with a wide turn at the northern bank, the Dundee side, the full length of the bridge over water was about two miles. Because of its width, however, the waterway was generally no deeper than about fifty feet, and no more than twenty feet of sand and gravel was believed to cover bedrock.

Until bridges are actually financed and begun, there usually remains a degree of uncertainty about the real conditions upon which physical foundations must rest, for there is seldom enough money to explore every square foot of bottom for a bridge that is as yet only some engineer's dream. In the case of the Tay, excavation was to take place within large cylindrical caissons. Unlike those of the St. Louis Bridge, these caissons were not pressurized with air, and workers protected only by diving bells prepared the bottom for the construction of heavy brick piers. Once work was begun, however, it was soon evident that the riverbed conditions were not so substantial as test borings had indicated, and Bouch redesigned the piers to consist of groups of cast-iron columns on a wider base. This reduced the pressure on the sandy bottom to less than half of what was originally intended. The girders for the superstructure were fabricated near the shore, floated next to piers, and jacked into place. After about six years of work, in September 1877, the first train passed over the bridge.

The Tay Bridge consisted of eighty-five individual spans, the eleven greatest of which were 245 feet in length and known as the "high girders," so designed to allow trains to pass through rather than over, and thereby to offer the least impediment to shipping. While no one of its girders was anywhere near record length, the Tay Bridge was overall the longest bridge in the world by far when it opened officially on June 1, 1878. A year later, Queen Victoria crossed the bridge and knighted Bouch for his accomplishment. A half year after that, during a fierce storm, the high girders of the bridge fell or were blown into the Tay, carrying with them the evening train from Edinburgh to Dundee and all of its seventy-five passengers. There were no survivors.

Though Bouch maintained that the "capsizing" of some of the cars of the train off the track and into the sides of the high girders brought them down, a court of inquiry discovered major flaws in the design and construction of the Tay Bridge. It was found, for example, that Bouch had grossly underestimated the effect of the strong winds that could develop along the firth. Upon being questioned about this during the inquiry, he gave no sign of having reconsidered the issue:

The Tay Bridge after the collapse of its high girders, on December 28, 1879

Q: Sir Thomas, did you in designing this bridge, make any allowance at all for wind pressure?
A: Not specially.
Q: You made *no* allowance?
A: Not specially.
Q: Was there not a particular pressure had in view by you at the time you made the design?
A: I had the report of the Forth Bridge.

The Forth was, of course, the other firth that had to be crossed by a bridge, and early in his design considerations Bouch had learned indirectly from Astronomer Royal Sir George Airy how great a wind force might be expected to push against each square foot of bridge area that might be thrown across the Firth of Forth. Airy, writing from the Greenwich Observatory, acknowledged that "for very limited surfaces, and for very limited times, the pressure of the wind does amount to sometimes 40 lb. per square foot, or in Scotland to probably more." He suggested that, on average, the entire bridge would experience a pressure of ten pounds per square foot, which led Bouch to believe he could for all practical purposes ignore the effects of the wind on any firth, even though French and American engineers of the time were assuming wind forces five times Airy's average amount and accordingly designing bracing and connections to resist them. The Tay inquiry concluded that "the fall of the bridge was occasioned by the insufficiency of the cross bracing and its fastenings to sustain the force of the gale."

In addition to the inadequacy of the superstructure of the Tay Bridge to resist the wind, its piers were found to have been poorly constructed. After the accident, the iron was discovered to have been badly cast. It had tapered boltholes, resulting in loose connections; it had uneven thicknesses, causing unintended variations in strength; and it had large voids left by cast-in air pockets that were subsequently filled in at the Wormit Foundry with mixtures known there as Beaumont Egg, consisting of "beeswax, fiddler's rosin, and the finest iron borings melted up, and a little lamp black." The filler material was smoothed down and painted over—resulting, of course, in weak spots that were merely cosmetically sound.

As authorized by the Regulation of Railways Act of 1871, the Board of Trade had appointed a court of inquiry comprising three members: William Henry Barlow, president of the Institution of Civil Engineers; Colonel William Yolland, chief inspector of railways; and Henry Cadogan Rothery, wreck commissioner. Only the first two were engineers, and professional loyalties appear to have surfaced when the time came to draft the final report "upon the circumstances attending the fall of a portion of the Tay Bridge." Though there seemed to be no substantial disagreement as to contributing factors to the failure, the two engineers could only conclude that they had "no absolute knowledge of the mode in which the structure broke down," and hence were reluctant to place blame too squarely and explicitly on Thomas Bouch. Rothery, on the other hand, had no such uncertainty, and he chose to write a separate report. (The official report, which was issued in June 1880 and circulated "to both Houses of Parliament by Command of Her Majesty," consisted of both views.) Rothery felt it was his duty to call it as he saw it:

> We find that the bridge was badly designed, badly constructed and badly maintained and that its downfall was due to inherent defects in the structure which must sooner or later have brought it down. For these defects both in the design, the construction, and the maintenance, Sir Thomas Bouch is, in our opinion, mainly to blame. For the faults of design he is entirely responsible. For those of construction he is principally to blame in not having exercised that supervision over the work, which would have enabled him to detect and apply a remedy to them. And for the faults of maintenance he is also principally, if not entirely, to blame in having neglected to maintain such an inspection over the structure, as its character imperatively demanded.

Sir Thomas went into seclusion and died four months later, at fifty-eight. He would be remembered in history not for his legacy of three hundred miles of railway functioning properly in England and Scotland but for the failure of his Tay Bridge. Not only a successful bridge across the Tay, but also one across the Firth of Forth were the legacies Bouch must have dreamed about. Indeed, his design for a bridge across the deeper Firth of Forth had been under construction for over a year when the Tay fell. Bouch's scheme for a suspension bridge of two great spans of sixteen hundred feet each might have resulted in an accomplishment that would have overshadowed the Brooklyn Bridge, then under construction in New York. With the disaster at the Tay, however, work at the Firth of Forth was suspended, and there was a general loss of confidence in the project, especially when it came out that Bouch had designed his bridges, on average, for only ten pounds of wind pressure. Even then, Bouch's design was not formally abandoned by the Railway Board until a full year after the Tay disaster, and six months after the court of inquiry's report, in part because there was still a great desire for bridges across the Scottish firths.

The Tay Bridge had, after all, operated successfully for more than a year, and it had demonstrated the great economic value of bridges across the firths. It should not be surprising, then, that a new Tay Bridge, which would have two tracks rather than one and would accordingly be wider and more substantial, was soon proposed. Less than eighteen months after the first Tay Bridge collapsed, plans for a new one were submitted for parliamentary approval, which was required for all civil-engineering works. The new undertaking was overseen by the firm of Barlow, Son & Baker, and the engineer was to be William H. Barlow himself, with his son, Crawford Barlow, as his assistant. The second Tay Bridge was to be constructed sixty feet upstream from the first, with piers spaced the same distance apart so that the sound girders, which did not fall, could be easily reused. The high girders were redesigned to much more substantial standards than Bouch's, including the ability to resist a wind pressure of fifty-six pounds on every square foot, and all cylinders of the supporting piers were to be tested well beyond the load they were expected to carry. There was naturally great local interest in the design of the replacement structure, and the piers of the "New Viaduct" and the "Old Bridge" were contrasted in the October 18, 1881, issue of the Dundee *Advertiser* as follows:

> The massive character of the new structure as compared with the old is obvious at a glance, especially (1) the greater lateral stability from the substitution of twin piers for the single pier below, and the increased width for the double line of rails above; and (2) the greater vertical stability from the diminished height of the superstructure and the arched formation at the upper junction of the piers.

As in the wake of all major failures, the design of the new structure had improved features that far exceeded correcting the deficiencies of the old. With favorable public opinion thus assured, tenders were invited, and that of William Arrol & Company of Glasgow was accepted late in 1881. Construction began the next year and was completed in 1887. The stumps of the original bridge piers remain in place today, serving as tidal breakwaters for the much more substantial piers of the second Tay Bridge, and as a stark reminder of the accident and its victims.

2

When the time came, in 1881, to have a new bridge designed for the Firth of Forth, proposals were invited by the Railway Board from its consulting engineers, Sir John Fowler, William Barlow, and Thomas E. Harrison, who had been set up as a panel to reexamine Bouch's scheme. John Fowler, then

The high girders of the rebuilt Tay Bridge, as they stand today, with the stumps of the original bridge still visible in the water

approaching sixty-five years of age, was younger by almost five years than Barlow and by almost ten years than Harrison. Fowler had begun his training as a civil engineer at sixteen, given evidence before Parliament by the time he was twenty-one, and been in charge of a railway-construction project at twenty-two. In his mid-forties, he had been involved in seventy to eighty "major schemes" a year, and it has been estimated that he must have been involved in over a thousand jobs over the course of a professional career that spanned more than sixty years, working with at least fifty different assistants. Talented assistant engineers were clearly essential to someone like Fowler, and his assistant on the Firth of Forth project, Benjamin Baker, was among the best. Baker, thirty years Fowler's junior, began work in his London office on the Metropolitan Railway project, the first link in the London underground when it opened in 1863, but preferred designing long-span bridges.

A truss- or girder-bridge design was not appropriate for the Firth of Forth, because the many piers that would have had to be sunk in deeper water would have presented an engineering challenge and an unwanted expense. Besides, it was a girder bridge across the Tay that had failed, so adverse public opinion would have had to be overcome. Suspension bridges had long been suspect in Britain for rail traffic, but John Roebling's successful one over the Niagara Gorge had put the form in a new light. However, the problems with wind, and the fact that the now abandoned design of Bouch had been of the suspension type, again cast it in disfavor. Fowler and Baker were thus inclined, for both technical and nontechnical reasons, to look to different bridge forms entirely.

The location that Bouch had identified was ideal, in that the firth was relatively narrow there, albeit relatively deep, and in that about midway between the shores, at Queensferry and South Queensferry, was an island, or "garvie" in Scots. It was said to be named Inchgarvie because its representation on a scale map was an inch long; coincidentally, its shape also resembled that of a small herringlike fish called a garvie. According to the engineer Baker, who later lectured on the design of the bridge, the area "should be well enough known to every reader of fiction," for it was the setting of Robert Louis Stevenson's *Kidnapped*, whose hero was taken "at the very spot where the bridge crosses." However, what presented themselves here to Baker and Fowler were not fictional settings but physical conditions for a bridge with piers only on the island and on or near the shores, which thus required two free spans each on the order of seventeen hundred feet. No such bridge had ever been built anywhere in the world.

What was being constructed in recent years in Europe and America, however, was a somewhat new type of bridge that was being used to span increasingly greater distances with as few supports in the water as possible. One of the earliest of these bridges was completed in 1869 over the River Main at Hassfurt, Germany, by the Bavarian engineer Heinrich Gerber, who a few years earlier had been granted a patent for the design, which came to be known as a Gerber bridge. In his bridge, which looked not unlike the high dark bridge that today carries the Pulaski Skyway over New Jersey marshland, the girder depth varied along the length of the bridge, which was articulated at strategic locations in order to simplify design calculations and to allow for minor settlement of the piers without exerting undue stress on the superstructure.

Gerber's concept had considerable appeal to engineers, and many other "Gerber bridges" were built according to similar principles, in part because bridge engineering generally had evolved to the point where this bridge type was a natural solution for supporting the increasingly heavy loads of commerce. Thomas Bouch had, in fact, designed and built one of the first such bridges in England in 1871, at Newcastle. This was apparently forgotten when the Tay disaster occurred; otherwise Bouch's involvement might have given this genre a bad name as well, at least among the general public and politicians.

In America, the expanding railroads presented many opportunities to build new bridges, and by 1877 a Gerber-type bridge with three 375-foot spans was carrying the Cincinnati Southern Railway over the Kentucky River. The engineers of this bridge were Louis G. F. Bouscaren, who had been an assistant engineer on the Eads Bridge, and Charles Shaler Smith, who was responsible for the Illinois approach to that bridge. Shaler Smith was born in Pittsburgh in 1836 and was to die in St. Louis fifty years later. As a child, he attended private school, but he apparently had no formal engineering training. Nevertheless, his work with Gerber-type bridges was instrumental in introducing the form to America, and his writings on comparative analyses of different truss types and on wind pressure on bridges were important contributions to bridge design in America.

In 1883, as construction was beginning on the Firth of Forth bridge, a Gerber-type with a 495-foot span was completed, built for the Michigan Central and Canada Southern railways. It was almost 240 feet above the Niagara Gorge and just south of Roebling's suspension bridge, also known as the Grand Trunk Bridge. The chief engineer was Charles Conrad Schneider, born in 1843 in Saxony, where he was trained and practiced as a mechanical engineer before coming to America in 1867. He began work here, as many immigrant engineers then did, as a draftsman. His early work with the Rogers Locomotive Works in Paterson, New Jersey, led to his involvement with railroad companies, and before long he was in charge of engineers in the New York office of the Erie Railroad, whose chief engineer was Octave Chanute. Born in Paris in 1832, Chanute moved to America with his family when he was eight years old, attended private schools in New York, became an assistant chainman on the Hudson River Railroad while still a teenager, and worked his way up the railroad ladder. In the process, he had gained enough experience in matters of the maintenance of ways to become involved in extending existing railroads. From 1867 to 1869, Chanute designed and oversaw construction of the first bridge over the Missouri River, at Kansas City. Later in life, he became interested in the nascent field of aviation, which led ultimately to the association of his name with an air-force base in Illinois.

In Chanute's office, Schneider had the responsibility for checking the bridge plans submitted by bridge companies, a practice not generally carried out by railroads at that time. In particular, Schneider was responsible for checking the strain sheets, which showed the portion of the load that each member of a bridge was designed to carry. It was a natural development for bridges also to come to be designed in railroad offices. Once designs became specified by the railroad, rather than as part of a lump-sum contract that included everything from design to construction by a bridge-building company, the Erie Railroad began to buy its bridges by weight, for

John Roebling's Suspension Bridge and the cantilever bridge over the Niagara Gorge, with Whirlpool Rapids in the foreground, in an etching from one of the many late-nineteenth-century tour guides of the area

the price of materials was the chief determinant of cost. The practice of letting bridge contracts by the pound soon became widespread.

The invitation to the opening of the bridge below Niagara Falls was signed by Chanute's protégé, Schneider, as chief engineer of the project. The engraving on the invitation shows Schneider's bridge in the foreground and the famous suspension bridge in the background—symbolic and prophetic of the relative positions the two bridge types were to hold, in the minds of some at least, for the next three decades or so. Also indicative of the climate in which bridge building was taking place in the closing decades of the nineteenth century, the 1883 invitation associated no personal name with the great structure, which was called simply, directly, and technically the "Canti-lever Bridge below Niagara Falls."

The hyphenation of the word "canti-lever" attests to how newly coined it was, at least with reference to bridge building; it required explanation, especially when applied to Fowler and Baker's bridge, under construction across the Firth of Forth, less than fifty miles south of where the Tay Bridge had collapsed. Though Eads had in fact begun his explanation of the principle of his arch bridge with a discussion of a canted lever, that fifteen-year-old report to the Illinois and St. Louis Bridge Company had been generally forgotten. In fact, the cantilevered method of construction used in erecting the Eads Bridge, once the most striking visible feature of the project, was now seldom referred to.

With the growing publicity in both Britain and America surrounding the supposedly new type of bridge, the hyphen was quickly dropped—but not the curiosity about the form. A reader of *Engineering News*, who wrote to the editor in late 1887 from an engineer's camp near Danielsville, Georgia, asked, "Whence comes the term 'Cantilever' as applied to bridges; or in other words what is a Cantilever bridge?" The editor responded:

> This is a question quite frequently asked and we might as well answer once for all. The term, as applied to a bridge, is of comparatively recent origin, but the principle is as old as the Hindoos and the art of building itself. It has been applied to wooden bridges for centuries, and it is only its later scientific solution by modern builders of steel and iron bridges that has brought it forward again prominently. Its advantage over other forms of truss construction is, that by a proper method of anchoring or balancing and the arrangement of its tension and compression members, it can be erected over space without supporting false-work. The Niagara and the Forth bridges are the latest examples of its application to a site where the conditions made false works impossible, or very expensive.

The idea of a cantilever played a central role in a remarkable lecture that Baker delivered at the Royal Institution in 1887, in which he also gave some indication of the public scrutiny under which the new bridge was being built. Eight years after the Tay Bridge disaster at Dundee, he felt it necessary to preface his remarks with a declaration that the Forth and Tay bridges were quite different structures, albeit still confused in the public's mind. He related an exchange he fully expected to have with every second Britisher: " 'How are you getting on with the Tay bridge?' I suggest 'Forth Bridge,' and the correction is generally accepted as a mere refinement of accuracy on my part." Not surprisingly, Americans were no more sure about Scottish geography. A report in *Scientific American* in early 1888 on the progress of construction on the cantilever bridge at Poughkeepsie, New York, misidentified the setting of the Forth cantilever as "between England and Scotland."

It was not just the location of the bridge that Baker was at pains to explain; he had also to convey a sense of its great size, each span being almost four times as long as the longest tubes in Stephenson's famous Britannia Bridge. Baker appealed to common points of reference to enable his London audience to appreciate the size of the Forth's spans:

> To get an idea of their magnitude stand in Piccadilly and look towards Buckingham Palace, and then consider that we have to span the entire

Benjamin Baker, circa 1890

> distance across the Green Park, with a complicated steel structure weighing 15,000 tons, and to erect the same without the possibility of any intermediate pier or support. Consider also that our rail level will be as high above the sea as the top of the dome of the Albert Hall is above street level, and that the structure of our bridge will soar 200 ft. yet above that level, or as high as the top of St. Paul's. The bridge would be a startling object indeed in a London landscape.

The image of the bridge amid familiar projections into the London skyline, as well as transported landmarks such as the Great Pyramid at Giza, St. Peter's in Rome, the Cathedral at Chartres, and a host of others, was to be made real in a mural in the South Kensington Museum. But Baker wished to convey to his audience not only the monumental size of his bridge but also the principles by which it stood. He asserted that it had "excited so much general interest" in part because it was "of a previously little known type." He would "not say novel, for there is nothing new under the sun," but made no references to Gerber's bridge, the American examples, or the cantilever method of construction of the Eads Bridge.

Baker did acknowledge that a visitor being shown about the construction site had suggested a Chinese precedent, to which the engineer replied, "Certainly." He went on to elaborate:

The Forth Bridge drawn to scale before familiar structures and landmarks

> Indeed, I have evidence that even savages when bridging in primitive style a stream of more than ordinary width, have been driven to the adoption of the cantilever and central girder system, as we were driven to it at the Forth. They would find the two cantilevers in the projecting branches of a couple of trees on opposite sides of the river, and they would lash by grass ropes a central piece to the ends of their cantilevers and so form a bridge. This is no imagination, as I have actual sketches of such bridges taken by exploring parties of engineers on the Canadian Pacific and other railways, and in an old book in the British Museum, I found an engraving of a most interesting bridge in Thibet upwards of 100 ft. in span, built between two and three centuries ago and in every respect identical in principle with the Forth Bridge. When I published my first article on the proposed Forth Bridge some four years ago, I protested against its being stigmatised as a new and untried type of construction, and claimed that it probably had a longer and more respectable ancestry even than the arch.

Baker seemed willing to acknowledge the ancient roots of his bridge, but nowhere in his lecture did he emphasize the precedents of his near contemporaries, Gerber, Shaler Smith, or Schneider. Indeed, after establishing the time of his first publication on the topic, Baker made remarks that cause the modern reader to wonder whether he read nothing but old books or simply did not wish to acknowledge the competition. Charles Schneider's fifty-page article on a cantilever bridge at Niagara Falls, followed by

An Asian cantilever bridge with a central girder span, which Baker found identical in principle to the Forth Bridge

even more pages of discussion, had appeared in the *Transactions of the American Society of Civil Engineers* two years before Baker's lecture. However, at the time, British engineers generally seemed little interested in recognizing or acknowledging American precedents. In a passage remarkably reminiscent of Eads's feud with Roebling in the pages of *Engineering*, Baker continued:

> The best evidence of approval is imitation, and I am pleased to be able to tell you that since the first publication of the design for the Forth Bridge, practically every big bridge throughout the world has been built on the principle of that design, and many others are in progress.

Interestingly, Baker avoided the use of the term "cantilever" in this passage, perhaps to make his assertions literally correct. Earlier in his lecture, however, he had introduced the new term with some elaboration, and reservation as to whether his bridge was a true cantilever:

> One of the first questions asked by the generality of visitors at the Forth is—why do you call it a cantilever? I admit that it is not a satisfactory name and that it only expresses half the truth, but it is not easy to find a short and satisfactory name for the type. A cantilever is simply another name for a bracket, but a reference to the diagram will show that the 1700 ft. openings of the Forth are spanned by a compound structure consisting of two brackets or cantilevers and one central girder.

The anthropomorphic model that Baker used in his lectures on the Forth Bridge

Baker then described how, in preparing for his lecture, he "had to consider how best to make a general audience appreciate the true nature and direction of the stresses on the Forth Bridge, and after consultation with some engineers on the spot, a living model was arranged." The elusive Baker did not make clear whether he, the engineers, or all together came up with the idea, but the striking anthropomorphic model was very effective and often reproduced, both literally and visually, at the time:

> Two men sitting on chairs extended their arms and supported the same by grasping sticks butting against the chairs. This represented the two double cantilevers. The central girder was represented by a short stick slung from one arm of each man and the anchorages by ropes extending from the other arms to a couple of piles of brick. When stresses are brought on this system by a load on the central girder, the men's arms and the anchorage ropes come into tension and the sticks and chair legs into compression. In the Forth Bridge you have to imagine the chairs placed a third of a mile apart and the men's heads to be 300 ft. above the ground. Their arms are represented by huge steel lattice members, and the sticks or props by steel tubes 12 ft. in diameter and 1¼ in. thick.

Nor did Baker mention who represented the "load on the central girder" in the human model. In fact, it was Kaichi Watanabe, a young Japanese engineer, apparently among the first from his country to study in Britain, who was sitting, hands to his sides, grasping what looks like a very narrow swing-like seat representing the central span of the bridge. He was a student of Fowler and Baker and "was invited to participate in the human model of the cantilever to remind audiences of the debt the designers owed to the Far East where the cantilever principle was invented." In fact, New York's *Engineering News* actually reproduced the "ingenious illustration of the cantilever bridge principle" a full six weeks before it appeared in London's *Engineering* and called the model "a Japanese idea, as may be suspected from the central figure," Watanabe. The American journal also reported that the model was "received with loud and general applause" during Baker's lecture.

The report in *Engineering News* was no anomaly, for in America there was considerable interest in the design and construction of the Forth Bridge and large and unique engineering projects generally, even if their American

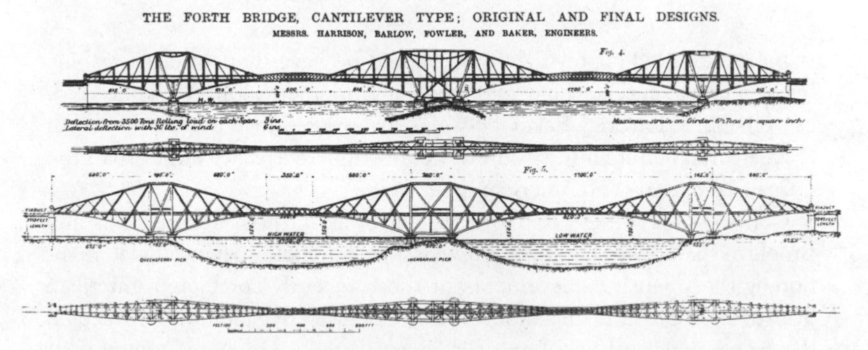

Two of many possible ways proposed to bridge the Firth of Forth, with the accepted design below

precedents were not acknowledged. Engineers and information could and did travel freely and relatively quickly by ship across the ocean, as Eads and Roebling knew, and the transatlantic cable had been operational for two decades. In the case of the human model of the Forth Bridge, *Engineering News* acknowledged its indebtedness to Thomas C. Clarke, a former director and future president of the American Society of Civil Engineers, for the use of the original photograph from which the illustration could be reproduced only weeks after Baker's speech. This makes Baker's silence about the existence of not insignificant contemporary cantilever bridges in America all the more inexcusable, and it reinforces the judgment that he at least, if not the entire British engineering profession, was unwilling to acknowledge that there was much to be learned from contemporaneous American experience. If anything, Baker offered a bit of gratuitous ridicule of the eighteen-hundred-foot cantilever bridge that the American Thomas Pope had proposed in 1810 to join New York and Brooklyn. He called his bridge a "flying lever pendant" and, reminiscent of a more famous Pope, described it in heroic couplets. Baker quoted from these in his lecture, no doubt because they gave so apt a description of his own project:

Each semi-arc is built from off the top,
Without the aid of scaffold, pier, or prop;
By skids and cranes each part is lowered down,
And on the timber's end grain rests so sound.
Sure all the bridges that were ever built,
Reposed their weight on centre, pier, or stilt;
Not so the bridge the author has to boast.
His plan is sure to save such needless cost. . . .

Baker admitted approvingly that, should he have thought of describing his own bridge in verse, he would have "appropriated bodily Mr. Pope's lyrical version." However, Baker chose "sober prose" to describe the Forth Bridge, which after all was soon to be a completed reality rather than the unrealized dream of an American poetaster.

Construction of the Forth Bridge began, as did that on the Eads and Brooklyn spans, with the sinking of caissons, and Baker again avoided mentioning the singular achievements of those recently completed American bridges except to note that almost one out of every five men who worked on the foundations of the St. Louis Bridge was attacked by some form of caisson disease, with sixteen deaths, whereas his bridge had "no deaths directly attributable to the air pressure." There were accidents, of course, especially among workers assembling the superstructure. Baker closed his speech by speaking of the risk that the "zealous and plucky workmen" performed high above the firth. Speaking on behalf of the engineers, he said, "we never ask a workman to do a thing which we are not prepared to do ourselves, but of course men will, on their own initiative, occasionally do rash things." He concluded, rather insensitively:

> Happily there is no lack of pluck among British workers; if one man falls another steps into his place. Difficulties and accidents necessarily occur, but like a disciplined regiment in action we close up the ranks, push on, and step by step we intend to carry on the work to a victorious conclusion.

The Forth Bridge under construction

Construction of the Forth Bridge was under the personal direction of the contractor William Arrol, whose company built the second Tay, but here he was in partnership with Joseph Phillips, who had considerable experience building large bridges, and with Sir Thomas Tancred and Travers H. Falkiner. Work was concluded in 1890, and early that year the definitive and comprehensive account of the project was written by Wilhelm Westhofen, the engineer who had supervised the building of the central section. Among the "chief desiderata," according to Westhofen, was the need for the "maximum attainable amount of rigidity," not only vertically, under railroad trains, but also horizontally, against wind pressure, so that the completed bridge "may by its freedom from vibration gain the confidence of the public, and enjoy the reputation of being not only the biggest and strongest, but also the stiffest bridge in the world." This was necessary, of course, because the memory of the decade-old Tay disaster was still fresh. It might not have been necessary, however, for Westhofen to remind his readers that Bouch's "original suspension-bridge design complied with none" of the desiderata, which also included assurances of fully tested materials, facility of erection (during which time the incomplete bridge was expected to be as safe against the wind as the completed structure was to be), and maximum economy consistent with safety.

The Eads Bridge is generally acknowledged as the first to use steel, and its success may have led to the Board of Trade's 1877 lifting of its ban on the use of steel in British bridges. The Forth was thus the first major bridge to be made fully of steel, which was manufactured using the Siemens-Martin open-hearth process. The bulk of the material was supplied by the Steel Company of Scotland, with some also coming from Wales. The Clyde Rivet

The completed Forth Bridge, showing Inchgarvie under the central pier and the South Queensferry landing

Company in Glasgow supplied the forty-two hundred tons of rivets required. Although steel was 50 percent stronger than wrought iron, Baker assured his audience at the Royal Institution that the material used in the bridge was "not in any sense of the word brittle, as steel is often popularly supposed to be, but it is tough and ductile as copper." He went on to make the point in more familiar terms: "You can fold ½ in. plates like newspapers, and tie rivet bars like twine into knots. The steel shavings planed off form such long, true and flexible spirals, that they are largely used for ladies' bracelets when fitted with clasps and electro plated."

For all the flexibility of the steel shavings, the bridge certainly was stiff—even during construction, when its great cantilever arms grew out symmetrically from each pier. It looked not unlike the Eads Bridge during its construction, but whereas the towers and cables that held up the Eads could be removed once the arches were complete, there would be no such extraneous falsework to be removed once the Forth Bridge was completed. Only the riveting cages and cranes would have to be taken off the finished bridge. Baker had predicted that each 1,710-foot span would deflect no more than four inches under the heaviest train, and measurements on the finished bridge showed an actual deflection of only three and a half inches.

The bridge was built "straddle legged" not only to achieve a great stiffness against the wind but also to look as if it did. The columns over the piers are as much as 120 feet apart at the base but only thirty-three feet apart at the top, giving the bridge the appearance of being able to withstand the severest blow. The bridge has been referred to as having a "Holbein straddle," after the stance that characterized male portraits by the German artist Hans Holbein. Fowler was apparently quite aware of this, reportedly having once remarked to the Scottish-born mechanical engineer James Nasmyth, at a London exhibition of Holbein's work, that the Tay might not have fallen had its piers had such a straddle.

The question of the aesthetic qualities of cantilever bridges in general and the Forth Bridge in particular was a much-discussed subject in the late 1880s. In a letter to *Engineering News*, F. J. Amweg, a civil engineer from Philadelphia, admitted that "the 'cantilever fever' is prevalent at the present time." However, he defended the design of his city's Market Street Bridge, in response to an article by the Pittsburgh engineer Gustav Lindenthal, who alluded to the bridge as just "another ugly looking cantilever bridge." Such an aversion to the form foreshadowed the career of Lindenthal, who was to play a dominant role in late-nineteenth- and early-twentieth-century American bridge building, but that is a topic for the next chapter.

In another of his many lectures on the Forth Bridge—this one in 1889, before the Library Institution of Edinburgh—Baker contended that the beauty or ugliness of the bridge had to be considered in the context of its size. He was reported to have said that "it was useless to criticise the design on paper, because the mental emotion arising from its enormous size was absent." His remarks were in response, at least in part, to growing criticism from the likes of William Morris, who had recently said that "there would never be an architecture in iron, every improvement in machinery being uglier and uglier until they reach that supremest specimen of ugliness—the Forth Bridge." Baker argued in his lecture that it was not possible to judge an object's beauty without knowing the functions it was designed to satisfy. Though the marble columns of the Parthenon were beautiful in their place, for example, they would cease to be so if bored through and used as funnels for an Atlantic steamship; he admitted, however, that, "of course, Mr. Morris may think otherwise." Baker stated that he and Fowler had indeed considered aesthetics in their design.

Other critics of the Forth Bridge faulted it for its structural excess and its consequent great cost. Not the least among these critics was Theodore Cooper, who shortly after the great Scottish bridge was finished would make comments upon it that years later would come back to haunt him:

> You all know about the Firth of Forth bridge, the clumsiest structure ever designed by man, the most awkward piece of engineering in my opinion that was ever constructed, from the American point of view. An American would

> have taken that bridge with the amount of money that was appropriated and would have turned back 50 [percent] to the owners, instead of collecting when the bridge was done, nearly 40 [percent] in excess of the estimate.

The Forth Bridge has continued to have its aesthetic and structural champions and detractors, which is not surprising considering the scale and unusualness of the structure. To give a sense of the bridge's scale, *Engineering* published in 1889 a diagram that compared the size of the Forth Bridge with that of the Eiffel Tower, then recently completed in Paris. Two Eiffel Towers were drawn on their side, balancing each other foot to foot, with two equally large half-spans of the Forth Bridge superimposed. Not incidentally, this comparison with the tower widely known to have been designed to resist principally the force of the wind, could do little but reinforce the impression of a sturdy and safe bridge. However, neither this exhibition of scale or stance nor the human model demonstrating the structural principles involved, represented the bridge in full. Whereas the Eiffel Tower diagram showed only one pier and two half-spans, and whereas the human model showed two piers and their respective four half-spans, plus a suspended span, the actual Forth Bridge has three piers, six cantilever arms, and two suspended spans. Were the three towering column structures not such a dominant visual feature of the bridge, it might have been criticized as having an even number of full spans, something many bridge engineers believe makes for an inferior visual composition. However, in the Forth Bridge, the light suspended spans are not so much the focus of attention as are the heavy, straddling towers and their symmetrically cantilevered arms, so the bridge appears to bound across the firth in an odd number of leaps that works both visually and structurally.

The completed bridge was tested in January 1890, during a severe gale, with two trains, each comprising three seventy-three-ton locomotives pulling fifty cars full of coal, with another three locomotives bringing up the rear. The total weight on the bridge was eighteen hundred tons, under which the ends of the cantilevers deflected only about seven inches. Though the trains ran side by side on the two-track bridge, they were not allowed to cross it completely—that first full crossing would not occur until January 24, when it was achieved by a special passenger train carrying the chairmen of the railroad companies. The formal opening of the Forth Bridge occurred on March 4, when the Prince of Wales, accompanied by his son, Prince George, and the dukes of Edinburgh and Fife, rode across in a train and declared the bridge opened. In a ceremony on the occasion, the prince announced that knighthood had been conferred by Queen Victoria upon Benjamin Baker and William Arrol—Sir John

Fowler had already been so honored—for their work on the Tay and Forth bridges, which signaled the progress that was being made in bridge building worldwide.

The cover of the souvenir program showed a North British Railway locomotive labeled "Progress" pulling a through passenger carriage labeled "Aberdeen to New York, via Tay Bridge, Forth Bridge, Channel Tunnel, and Alaska." Dreams are always in advance of reality, however, and a Channel tunnel, the boring of which had in fact begun years before, would not be completed for more than a hundred years. Prior to the 1990 centennial of the Forth Bridge, it was given a thorough inspection and declared to be in fine shape, a result of its having been conscientiously maintained throughout its first hundred years; and it was predicted that "given reasonable care and maintenance [the bridge] will last for another 100 years." Thus the Aberdeen–to–New York through train might still someday be achieved, for a trans-Siberian railway was officially completed in 1991, and a bridge connecting Siberia and Alaska across the Bering Strait was for some time the dream of such engineers as Joseph Strauss and Tung-Yen Lin. Though none of these would realize their dreams, that is not to say that their successors will not take the project up; but it will be one for another century.

3

While the Forth Bridge was being built in Scotland, American bridge construction was also continuing to advance. Octave Chanute had overseen the building of the two-thousand-foot-long, three-hundred-foot high Kinzua Viaduct, carrying the Erie Railroad's new division into the mining region of northwestern Pennsylvania. Originally constructed out of wrought iron in 1882 by the Phoenix Bridge Company of Phoenixville, Pennsylvania, the structure would be replaced with a stronger steel design in 1900. (It now stands unused in Kinzua Bridge State Park.)

The Brooklyn Bridge was completed in 1883, under the supervision of Washington Roebling and his wife, Emily Warren. With a clear span of 1,595 feet between stone towers with Gothic arches, it was famous for being the longest suspension bridge in the world, but it was soon so congested with traffic that talk of other bridges across the East River began almost immediately upon its opening. Furthermore, the bridge, designed for the lighter trolley and road traffic of almost twenty years earlier, was not structurally able to carry heavy locomotives, and for years elaborate schemes of switching cable cars at the New York and Brooklyn terminals would be an almost constant topic of discussion in the pages of *Engineering News*.

The Kinzua Viaduct, as it looked in the late nineteenth century, in northwestern Pennsylvania

Expanding cities and railroads needed more and more bridges, and the steadily increasing volume and weight of traffic they were being required to carry made bridge design an ever-challenging endeavor, whose practitioners welcomed new and better ways of determining the loads that various kinds of vehicles and trains imposed on a structure. There is seldom a single method that is all things to all engineers, and different individuals at a given time often have different views of how best to design a bridge. In the late nineteenth century, a technical paper published in the *Transactions of the American Society of Civil Engineers*, which was a principal forum for such discussions, was often followed by at least as many pages of discussion by a dozen or more of the author's prominent contemporaries (as had Schneider's 1885 paper on the Niagara Falls cantilever). The pages of *Engineering News* and its successor, *Engineering News-Record*, also often contained exchanges between engineers, like those that had appeared in *Engineering* between Eads and Roebling over the design of caissons, but these tended to be more one-on-one serial debates and could be less dignified professionally.

The bulk of what the likes of James Eads and the Roeblings wrote about bridge design appeared in the form of reports to those from whom they wanted initial or continuing financing. They published relatively little about bridge building in professional journals; therefore, with a few notable exceptions, their theoretical outlook tended not to be so influential as their practical achievements in construction. Those engineers who wrote more constantly and openly about bridge design and construction tended to have a much more immediate influence on the nature of design as practiced within the profession. Having technical papers noticed and discussed in the *Transactions of the American Society of Civil Engineers* was a guarantee for gaining recognition among engineers, but an alternative way to disseminate one's ideas and methods was book publication.

Theodore Cooper was a frequent contributor of papers to meetings and journals of the American Society of Civil Engineers, and he twice received that society's prestigeous Norman Medal for his contributions. His 1880 medal-winning paper, "The Use of Steel for Railway Bridges," showed him to be innovative, for no major bridge had yet been built entirely of the relatively new material. In the mid-1880s, he first published his book, *General Specifications for Iron Railroad Bridges and Viaducts*, which has been described as comprising "the first authoritative specifications on bridge construction that had been published and circulated." The title soon expanded to include steel bridges. Cooper's book was issued in its seventh edition in 1906.

Among the things that made Cooper's name most well known among engineers was his system of accounting in the design process for the loads of railroad trains on a bridge structure. He represented the heaviest locomotive then known by means of the forces it exerted through its driving wheels, and represented the train it pulled as a single, uniformly distributed load related to those forces. This made it convenient to modify earlier bridge designs as locomotives and the cars they pulled became heavier, which they seemed constantly to do. Cooper's system was widely adopted, and by the early twentieth century had become "the almost universal standard for railway bridge design in America." As convenient as his method was for dealing with trains of increasing weight, Cooper also strongly advocated the more accurate method of representing the load of a train on a bridge by the individual loads transmitted through each wheel. He published tables that made it possible for design engineers to carry out such analyses rapidly and conveniently. All such refinements in calculation meant, of course, that railroad bridges could be designed more accurately, and therefore more economically. No undue iron or steel needed to be added because of uncertainties as to how the bridge might be loaded by a heavy train moving across it.

Cooper's 1889 paper on American railroad bridges constituted a concise history, beginning with seventeenth-century wooden bridges and concluding with a section on the failure of bridges, then a matter of increasing concern to the railroads and their passengers. What the disaster of Tay Bridge was to Britain, the collapse of the Ashtabula Bridge almost three years to the day earlier, on December 29, 1876, had been, even more immediately, to American bridge building. The bridge had been erected in the mid-1860s across a deep gorge at Ashtabula, Ohio, about fifty miles east of Cleveland on Lake Erie, on the Lake Shore & Michigan Southern Railroad. It was basically what is known as a Howe-type truss design, but with additional diagonals, and with cast iron replacing wooden members. The design had been modified by Amasa Stone. Born in Charlton, Massachusetts, in 1818, Stone began his career as a carpenter and, with his brother-in-law, William Howe, the inventor of the truss type, acquired a contract to build the first railroad bridge over the Connecticut River, in 1840, and with a partner, acquired patent rights to the Howe truss in 1842. He went on to a remarkable railroad career, which included being president of the Cleveland, Painesville & Ashtabula Railroad, which he had built, and which had merged with the Lake Shore & Michigan before the bridge collapsed. The 165-foot span was being crossed by a train of two locomotives pulling eleven railroad cars westbound at fifteen miles per hour, when the driver of the first locomotive felt the bridge sinking. He opened his throttle wide and got safely across, but the other locomotive and all the cars went down with the bridge. At the time, it was still snowing after a recent blizzard; eighty people died in the icy wreck.

According to some accounts, Stone had been warned by engineers that the structure would be unsafe, and it was generally assumed that the bridge was cheaply built. However, after the accident it was found that, at $75,000, the bridge had actually been relatively costly to build. The engineer responsible for inspecting the structure became so disturbed by the blame heaped upon him that he committed suicide. Although Stone and his design were eventually condemned by the American Society of Civil Engineers, the exact cause of the failure is still not known with absolute certainty. One possible explanation is that a car or cars became derailed, perhaps when the driver lurched his locomotive forward because he thought the bridge was falling, and the truss was so damaged by the impact of the wheels that it could no longer support the unusual load. Whatever the cause, increased attention was devoted to bridge building in America after 1876.

Cooper wrote, more than ten years after the Ashtabula accident, that it "not only alarmed the general public, but also shook the blind confidence of the railroad companies in their existing bridges," and he outlined how the situation had changed in the meantime. Subsequent inspections of bridges and procedures uncovered weaknesses, not only in existing bridges but also in existing ways of designing and building them. Cooper was especially critical of designs that did not balance careful analysis with judgment about constructability. He documented how the loads that railroad trains exerted on longer bridges had about doubled from 1873 to 1889, and he called for more precise methods of analysis. In particular, Cooper advocated the adoption of his method based on locomotive wheel loadings, which he acknowledged had been "worked out independently but simultaneously by Mr. Robert Escobar, C.E., of the Union Bridge Company," in 1880.

Cooper called attention to an 1878 publication of his own as being "the first paper on bridge construction in which that relic of ignorance, the 'factor of safety,' was entirely omitted." He recognized that bridges should be built with a considerable measure of strength beyond that calculated, but he believed that safety should be based upon rational calculation coupled with careful material testing and inspection, rather than upon an arbitrary numerical multiplier. He opposed "stringent rules," what today would be called codes of practice, that placed mandatory numerical requirements on such things as factors of safety. Good bridge building, he maintained, could not be based "merely upon a theory of stresses." Rather, it "must provide for all those requirements of stability and solidity which are instinctively recognized by the practical engineer, and which cannot be complied with by merely using a large factor of safety." Cooper could not have known at the time how ironic his own words would become.

Toward the end of his authoritative paper, Cooper summarized the benefits that had accrued from "the American system of competitive bridge construction," which, according to him, involved aspects of evolution of truss designs, economy of details, and advances in theory absent from British practice. But Cooper also acknowledged some limitations of the competitive system, including "the American idea of building cheap railroads far in advance of the immediate demands of the regions through which they run, to settle those districts and build up a future paying traffic," which "compelled the use of cheap and light bridge structures," thus "favoring the lowest bidder." He insisted, however, that bridges used for the traffic originally designed were safe, and that it was as irresponsible for engineers to build excessively strong bridges as excessively weak ones. He approvingly quoted a Professor Unwin:

> If an engineer builds a structure which breaks, that is a mischief, but one of a limited and isolated kind, and the accident itself forces him to avoid a repetition of the blunder. But an engineer who from deficiency of scientific knowledge builds structures which don't break down, but which stand, and in which the material is clumsily wasted, commits blunders of a most insidious kind.

In other words, Unwin, like Cooper, believed that bridges should be made strong enough to perform their function but not so strong that they are heavier and more expensive than they have to be. This view has always been and always will be shared by the best of engineers, for they recognize that in the final analysis engineering is part of a much larger social enterprise, and money spent unnecessarily for civil-engineering structures becomes unavailable for other civic endeavors, or for initiatives of a humanitarian kind. The line between too little and too much safety is not always very clear and distinct, however, and that is what makes the best engineering also the most difficult. It is also what can lead to the worst engineering.

Cooper's reputation was so solid in 1894 that he was appointed by President Grover Cleveland to a commission of five expert engineers that was to recommend the length of span which would be safe and practical for a bridge across the Hudson River at New York City. Two competing bridge companies had proposed two competing bridge types, one a cantilever and one a suspension bridge. The cantilever would have had a span greater by three hundred feet than the recently completed Forth Bridge, and the suspension bridge, with a thirty-one-hundred-foot main span, would have been almost twice as long as the Brooklyn Bridge. The commission's rejection of a cantilever in favor of a suspension bridge for this dramatic site touched off a debate that was to last for years, and would include considerations relating to the interference with shipping, the length of the maximum possible suspension bridge, and the alternative of tunnels under the Hudson. That saga will take the entire next chapter to relate.

Engineering News, in part because it was based in New York, took considerable interest in the issue of bridging the Hudson, and the extended discussion of cantilever bridges in its pages prompted one reader, perhaps unfamiliar with the discussion of the word that had taken place years earlier, to inquire as to why that spelling was used, rather than "cantaliver," as recommended by the *Century Dictionary*. *Engineering News* admitted that, whereas it followed that "eminent authority" in such Americanized spellings as "center" and "gage," it preferred the "cantilever" spelling "as being more euphonious and as least upsetting long-established and well-nigh universal usage." The editors went on to reopen the question of etymology, admitting that "the origin of the word is uncertain," an opinion to be echoed by the ponderous and authoritative *Oxford English Dictionary*, in which "no satisfactory suggestion could be offered" for the word that had become ever more familiar to many engineers and laypersons alike. According to *Century*, the editors reported, the "probable original . . . was 'quanta libra,' of what weight or balance," and this led to the dictionary's spelling preference. Nevertheless, the *Universal Dictionary*, copyrighted in 1897, was to relegate the spelling "cantaliver" to "unusual" status. Under the definition of the word relating to bridge building, this dictionary credited, perhaps because of the popularity of Benjamin Baker's lecture and its human model, the Japanese with "the earliest known application of the principle," and described quite precisely the already "celebrated Forth Bridge" as comprising a "double cantilever (of 1,360 ft. length) . . . connected by girders 350 ft. long," thus adding up to the remarkable 1,710-foot spans that impressed Cooper and so many of his contemporaries.

The Forth Bridge had clearly become *the* cantilever of the world, to the virtual exclusion of its progenitors. Under the entry "cantilever bridge," for example, the *Universal* again was quite technically correct in describing such a structure as one "constructed on the cantilever system, the two sides being pushed out towards the centre and supported by a greater weight on land, until they meet and are joined at the centre." Though the dictionary acknowledged that "numerous important bridges" were built on the principle, none were named in the entry. The temporary superstructure of the Eads Bridge in particular was no more mentioned in this American dictionary than it was in Baker's lecture. Yet it was Eads who, almost three decades earlier, in his argument for the advantages of the arch over the truss, had so convincingly employed the bent, angled, or canted lever to make his point for his design for a bridge across the Mississippi. Indeed, according to Cooper, who could take personal pride in its construction, the St. Louis bridge was "the first practical solution of the cantilever principle on a large scale." He went on to observe that "the erection of two balanced cantilevers, each over 250 feet, with ease, safety and economy, made clear to the mind of engineers that the cantilever was the economic method of erecting long spans over deep gorges or rivers, where ordinary methods of scaffolding would be too expensive, or subject to great risks, or where navigation forbids the obstruction of the waterway." In fact, to the casual observer, photographs of the Eads Bridge under construction can be easily mistaken for those of the Forth Bridge. I once assembled a series of slides for a lecture by squinting at them before a light bulb and was later embarrassed to find that the Eads Bridge under construction had been projected on the screen while I was describing the Forth.

An early proposal for a cantilever bridge across the St. Lawrence River at Quebec

Cooper would not have made such a mistake. He did, however, near the end of his career, supervise the construction of a great cantilever from such a distance as to lose sight of some of the details that earlier he had written about so authoritatively. The history of his bringing to reality the idea for a bridge across the St. Lawrence River, near the city of Quebec, is, like the history of virtually all great bridges, long and arduous. A span was proposed as early as 1852 by Edward Wellman Serrell, who had by then designed and built the Lewiston & Queenston Suspension Bridge over the Niagara River between New York and Canada. That bridge was advertised as the "Largest in the WORLD!!!" because its stone towers atop the bluffs were 1,040 feet apart, but its deck was only 849 feet long because of the manner in which it joined the shores below the bluffs. Guy wires, mimicking those on Roebling's nearby Niagara Gorge Suspension Bridge, were added after a storm in 1855, but they were detached in early 1864, when they were threatened by an ice jam that had formed in the river. The guys were still unfastened when a gale struck on the first day of February of that year, and the bridge was destroyed. Serrell's proposal for a railway and highway bridge at Quebec was not acted upon at the time, but the site he identified was to become the location for another ill-fated bridge more than forty years later.

4

The Quebec Bridge Company was incorporated in Canada in 1887, with the authority to issue bonds and the right to build a railway bridge that might also serve for pedestrians and vehicles to cross the St. Lawrence River. Construction was to begin within three years, but three extensions were granted by Parliament, with the last due to expire in 1905. Though other legislation changed the name of the organization to the Quebec Bridge & Railway Company, it continued to be known by its shorter name. Legislation also declared the bridge to be for the general advantage of Canada, and so subsidies were granted to allow the work to begin in earnest.

In 1897, E. A. Hoare, chief engineer of the bridge company, wrote to the president of the Phoenix Bridge Company in Pennsylvania and asked that any of its engineers planning to attend the annual meeting of the American Society of Civil Engineers in Quebec that June stop in to see him regarding a bridge project. Among those attending the meeting was John Sterling Deans, chief engineer of Phoenix, and he and many other visiting engineers were taken to the bridge site. Also in the group was Theodore Cooper, and within about a week Deans had written to Hoare that Cooper would be happy to lend his experience to the project, which called for a bridge with piers sixteen hundred feet apart. When invitations to bid were issued in 1898, specifications for a cantilever bridge were included; bidders proposing any other kind of bridge would have to provide their own specifications. Among the tenders received were four cantilever and three suspension-bridge designs, including a cantilever from the Keystone Bridge Company and both a cantilever and a suspended span from the Phoenix Bridge Company. Since any type of bridge at Quebec would necessarily be on a massive scale, the advice of an experienced consulting engineer was sought, and Theodore Cooper agreed to review all the plans and tenders. In spite of his prior relations with Phoenix, Cooper's integrity as an engineer was believed to be above favoring their design for any but sound technical and economic reasons.

Cooper preferred the cantilever designs, because he believed them to be realizable at a lower cost than suspension-bridge alternatives. In mid-1899, he reported to the Quebec Bridge Company his conclusion that the Phoenix cantilever design was indeed the "best and cheapest" overall. (In fact, Keystone's design was considerably less expensive per ton of steel, but it was more costly overall, since the bridge itself would have been much heavier.) Along with his recommendation, Cooper called for further exploration of the riverbed, in order to establish the final position of the bridge's foundations and thus set a final determination of its length. Deeper foundations, for example, could require more time and money to construct than would a longer span. In early 1900, Cooper had the additional information he had requested; after studying the situation for three months, he concluded that the original pier locations, sixteen hundred feet apart, would take a year longer and be accompanied by more "real and imaginary contingencies" than shallower piers farther apart, and he recommended increasing the main bridge span to eighteen hundred feet—thus proposing a cantilever bridge with a span longer than any in the world.

In the meantime, negotiations were going back and forth between the Quebec and Phoenix companies, with the latter concerned about the financial status of the former. Detailed design work did not progress very quickly under such circumstances, for Phoenix was not assured of being paid for its services. It was not until 1903, when plans for a National Transcontinental Railway project were revealed, that a bridge at Quebec became such a necessity that government backing was assured. By then the government had also become more interested because of planning for the Quebec Tercentenary in 1908, and it was intimated that the bridge should be ready for the celebration. Thus the pace of design work was suddenly accelerated in 1903—with consequences that were only to be realized years later.

Cooper's changes relevant to an eighteen-hundred-foot span were sent to Phoenix, where Peter L. Szlapka, the company's design engineer, raised some questions about the degree to which some of the steel was stressed. In the final analysis, however, the exceptional magnitude of the structure was invoked as justification for the exceptional loading of its parts. The new specifications were submitted to Collingwood Schreiber, chief engineer of the Department of Railways and Canals, for the required government approval, and Schreiber proposed to his superior that the department "employ a competent bridge engineer to examine from time to time the detailed drawings of each part of the bridge as prepared, and to approve of or correct them," and submit them to Schreiber for final approval. When Cooper learned of this, he wrote to Hoare at Phoenix: "This puts me in the position of a subordinate, which I cannot accept. . . . I have written to Mr. Schreiber that I do not see how such an engineer could facilitate the progress of the work or allow me to take any responsible steps independently of his consent." In other words, Cooper wanted to have the last word on, and the full credit for, the design of the longest bridge in the world. A couple of weeks later, Cooper went to Ottawa to meet personally with Schreiber, after which the minister of railways and canals was advised that, "provided the efficiency of the structure be fully maintained up to that defined in the original specifications attached to the company's contract, the new loadings proposed by their consulting engineer be accepted." Though all plans were to continue to be submitted to Schreiber, Cooper was, for all practical purposes, to be the final authority.

Theodore Cooper, like Benjamin Baker, never married; their bridges were their children. As the Forth Bridge was Baker's magnum opus, so the Quebec was to be Cooper's. Cooper had few equals in America, but at the time of his involvement in the Quebec Bridge he was an elderly man in poor health, which pretty much kept him confined to New York. Though he had visited the Quebec site on several occasions while the piers were being constructed, he never once went to Quebec during the erection of the steel superstructure. Cooper may have been "*de facto*, chief engineer," and thus ultimately responsible for checking every aspect of the bridge design, but he had no staff in New York to assist him. The design of the bridge, as far as selecting the sizes of members and checking that they were not overloaded, was done to Cooper's specifications and modifications at the Phoenix Bridge Company by Szlapka, a German-trained engineer who over the course of twenty-odd years with Phoenix had worked on many major projects. Szlapka was, however, a desk engineer, without experience in the erection phase of bridge building, and so was not necessarily in a position to judge the structure itself on his own visits to the construction site. Yet Cooper, who was known for his hypercritical disposition, had full confidence in Szlapka's work, accepting it on faith when Cooper could not study it thoroughly himself; he had little concern in 1907 that the bridge was progressing in any but a normal fashion.

The south arm of the Quebec Bridge had been cantilevered out about six hundred feet over the St. Lawrence River by early August 1907, when it was discovered that the ends of pieces of steel which had been joined together were bent. Cooper was notified, by letter, by Norman R. McLure, a 1904 Princeton graduate who was "a technical man" in charge of inspecting the bridge work as it proceeded, who suggested some corrective measures. Cooper sent back a telegram rejecting the proposed procedure and asking how the bends had occurred. Over the next three weeks, in a series of letters back and forth among Cooper, chief engineer Deans, and McLure, Cooper repeatedly sought to understand how the steel had gotten bent, and rejected explanation after explanation put forth by his colleagues. Cooper alone seems to have been seriously concerned about the matter until the morning of August 27, when McLure reported that he had become aware of additional bending of other chords in the trusswork and, since "it looked like a serious matter," had the bends measured; he explained that erection of additional steel had been suspended until Cooper and the bridge company could evaluate the situation.

Yet, even as McLure went to New York to discuss the matter with Cooper, Hoare, as chief engineer of the Quebec Bridge Company, had authorized resumption of work on the great cantilever. As soon as McLure and Cooper had discussed the bent chords, Cooper wired Phoenixville: "Add no more load to bridge till after due consideration of facts." McLure had reported that work had already been suspended, and so contacting Quebec more directly was not believed to be urgent, but when McLure went on to Phoenixville, he found that the construction had in fact been resumed. Some conflicting reports followed, thanks in part to a telegraph strike then in progress, as to whether Cooper's telegram was delivered and read in time for Phoenixville to alert Quebec.

In any event, the crucial telegram lay either undelivered or unread as the whistle blew to end the day's work at 5:30 p.m. on August 29, 1907. According to one report, ninety-two men were on the cantilever arm at that time, and when "a grinding sound" was heard, they turned to see what was happening. "The bridge is falling," came the cry, and the workmen rushed shoreward amid the sound of "snapping girders and cables booming like a crash of artillery." Only a few men reached safety; about seventy-five were crushed, trapped, or drowned in the water, surrounded by twisted steel. The death toll might also have included those on the steamer *Glenmont*, had it not just cleared the bridge when the first steel fell. Boats were lowered at once from the *Glenmont* to look for survivors, but there were none

The south arm of the Quebec Bridge, as it appeared just before its collapse on August 29, 1907

to be found in the water. Because of the depth of the river at the site, which allowed ocean liners to pass, and which had demanded so ambitious a bridge in the first place, the debris sank out of sight, and "a few floating timbers and the broken strands of the bridge toward the . . . shore were the only signs that anything unusual had happened." The crash of the uncompleted bridge "was plainly heard in Quebec," and the event literally "shook the whole countryside so that the inhabitants rushed out of their houses, thinking that an earthquake had occurred." In the dark that evening, the groans of a few men trapped under the shoreward steel could be heard, but little could be done to help them until daylight. The sounds of the bridge falling and the moans of the lives it claimed would echo around the world for days, weeks, and years to come.

Within days, the story of the unread telegram was reported as part of the tragedy. According to one version, Cooper filed it in New York before noon and, though delayed by the strike, the wire did reach Quebec in the middle of the afternoon—in time to save the men, if not the bridge. Cooper was first reported to have said that the message included an admonition to get off the bridge at once, but in fact it called only for halting the addition of any more steel to the structure. The less-than-urgent-sounding message thus lay on Deans's desk until he returned to his site office at about 5 p.m., shortly before McLure himself returned from New York. Even if it had not been too late to clear the structure, no warning might have been given, for the halt asked by Cooper was in fact to buy time to analyze the anomalies that had developed in the structure. It is not clear that Cooper or anyone else believed that collapse was imminent.

Shortly after the accident, amid speculations as to its causes and as to whether the men might have been saved, Cooper was reported to have reproached himself "for not having visited the work in two years," and confessed that he had "tried to obtain his release from the responsibility of serving as the consulting engineer" on the Quebec Bridge because of his poor health, but "the builders would not listen to that." With regard to the ill-fated telegram, Cooper qualified earlier reports regarding its message and pointed out that, serving officially only as consulting engineer to the project, he had no authority to order the workmen off the structure.

A royal commission was formed immediately to inquire into the cause of the collapse of the Quebec Bridge. The commission comprised: Henry Holgate, a civil engineer from Montreal; John G. G. Kerry, a civil engineer from Campbellford, Ontario; and John Galbraith, dean of the Faculty of Applied Science and Engineering at the University of Toronto. The site was visited the day after the accident, and the taking of evidence commenced in Quebec within two weeks. Cooper himself, who remained in New York, was interviewed by the commission there for a week in mid-October. After the confusing and conflicting newspaper reports concerning what he had said about his telegram in the days immediately following the accident, Cooper had remained silent on the matter, until the commission visited "the Nestor of American bridge designers," as *Engineering News* identified him in its report on the visit, thus raising his reputation to mythic proportions. Although he was actually sixty-eight at the time, he was described as "now 70 years of age" and as having been "in poor health for several years." The trade journal's sympathetic portrait of the engineer was no less deferential than the commission's treatment of him:

> The Canadian Commission, therefore, which spent the week from Oct. 14 to Oct. 19 in New York City, visited Mr. Cooper at his residence, conferred with him concerning the matter on several successive days, as his strength permitted, and finally formulated a series of questions covering the matters pertinent to the inquiry. Replies to these questions were dictated by Mr. Cooper at his leisure, and this testimony, after its reduction to writing, was fully reviewed and revised by the Commission and Mr.

Cooper in further conference, until it represented as completely as the Commission could determine, the full testimony which Mr. Cooper was able to give.

By the end of the month, the questions of the commission and Cooper's responses were reprinted in full in *Engineering News*. The transcript showed that the commission was looking into the entire history of the bridge project, the involvement and interrelationship of Cooper, the Phoenix Bridge Company, and the Quebec Bridge Company, and the nature of the various organizations and their respective involvement in the design and oversight of the work. Among the leading questions posed to Cooper was whether the plans were approved to his satisfaction or whether he would have given them further study had he been able to do so. His reply was that of a man seeking sympathy:

> I should have been glad to have had the physical strength and the time allowed me to have given further study to many parts of this structure, but in my physical condition I have been compelled, and must accept the responsibility for the same, to rely to some extent upon others. I had and have implicit confidence in the honesty and ability of Mr. Szlapka, the designing engineer of the Phoenix Bridge Co., and when I was unable to give matters the careful study that it was my duty to give them, I accepted the work to some extent upon my faith in Mr. Szlapka's ability and probity.

Engineering News, in a prefatory editorial to its printing of Cooper's testimony, cautioned engineers to "maintain a judicial attitude in considering the serious question how responsibility should be apportioned for faults in design, construction and erection." Indeed, the trade publication reminded its readers, "every engineer will recognize the fairness of suspending judgment as he reads Mr. Cooper's statement until the statements of the Phoenix Co. engineers are presented."

After taking Cooper's testimony in New York, the commission traveled to Phoenixville and Philadelphia to collect further information and take testimony from the Phoenix Bridge Company and its officers and engineers, including Szlapka. Here the commission heard "vigorous language directed against Theodore Cooper," which included charges that he had played down concerns over the incomplete bridge's structural behavior when it was questioned by the company's engineers, that he allowed more stress on the materials in this bridge than in any previous structure, that he ordered the main span increased to eighteen hundred feet, and that he refused to visit the Phoenixville plant where the first parts made for the bridge were being assembled.

The commissioners delivered their report, to which was appended a "Report on Design of Quebec Bridge" by C. C. Schneider, within six months of the accident. Among the main findings of the inquiry were that the collapse was initiated by the inability of the lower chords near the main pier to withstand the high though not unexpected compression loads to which they were subjected. Szlapka had designed these chords, and Cooper had examined and approved them, and their failure "cannot be attributed directly to any cause other than errors in judgment on the part of these two engineers." The report continued:

> These errors of judgment cannot be attributed either to lack of common professional knowledge, to neglect of duty, or to a desire to economize. The ability of the two engineers was tried in one of the most difficult professional problems of the day and proved to be insufficient for the task. . . . A grave error was made in assuming the dead load for the calculations at too low a value and not afterwards revising this assumption. . . . This erroneous assumption was made by Mr. Szlapka and accepted by Mr. Cooper, and tended to hasten the disaster.

In short, what Szlapka had done was to let stand an educated guess as to the weight of steel that the finished bridge would contain. Such guesses, guided by experience and judgment, are the only way to begin to design a new structure, for without information on the weight of the structure the load that the members themselves must support cannot be fully known. When the loadings are assumed, the sizes of the various parts of the bridge can be calculated, and then their weight can be added up to check the original assumption. For an experienced engineer designing a conventional structure, a final calculation of weights only serves to confirm the educated guess, and so such a calculation may not even be made in any great detail. In the case of a bridge of new and unrealized proportions, however, there is little experience to provide guidance in guessing the weight accurately in the first place; a recalculation, or a series of iterated recalculations, is necessary to gain confidence in the design. (The situation is not unlike that of a veteran weight guesser at a carnival, who might be expected to predict quite well the weights of normal-sized fairgoers but not the weights of dwarfs or giants, who fall outside the range of even sideshow experiences.) According to the findings of the commission, "the failure to make the necessary recomputations can be attributed in part to the pressure of work in the designing offices and to the confidence of Mr. Szlapka in the correctness of his assumed dead load concentrations. Mr. Cooper shared this confidence." Since Cooper was well known to have a "faculty of direct and unsparing criticism," his confidence in Szlapka's design work went unquestioned.

Just as Cooper had confidence in Szlapka's work, so did the resident engineer at the construction site have confidence in the work of them both. When a construction foreman expressed serious concern over the condition of the fatal member, the resident engineer thought the matter of little importance, telling the foreman, "Why, if you condemn that member, you condemn the whole bridge." After the collapse, it was reported that the resident engineer "had confidence in that failing chord because it was to him unbelievable that any mistake could have been made in the design and fabrication of the huge structure over which able engineers had toiled for so many years." As a result of the accident, however, *Engineering News* reported that mistakes of all kinds had become more believable.

The underestimation of the true weight of the bridge had actually come to Cooper's attention earlier in the design process, but only after considerable material had been fabricated and construction had begun. At this time, a recalculation of the stresses in the bridge led Cooper to consider that the error had meant that some stresses had been underestimated by 7–10 percent. All structures are designed with a certain margin of safety; he felt the error had reduced that margin to a small but acceptable limit, and so the work was allowed to proceed. In fact, some of the effects of the underestimated weights were, in the final analysis, of the order of 20 percent, and this was beyond the margin of error that the structure could tolerate.

In its discussions of the various bridge-building organizations involved and their respective faults, the inquiry commission was clear regarding the sense of hubris and overconfidence that success can bring to an organization. In this regard, the Royal Commission anticipated in some ways by eight decades what the Presidential Commission would find in its investigation of the space shuttle *Challenger*'s accident. Although there do not seem to have been too few assistants in that more recent accident, there certainly seem to have been too many overconfident bosses, or at least too many bosses willing to make compromises for other than purely technical ends. According to the Royal Commission, reporting in 1908:

> Mr. Cooper states that he greatly desired to build this bridge as his final work, and he gave it careful attention. His professional standing was so high that his appointment left no further anxiety about the outcome in the minds of all most closely concerned. As the event proved, his connection with the work produced in general a false feeling of security. His approval of any plan was considered by every one to be final, and he has accepted

> absolute responsibility for the two great engineering changes that were made during the progress of the work—the lengthening of the main span and the changes in the specification and the adopted unit stresses. In considering Mr. Cooper's part in this undertaking, it should be remembered that he was an elderly man, rapidly approaching seventy, and of such infirm health that he was only rarely permitted to leave New York.
>
> Mr. Cooper assumed a position of great responsibility, and agreed to accept an inadequate salary for his services. No provision was made by the Quebec Bridge Company for a staff to assist him, nor is there any evidence to show that he asked for the appointment of such a staff. He endeavoured to maintain the necessary assistants out of his own salary, which was itself too small for his personal services, and he did a great deal of detail work which could have been satisfactorily done by a junior. The result of this was that he had no time to investigate the soundness of the data and theories which were being used in the designing, and consequently allowed fundamental errors to pass by him unchallenged. The detection and correction of these fundamental errors is a distinctive duty of the consulting engineer, and we are compelled to recognize that in undertaking to do his work without sufficient staff or sufficient remuneration both he and his employers are to blame, but it lay with himself to demand that these matters be remedied.

The issues raised in the report were to reverberate throughout the engineering profession for many years to come, and in some form remain as issues today. Engineering work, especially relating to novel and untried projects, requires considerable time for thinking and rethinking about assumptions and tentative solutions, often among a broad range of colleagues and even in public forums. In cases like the Forth Bridge, the time and openness have been repaid in structures that stand as monuments. The very success of once bold endeavors like the Forth Bridge, however, can lead engineers like Cooper into a sense of security concerning ostensibly similar designs that may not be warranted. Incidents like the Quebec Bridge collapse provide rude awakenings, as *Engineering News* reported within weeks of the accident: "The Quebec Bridge collapse has been an object lesson to every structural designer; and we risk nothing in saying that in a thousand offices, stress computations are being checked over and details of design are being investigated and discussed with greater care and thoroughness than ever before."

The collapse of the Quebec Bridge, like that of the Tay, did not remove the need for a bridge at the location. Indeed, one could almost say there was renewed resolve to show that it could be done—and done right. The new design that was finally chosen was described at the time as "commonplace in appearance and costly to build." This should not have been surprising, for matters of aesthetics and economy, so important when bridges are first planned, come to appear almost as luxuries in the wake of a tragedy of the magnitude of that which occurred on the St. Lawrence River in 1907. The bridge that was finally built at the site of the wreckage reinstated the look of straight bottom chords, which Szlapka had testified had been changed to curved ones "for the sake of artistic appearance." Its outline did make the structure easier to analyze for load and stress distribution, but what was more significant about the new design was that it was a heavier and more substantial-looking bridge. If Cooper's Quebec seemed to have the lightness of the Tay Bridge, the redesigned Quebec would appear to have the firmness of the Forth.

5

The collapse of the first Quebec Bridge in 1907 had a profound and immediate effect on the direction of bridge building worldwide. New York's Queensboro Bridge, whose more lacy and graceful cantilever design with a maximum span of almost twelve hundred feet is often mistaken for that of a suspension bridge, was under construction when the Quebec collapse occurred. The Queensboro was completed in 1909 amid considerable protest and concern over its safety in particular, as well as over the safety of the entire genre. A second mishap in Quebec—which would occur in 1916, when the closing span of the redesigned structure fell to the bottom of the river while being hoisted into place, would reinforce reservations about the form. In spite of the resolve of the Canadian government to complete a successful cantilever design across the St. Lawrence River and thus vindicate the original decision, no other major cantilever bridge would be completed until the 1930s. To this day, none but the Forth comes within a hundred feet of the eighteen-hundred-foot span of the Quebec.

The incidents at Quebec were naturally the subject of doubting editorials in newspapers and trade journals alike, for both the public and the profession took a keen interest in record-setting bridge building. Nevertheless, according to *Engineering News-Record*, "Twice the hopes of success have been dashed, but never in the heart of the true engineer was there doubt that the enterprise would be brought to a successful conclusion." As with all failures, there were lessons learned in the collapse of the first bridge and the subsequent embarrassment during the final stages of the second, and it was the knowledge contained in these lessons that gave engineers the understanding to attack the problem of bridging the St. Lawrence with re-

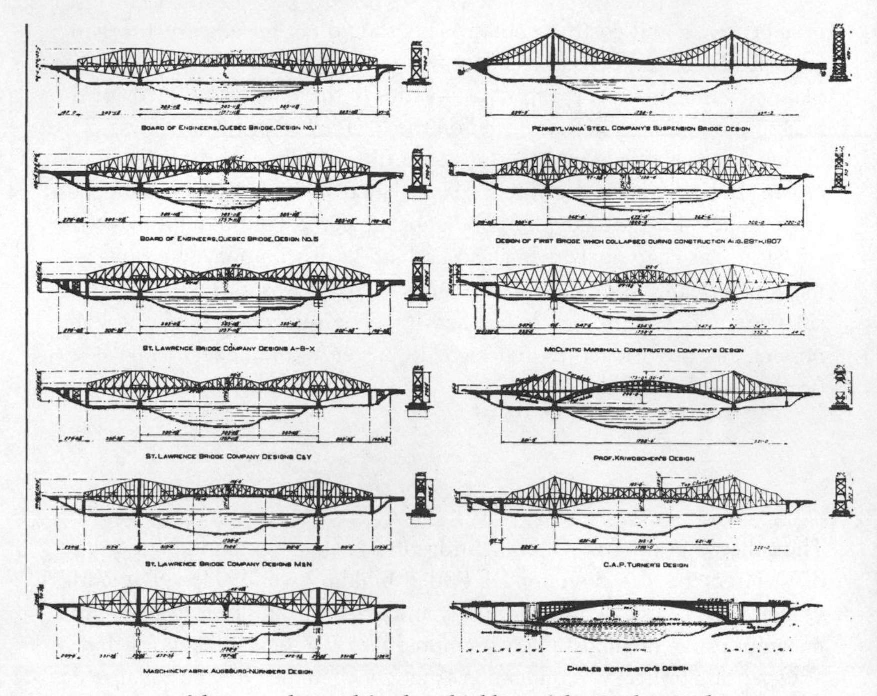

A variety of designs submitted for the rebuilding of the Quebec Bridge

newed confidence even in the wake of defeat, and enabled them in the end to "have vindicated the profession before a doubting world."

Years before the second Quebec accident, some members of the profession had their own concerns and prejudices about the whole process of choosing a bridge design. Among the most prominent and vocal of these was Gustav Lindenthal, who at the time of the Quebec collapse was a consulting engineer in New York City. Lindenthal had prepared the suspension-bridge proposal that the Phoenix Bridge Company had submitted along with its winning cantilever design in the original Quebec Bridge competition. He had also prepared a modified suspension-bridge design in response to the invitation by the board of engineers constituted to design a new bridge. This board had come up with specifications and with its own official concept, a cantilever structure with a straighter and bulkier outline than that of the collapsed structure. However, when bids were invited from construction companies, they were also given the option of submitting their own designs, though a company that did so had to assume "the entire responsibility not only for the materials and construction of the bridge, but also for the design,

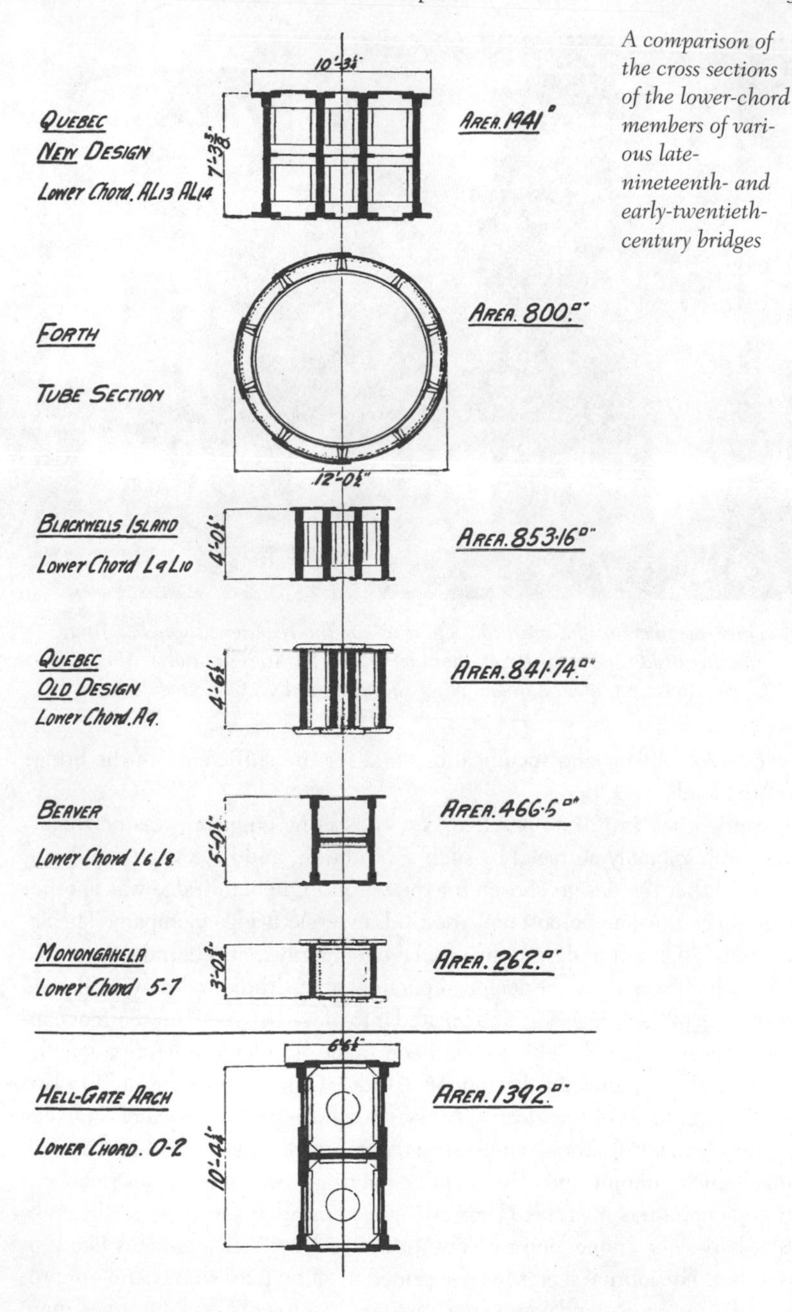

A comparison of the cross sections of the lower-chord members of various late-nineteenth- and early-twentieth-century bridges

Three members of the board of engineers for the redesigned Quebec Bridge standing inside one of the lower-chord members (left to right: *Ralph Modjeski; C. N. Monsarrat, chairman and chief engineer; and C. C. Schneider)*

calculations, plans and specifications and for the sufficiency of the bridge for the loads" specified.

Lindenthal, in private practice as a consulting engineer, was personally and professionally outraged by such a condition, and he was greatly disappointed that the design chosen for the second Quebec Bridge was another cantilever, this one proposed by the St. Lawrence Bridge Company. The accepted design differed from the official one mainly in its central span of 640 feet, which was to be fabricated separately about three miles away, floated on barges to the bridge site, and hoisted into place between the 580-foot cantilever arms. After the choice had been made, Lindenthal wrote a lengthy tract entitled "Notes on the Quebec Bridge Competition," printed in two successive issues of *Engineering News*, which was published every Thursday in New York City. Although it identified itself at the time as "a journal of civil, mechanical, mining and electrical engineering," *Engineering News* concentrated on matters relating to large civil-engineering projects, especially those that had some connection to New York or to New York engineers like Lindenthal. The journal seems to have prided itself on publishing details of proposed, in-progress, and completed engineering projects, and the larger, more visible, and more expensive the projects it could report on, the better.

The second Quebec Bridge accident, showing the central span buckling upon impact with the water

Lindenthal's notes on the Quebec Bridge competition had been introduced by an editorial, which had called it "the most important international competition for the design of a great engineering structure which has ever been held." The potential value of such a competition was, of course, that the various alternative designs emanating from the minds of creative engineers with a wide range of experience and vision provided an excellent opportunity for the comparison of the different proposed bridge types, and thereby the opportunity for understanding their relative strengths and weaknesses. A given engineer, especially within the limited time between the announcement of a design competition and the deadline for submission, may only be able to pursue in sufficient detail one or two designs. Decisions must necessarily be made at the outset as to whether to explore a cantilever or a suspension-bridge design or both, for example. If the engineer or the bridge company he is associated with is without work, considerable time may be available to be devoted to speculation on the competition, in the hopes of winning. If the engineer or his firm is already busy with other projects, a decision has to be made as to whether to hire additional engineers to work on those, so that a new design may be developed for the competition, or to try to find some spare time to think in new directions. In all cases, there is the question of how the time will be compensated for, and this was a point of major and emotional emphasis in Lindenthal's notes on the Quebec competition.

According to Lindenthal, the official design and the specifications upon which tenders were invited were prepared by the Board of Engineers "after much labor and time" over a two-and-a-half-year period at a cost to the Canadian government of about $500,000. The bidders, on the other hand, "were expected to make their competitive designs without compensation and in four months' time." Lindenthal believed that one way such a competition could work was as follows:

> If five or more of the world's bridge firms had been invited to prepare competitive designs, on a proper general specification, in six months' time, and for a compensation to cover expenses, then at 20% of the cost of the official design (which afterwards was ignored) and in less than one-quarter of the time there would have been a choice from a number of superior plans with tenders thereon, which would have represented the best practice and advance in the art.

Lindenthal went into some detail about the matter of compensation for engineering services for more than pecuniary, self-serving reasons. He had long been an outspoken advocate for structures transcending the utilitarian, and for more respect for the engineering profession. Under a section of his notes entitled "Causes of the Disaster," Lindenthal stated clearly that "the primary cause of both disasters"—namely, the 1907 collapse of the Quebec cantilever under construction and the 1879 fall of the high girders of the Tay Bridge—was simply "bad engineering." But, rather than going on to detail technical causes, Lindenthal wrote about matters that were to become increasingly the topic of letters, editorials, and articles in *Engineering News* and other professional publications of the time. In connection with the Quebec Bridge, he wrote, there was "a contributory circumstance of which it is difficult for engineers to speak without a feeling of humiliation." This circumstance was not only "the beggarly compensation for engineering services on a work of unprecedented magnitude," but also "the willingness of an engineer of high reputation and unimpeachable integrity to assume very important and laborious duties for a fee for which they could not possibly and seriously be met."

Lindenthal admitted that the financial conditions of a company "during the incubation of large work" could be weak, so that the engineers and other professionals "aiding in its promotion" might be quite satisfied to take only nominal compensation until financing was secured and they could be fully paid. His comments vis-à-vis the Quebec Bridge project in particular are well worth quoting at length, because, using Theodore Cooper's experience as representative, they set down issues that were on the minds of an increasing number of engineers in the early part of the twentieth century, and they provide insight into the practice of engineering generally:

> While the Quebec Bridge Co. was struggling along, it could not pay more than a small amount for engineering advice. It got its plans for the bridge from contractors for nothing. But after the money, with the aid of the Canadian Government, was assured, the large and difficult engineering work should have been thoroughly taken in hand, through an efficient engineering organization, properly compensated.

Mr. Theodore Cooper, who as Consulting Engineer had assumed the largest share of responsibility, had no such organization and could not afford to have it. The fee of $3,750 per year, which he received for his services, was hardly enough to pay office rent and a stenographer. Most unfortunately Mr. Cooper seemed unable to see the wrong he did to himself, to the profession and to his clients, when he did not advise and explain to the last-named that neither he nor any other engineer could conscientiously undertake the important duties without adequate facilities and a competent working staff, and the compensation should be made sufficient both for his services and theirs. His action was a grievous wrong to the engineering profession, as it tended to create the impression that responsible engineering service was of little account and could be had for next to nothing, provided contractors' plans were furnished.

It is most pathetic to notice from the testimony how Mr. Cooper endeavored to serve his employers faithfully, unselfishly but mistakenly. With a proper engineering organization it would not have been necessary to rely in any degree upon the office work and strain sheets of the contractors. Entirely independent computations could and should have been made by the responsible engineer, and the errors in the assumption of dead-load would have been discovered before construction began. Systematic study and analysis could also have been given to the contractor's design to determine whether and where modifications in form and details, as for instance in the compression members, must be made for greater safety.

These thoughts, as here mentioned with the kindliest spirit to Mr. Theodore Cooper, an old and valued friend, are more particularly intended to call attention to a most essential requirement of good engineering service on large work, and that is resolute and great executive ability, which is rarer even than great technical ability. The failure of everyone concerned to recognize the importance of that requisite in the engineer's work contributed greatly to the failure of the bridge.

The editors of *Engineering News* recognized that the objection could be raised that "Mr. Lindenthal was one of the participants in the competition and is therefore biased in his views," but they defended their publication of what was at times the embarrassing diatribe of a loser by appealing to the author's reputation, which by then had surpassed Cooper's. The editors were sure that it would be generally agreed that "in the entire engineering profession of this or any other country there is hardly an engineer who is so competent by experience and ability to deal with the problem of long-span bridge design than Mr. Lindenthal." Indeed, *Engineering News* could have its own objectivity questioned: it had for two decades advocated Lindenthal's design for a great suspension bridge to span the Hudson River at New York, a project for the likes of which engineers such as Lindenthal himself were not to be much compensated, if at all, unless the project reached fruition. Ironically, Cooper, whom Lindenthal so severely criticized, had, of course, been among the distinguished engineers in favor of a suspension bridge across the Hudson.

The completion of the Quebec Bridge finally occurred in 1917; the distractions of a world war may have partly accounted for the lack of publicity accompanying the opening of what has become a symbol of Canadian resolve. Theodore Cooper had retired the year the first Quebec bridge fell,

The completed Quebec Bridge

and he spent the final twelve years of his life in New York City, where he had set up his consulting practice in 1879. When Cooper died, in 1919, almost two years had passed since the Quebec Bridge had finally been completed, and there was less than a week till the twelfth anniversary of the collapse of the structure that was to have been his final work. His obituary in *The New York Times* declared that he "foresaw" the Quebec disaster, and reported that nearly one hundred lives would have been saved "had a telegram sent by Mr. Cooper been received and heeded." The obituary in *Engineering News-Record*, perhaps three times as long as that in the *Times*,

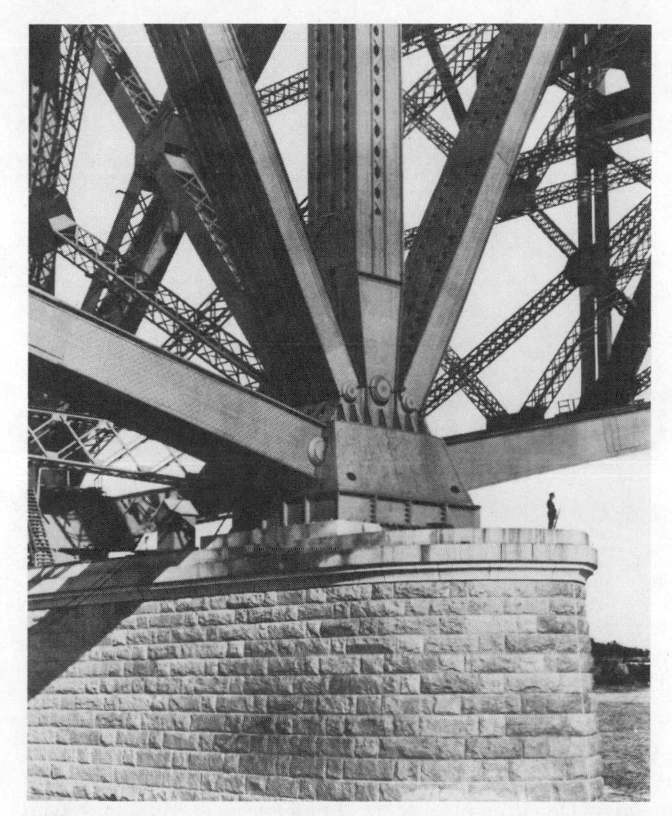

The scale of the Quebec Bridge shown by a guard posted during World War I

provided no such amelioration of Cooper's role in the Quebec accident but mentioned the bridge only in passing as among the many projects included in the "consulting work" of the "famous bridge engineer." Similarly, the memoir of Cooper that appeared two years later in the *Transactions of the American Society of Civil Engineers* treated his involvement in the Quebec Bridge project as only a passing credit, with no mention of the discredit it was in fact to his own and the profession's reputation. Since these sources provided the information on Cooper's life as presented subsequently in the *Dictionary of American Biography*, it too omits all mention of his role in the Quebec failure.

The obituaries and memoirs of Cooper were kind in their words of remembrance, but their generally short length belied their unqualified evaluation of his professional life, which encompassed so much of the latter part of nineteenth- and the early years of twentieth-century engineering. He was, after all, the 1858 civil-engineering graduate of Rensselaer Institute

Theodore Cooper, as pictured in an obituary

who then had begun his career working on the Hoosac Tunnel; who had entered the U.S. Navy at the outbreak of the Civil War and served on boats ranging from the *Chocura* out of Boston to the *Nyack* in the South Pacific; who had served as instructor at the Naval Academy in Newport, Rhode Island, and in the new Department of Steam Engineering at Annapolis; who had left the navy as first assistant engineer after Captain Eads appointed him inspector of steel being made for the bridge across the Mississippi; who had taken charge of the erection of the steel by the cantilever method for the great bridge at St. Louis; who had succeeded Eads as engineer of the bridge-and-tunnel company after Eads moved on to construct jetties and promote his dream of a ship railway; who had joined successively the Delaware Bridge Company and the Keystone Bridge Company, rising to assistant general manager of the latter; who had designed and built shops for the Mexican National Railroad; who had remodeled and rebuilt a plant for the Lackawanna Coal and Iron Company; and, with all this experience behind him at the relatively young age of forty, in 1879, had established himself as a consulting engineer in New York City, where twenty years earlier the iron magnate and philanthropist Peter Cooper, no relation to Theodore, had founded the Cooper Union for the Advancement of Science and Art.

Theodore Cooper's experience, coupled with the reputation he had established with his publications on bridge design and construction, especially with regard to the loadings to which increasingly heavy locomotives subjected steel bridges, had opened up many opportunities for him. In New York City alone, he had worked on projects involving a bridge over the Harlem River, on the first elevated railways, on the New York Public Library, as one of five engineers appointed by President Cleveland to determine the maximum span of a bridge proposed to cross the Hudson River, and as a member of the board of experts who evaluated a design of the Manhattan Bridge. In the context of such a distinguished and varied career, the Quebec Bridge may indeed have appeared to contemporary editors as an inappropriate focus for an obituary.

In spite of all that had been written of Cooper's inadequate compensation for his work on the Quebec Bridge, he did not die a pauper. His total assets amounted to about $180,000, the great majority of that in stocks and bonds, mostly in the American Telephone and Telegraph Company. Since Cooper never married, he left the bulk of his estate to a dozen nieces and nephews. The main beneficiaries were the two nieces, Alice and Mary Cooper, who had lived at the same West 57th Street address as their uncle, and who together received about a quarter of his estate.

Cooper's intangible and unspoken legacy was, however, the collapse of the Quebec Bridge. That event, no matter what its ultimate cause and who its agent, took the genre of the cantilever bridge from its previously high position of trust, which had been built up by Benjamin Baker's lectures in the late 1880s on the principle generally and on the Forth Bridge in particular, to its low position of doubt and distrust after 1907. The single event of the Quebec Bridge failure thus altered the course of bridge development, especially in America, from that set by Eads at St. Louis with his articulate argument for and achievement of an arch over a suspension design. Fowler and Baker's great cantilever over the Firth of Forth had created a further hurdle for proponents of suspension bridges to overcome in the late nineteenth and early twentieth centuries; that the Quebec Bridge was under construction as an even greater cantilever was in fact testimony to the growing competitiveness of the genre at the time. The collapse of the Quebec Bridge—Theodore Cooper's dashed dream—greatly influenced the bridgescape across our rivers and the bridgeline of our cities to become what we know today.

LINDENTHAL

At the peak of his career, Gustav Lindenthal would be hailed as "the Nestor" (as had Cooper before him) and also would become known as the "dean" of American bridge engineers, but his life seemed to be a constant striving to establish and maintain himself as precisely those things while holding fast to a dream that was never to be realized, even though he invested in it vast amounts of time and energy. Lindenthal was born in 1850 in Brünn, a manufacturing city in the Austro-Hungarian province of Moravia that, renamed Brno, became a part of Czechoslovakia after World War I. What appears to be incontrovertible about his background is that he was the oldest son of a large family born to a cabinetmaker and his wife, and that Gustav received a formal education through about age fourteen. The details of his further education have recently been revealed to be even more uncertain than one might gather from a close reading of standard biographical works like *The National Cyclopedia of American Biography*, where he is said to have been "educated at the Provincial College in Brünn and the Polytechnic schools in Brünn and Vienna." We may speculate that Lindenthal himself was the original source of such information, and, further, that his claim of having been "educated at" a school may have meant little more than having used the library and attended some public lectures there. However, it was natural to assume from the wording that the connection had been somewhat more formal.

The issue of Lindenthal's education was raised to a higher consciousness, however, in a 1991 article in *The New Yorker* that dealt mainly with

Lindenthal's masterpiece, the Hell Gate Bridge, upon which he was working at the same time that he wrote so authoritatively on the Quebec Bridge competition and related professional matters in *Engineering News*. The writer of the *New Yorker* article, Tom Buckley, revealed that none of the schools contacted in Brünn or Vienna could find any record of a Gustav Lindenthal's ever having been a student in the 1860s or early 1870s. According to a memoir of him published in the *Transactions of the American Society of Civil Engineers* five years after his death, Lindenthal was educated at the Polytechnicum College in Dresden, Germany, but that may have been merely an error based on his receiving an honorary degree from this institution in 1911. Long before then, his career had reached the point where the amount of formal education he had received mattered little in practical terms, for he had risen to the very top of his profession. But it may well have mattered to the man himself, or to his rivals.

What does seem certain is that young Gustav "received practical training from 1866 to 1870," for at the age of sixteen or so he "was put to work as a mason and carpenter," and "also worked in a machine shop." Perhaps he was forced to help support the family rather than enroll in school, but "the home soon became too confining for him," according to a tribute published in his hometown on the occasion of his eightieth birthday, and he ran away to Vienna "to start a life of his own." So young Lindenthal appears to have left home at about the age of twenty to make something of himself in Vienna, where he became an assistant in the engineering department of the Austrian Empress Elizabeth Railroad. Two years later, he joined the Union Construction Company, which was engaged in building an "incline plane and railroad," and the next year, 1873, he joined the Swiss National Railroad as a division engineer in charge of location and construction. Without a formal engineering degree, however, Lindenthal would have seen his future limited in Europe, and he emigrated to America, where self-educated engineers like James B. Eads and apprenticed ones like Octave Chanute could still, in the young profession, rise to considerable heights.

Among Lindenthal's first jobs in America was that of a journeyman stonemason, in which he found himself "working for several months on the foundation of the memorial granite building of the Centennial International Exhibition in Philadelphia" for the 1876 World's Fair. Lindenthal was to be remembered by his daughter as a man who "stood a little over six feet tall and was solidly built," and who "wore a mustache and a beard from the time he was a young man." His physical characteristics and his European experience, along with the self-determination of an immigrant wanting to make something of himself in the land of opportunity, no doubt helped him before too long become "an assistant engineer in the erection of the centennial exhibition permanent buildings in Philadelphia," a position that he would hold for the next three years. The judgment haltingly expressed by Buckley in *The New Yorker*—that "it appeared that the most eminent bridge designer of his time had been, in a sense, an impostor"—seems too harsh, for in the 1870s it was still possible to establish oneself as a professional on the basis of performance rather than college degrees. Indeed, the judgment seems to have been too harsh even for Buckley, who had just sung the praises of the engineer's Hell Gate Bridge, and who now seemed to want to soften the impact of the revelation:

> Lindenthal was neither the first immigrant to these shores nor the last to invent or to embellish his accomplishments or his ancestry—some did it to erase a criminal past, to free themselves from unhappy marriages, or simply to create new, more agreeable, and, perhaps, truer versions of themselves. His ersatz degrees doubtless opened doors, but he would have been quickly booted back onto the street if he had not been able to do the work. What his deception concealed, in fact, was the extraordinary intelligence, energy, and self-discipline that enabled him to teach himself mathematics, engineering theory, metallurgy, hydraulics, estimating, management, and everything else a successful bridge designer had to know—not to mention, in his case, English.

Whether Lindenthal ever proffered "ersatz degrees" may never be known, but it may indeed have been a "truer version" of himself that enabled him to become the engineer with the grandest dreams on the continent. These dreams were to be articulated in his adopted language in technical papers, prospectuses, tracts, letters, and a steady stream of words which belie the conventional wisdom that engineering and writing are alien endeavors. Indeed, Lindenthal, like virtually all great engineers before and after him, was a master of the pen and pencil as well as of bridge design, which should not be surprising. The dream of a bridge, which typically takes its first tangible shape in the form of a pencil sketch, would win no financial or political support were its engineer not able to flesh it out in words that convey not only the technical excitement of the project but also its benefit to the community of investors, merchants, politicians, and people generally.

After the Centennial Exhibition closed, Lindenthal began working for the Keystone Bridge Company on projects in Chicago and Pittsburgh. This experience, in turn, enabled him to become, in 1879, bridge engineer in Cleveland with the Atlantic & Great Western Railroad. Like many a young engineer of his time, Lindenthal thus followed a peripatetic career among the expanding railroads and bridge-building companies. However, shortly after he turned thirty, he decided to strike out on his own and returned to Pittsburgh to set himself up in private practice. There was plenty of work for a confident and competent engineer; many of the railroads needed help in carrying out surveys, designing and constructing new bridges, and replacing their old wooden-truss bridges with wrought-iron ones more capable of supporting the increasingly heavy locomotives that had come into use. Through such work Lindenthal came into contact with many of the most prominent engineers of the time.

I

Among the long-standing bridge needs in Pittsburgh was a crossing of the Monongahela River to reach the city's South Side. In the early nineteenth century, that need was provided by ferry service, but in 1810 a bridge charter was obtained, and by 1816 a fine covered wooden bridge was in place. This bridge was the work of Lewis Wernwag, whose earlier Colossus Bridge across the Schuylkill River in Philadelphia has been called "an American engineering superlative." Although the Colossus had a single clear span of more than 340 feet, which certainly contributed to its being referred to as "the most stunning and visually compelling engineering structure built in the early United States," the Monongahela bridge had eight much more modest spans of 188 feet each. Fire, however, the fate of many a covered bridge, destroyed the Colossus in 1838, and the wooden superstructure of the Pittsburgh bridge in 1845.

John Roebling thus was given the opportunity to build his first wire-suspension bridge to carry a road as opposed to a canal, and he was able to complete the structure especially quickly by employing the original masonry piers, which had not been harmed in the fire. Though Roebling's bridge was a great success at first, "in the course of time it became very shaky and loose, and its continuous swaying and creaking convinced every one that it was becoming unsafe for travel." In fact, the Pittsburgh suspension bridge was so flexible that at times of high water riverboat captains could arrange to have the headroom under one of its eight spans increased by a foot or two by hiring teamsters to position heavy wagons on the spans on either side of the one under which they wished to sail. As roadway traffic grew increasingly heavy, however, the large deflections and vibrations of the bridge became unacceptable, and in 1880 a new suspension bridge with larger spans was commissioned. After the new piers were under construction, the bridge company reconsidered its plans, and looked to something

The Colossus of 1812, a wooden bridge of uncommon span

other than a suspension bridge, which "would not be subject to undulations and would be capable of enduring the constantly increasing traffic without limitation of load or speed." Lindenthal himself may have pointed out the limitations of the suspension-bridge design, and he was invited by the directors and managers of the company to prepare alternative plans. He was subsequently awarded the commission for a new type of bridge of European design.

The Smithfield Street Bridge in Pittsburgh was Lindenthal's first important design project. Its principal structural form is now technically known as a lenticular truss, because it is lens-shaped, but was then called a Pauli truss, after its German inventor, Friedrich August von Pauli. The structural principle under which it functions is not unlike that of Isambard Kingdom Brunel's Saltash Bridge, built across the Tamar River in southwestern England in the 1850s, in which a top tubular member and a suspended chain act in opposing ways to produce a self-equilibrating truss, a variation on the bowstring girder that Eads described. Lindenthal's adaptation of the Pauli design was of a much lighter construction, however, because of the use of steel in some of its parts, and it showed "the triumph of architectural skill over the gross bulkiness that in the past was considered inseparable from an adequate amount of strength." In fact, Lindenthal's "use of steel instead of iron wher-

The original portal design of Pittsburgh's Smithfield Street Bridge

ever possible was based upon economy as much as anything," and the decision saved about 5 percent of the bridge's total cost of $458,000. As originally completed in 1883, the Smithfield Street Bridge carried a single roadway on two main spans of 360 feet each through towering portals with iron-fringed roofs. In 1891, as Lindenthal had made provision for in his original design, a second roadway was added on the already wide piers and a third set of lenticular trusses was erected, thus providing separate roadways for trolley and horse traffic. The original Victorian-style portal motif was retained after widening, though it was changed in 1915 to the less ornate dual-portal cast-

The widened Smithfield Street Bridge, with a less ornate portal

steel design that exists on the bridge today, and the Smithfield Street Bridge remains one of Pittsburgh's most significant landmarks.

An etching of the original portal design of the "new bridge at Pittsburg" dominated the front page of *Scientific American* for September 22, 1883, with a profile of the bridge relegated to a rather small inset engraving. Approaching a bridge like this one from Smithfield Street, or approaching Pittsburgh's Point Bridge, whose functional towers provided even more imposing portals to the main span, must have been an experience not unlike the one Victorian travelers encountered upon entering the crystal palaces that housed the world's fairs of the 1850s and 1860s. In fact, Lindenthal had been inspired—if not constrained, as all engineers are—by the style and technology of his own time, which in this case included the buildings for the Centennial fair in Philadelphia.

Among the curious features of the *Scientific American* story of the new bridge was the opening note that the cover engraving was made "from an excellent photograph by S. V. Albee, for a copy of which we are indebted to Mr. Alex. Y. Lee, C.E., of Pittsburg." That the engineer Lee's name was prefixed and suffixed in ways that Albee's was not suggests the status of the engineer, if not the profession itself, at the time, at least in *Scientific American*. All the more notable, therefore, is the fact that the "chief engineer, Mr. Gustavus Lindenthal," who was identified as the source of the particulars about the bridge, had no initials following his name. Evidently, at this early and important stage in his career, not only had Lindenthal not Americanized his given name but, more significantly, he seems not to have conveyed to the reporter that he held any such degree as C.E. Indeed, if anything, Lindenthal seems to have kept the reporter from making such an assumption—as well he might have. Lindenthal's engineering achievements were and would be his credentials.

In addition to his bridge over the Monongahela, Lindenthal also built one over the Allegheny River, at Seventh Street in Pittsburgh. This was a suspension bridge with four cables that employed not wire but chains composed of eyebars to hold up the roadway. The eyebar chains were suspended in pairs one above another from either side of the towers, and they were interconnected with bracing. Lindenthal may have been influenced in the design by the Point Bridge, completed in Pittsburgh in 1877, which also employed trussed chains to support its roadway. This scheme gave a considerable degree of stiffness to the chain structure, so that the relatively flexible roadway suspended from it was not subject to the degree of deflection and vibration that had been found unsatisfactory in wire suspension bridges. This preference for the use of eyebars rather than wire cables for suspension bridges was to be central in some of the debates Lindenthal would have with other engineers when he became involved in the design of bridges for New York City. Although Lindenthal's Seventh Street Bridge was to be replaced about a decade before the death of its engineer, it was, along with the Smithfield Street Bridge, one of the major structures erected in the 1880s in America.

Whereas the Brooklyn Bridge, which was completed in 1883, dwarfed Lindenthal's Pittsburgh bridges and thus captured the imagination of the wider public, his engineering reputation was firmly established, albeit principally in one locality. He had received the Rowland Prize from the American Society of Civil Engineers for his paper on the Monongahela bridge, which he read before the society in 1883, and he was well established as an engineer not only of bridges but also of unique forms of transportation like the inclined railroads used for moving wagons and streetcars along the steep slopes in and around Pittsburgh. Lindenthal, however, appears to have wanted to be more than an important engineer in Pittsburgh or among his colleagues in the American Society of Civil Engineers. One way of gaining wider recognition would be to design and build a bridge larger than the Brooklyn Bridge, then the largest anywhere in the world. If Pittsburgh did not need such large bridges, New York did, and spanning the Hudson River was a problem whose solution everyone would appreciate for its grand achievement. This would place its engineer in the category of a Roebling, if not higher.

2

According to his own account, almost fifty years after the incident, Lindenthal was approached in the fall of 1885 by Samuel Rae, assistant to the vice-president of the Pennsylvania Railroad, regarding the "practicability of a railroad bridge across the Hudson River." Being a "very able engineer with a penetrating and cautious mind," Rae also consulted other engineers over the situation at New York:

> There was keen competition among the railroad companies for Western traffic. The New York Central Railroad Company advertised a direct entrance, with four tracks, to the heart of Manhattan, while the Pennsylvania Railroad Company and the other railroads terminating in New Jersey were handicapped and had to transfer their passengers across the Hudson River by ferries. A tunnel under the Hudson River had been started at Hoboken, N.J., but it was intended only for small [railroad] cars and local traffic. A larger tunnel for locomotives and standard cars appeared objectionable because of the smoke, which was then a subject of daily complaint in the tunnels of the New York Central Railroad.

As Rae also noted, the situation favored an immense bridge with open-air railroad tracks. In the wake of the Tay Bridge failure, however, the Firth of Forth cantilever design of Fowler and Baker had succeeded the suspension-bridge design of the discredited Bouch, and so a cantilever had also been talked about for the Hudson River, which was about three thousand feet wide and very deep at New York. However, there were serious questions whether a pier would be allowed in the river, and whether the depth of the foundations might be practical. Having "given thought to the matter before," Lindenthal turned to a suspension-bridge design for New York, and he was convinced that it was technically possible. He reported as much to the Pennsylvania Railroad in the spring of 1886, but Lindenthal's integrated approach, which included a terminal plan, was prohibitively expensive for a single railroad to finance. Thus the North River Bridge Company, with Lindenthal as chief engineer, was organized in 1887 to seek financial support from several railroads, which would share the bridge and terminal facilities. This seemed like a very promising enterprise, for the otherwise uninterrupted transcontinental tracks then terminated in New Jersey, just across the river from New York, which was the ultimate destination of an enormous volume of passengers and freight. The closest bridge across the Hudson River was at Albany, over 150 miles north. Construction of the cantilever bridge at Poughkeepsie, about sixty miles upriver, had just begun, and ferry service between New Jersey and New York was slow, expensive, and subject to interruption by the weather. Furthermore, there was "annoyance and even danger to the landed passengers on the overcrowded and nasty streets" of New York City, which also housed the offices of *Engineering News*, the trade journal that was then poised to grow and expand its influence under the vision and energy of its new editor, A. M. Wellington.

Arthur Mellen Wellington, born in Waltham, Massachusetts, in 1847, was the son of a physician. He attended the Boston Latin School, and he learned engineering not through formal education but as an assistant in the Boston office of John B. Henck, himself a self-made engineer. Henck, born in Philadelphia in 1815, was self-educated until he entered Harvard, from which he was to graduate first in his class in 1840. He remained in Cambridge to serve as principal of Hopkins Classical School for a year before moving to the University of Maryland, where he spent a year as a professor of Latin and Greek. After five more years in a similar position at the Germantown Academy, and with a growing family, he returned to Boston to enter the employ of Felton & Parker, Civil Engineers. After a couple of years there, he left to join in a partnership with the engineer William S. Whitwell. Henck eventually set up his own consulting offices in Boston to do general engineering work, which included work on street railways, the Charles River Basin, and the development of Boston's Back Bay district. In 1865, he became head of the Civil Engineering Department of the newly established Massachusetts Institute of Technology, a position he held until 1881.

Though Wellington was introduced to engineering in Henck's firm in Boston, the young man did not wish to remain in that city. He took and passed an examination for assistant engineer in the U.S. Navy, but the war ended before he could assume such a position. Wellington then went to Brooklyn, New York, where he joined the Park Department under Frederick Law Olmsted, who with Calvert Vaux had laid out Prospect Park. Wellington apparently had wanderlust, however, and he began to work for a succession of railroads, beginning as a transitman on the Blue Ridge Railroad in South Carolina and working himself up from assistant engineer, principal assistant engineer, and locating engineer to engineer in charge. However, when railroad construction was suddenly stopped during the panic years of 1873 and 1874, Wellington found opportunities for engineers scarce. On his application for membership in the American Society of Civil

Engineers, he wrote: "1874–'78, was engaged in miscellaneous professional, business and literary occupation more interesting than lucrative and not always particularly interesting." According to one who knew him, however, he was later to refer to this "period of enforced idleness—so far as idleness was possible to a man of his restless energy—as a blessing in disguise."

One of the outlets for Wellington's energy was explicating his experience with railroad construction in books. His first dealt with the very important task of computing how much earth needed to be moved to construct a railroad, a key factor in its cost. The same year this book was published, 1875, Wellington began his "great work, and that by which his fame as an engineer would be established, *The Economic Theory of the Location of Railroads*." It was in this work, first published in 1876 as a series of articles in the *Railroad Gazette* and in 1877 as a book, that Wellington's famous definition of engineering appeared:

> It would be well if engineering were less generally thought of, and even defined, as the art of constructing. In a certain important sense it is rather the art of not constructing: or, to define it rudely, but not inaptly, it is the art of doing well with one dollar, which any bungler can do with two after a fashion.

Wellington's success as a writer brought him opportunities, and in 1878 he became principal assistant to Charles Latimer, chief engineer of the New York, Pennsylvania & Ohio Railroad. After three years at the "Nypano," as Latimer's company was known, Wellington went to Mexico to become first engineer in charge of location and surveys for that country's national railroad, and later assistant general manager and chief engineer in charge of location. He again grew restless, however, and in 1884 he returned to the United States to become one of the editors of the *Railroad Gazette*, a position for which his practical experience was invaluable. In January 1887, he left the *Gazette* to join *Engineering News* as one of the editors-in-chief and as part owner. According to another of the editors, "the influence of his energy and ability was at once seen in every department of this journal. Within two years its subscription list had more than doubled."

The juxtaposition of Wellington's name and the word "energy" was ubiquitous, and he seemed always to be looking for new challenges. In the summer of 1892, instead of taking his usual vacation, he stayed in New York and "devoted his leisure to working out some ideas in thermodynamics which had occurred to him years before," which led to the invention of an efficient engine whose development "became the all-absorbing work of his life, and in his earnestness and zeal all thought of care for his health was forgotten."

As he spent more and more time on his invention, he worked less and less on writing and editing, finally giving up work at *Engineering News* entirely in May 1894. He soon thereafter embarked on the European rest trip his physicians had advised earlier, but he became acutely ill while in Norway. Though he recovered sufficiently to return to America, his health again failed, and he died in April 1895, from overwork, according to those who knew him most closely.

Gustav Lindenthal could not have helped interacting with Wellington during the course of his editorship at *Engineering News*, for that journal was to follow closely the great bridge project of which Lindenthal dreamed. Among the first public mentions of a new plan for a New York City bridge over the Hudson River, which was then also known as the North River (as opposed to the South or Delaware River), appears to have been a letter that ran in mid-1887 in a newspaper in Philadelphia, the hometown of the Pennsylvania Railroad. According to *Engineering News*, and most probably according to editor Wellington, it was "from a man whom [*sic*] we happen to know is eminently qualified to discuss the subject on the great question of how to eliminate the Hudson river from the New York terminus problem." Though the letter-writer was not identified, it was almost certainly Lindenthal, whose name in 1887 was quite unlikely to be meaningful to New Yorkers generally, but who may very well have himself fed the letter to Wellington. The subject of the letter was itself of immense interest, however, and it was quoted at length. As in most preliminary reports of engineers setting out a complicated problem and a proposed solution, both were stated concisely:

> Are the proposed tunnels under the river the proper remedies for the present inconveniences? The projected tunnel is estimated to cost $11,000,000 for two tracks. But two tracks would not begin to accommodate the passenger business of a single railroad, much less all that now terminate on the Jersey side. The Pennsylvania Railroad alone would require four tracks for its steadily increasing business. There should be not less than six tracks, requiring six tunnels. . . . Six mud tunnels for necessarily slow trains with noisy, cramped terminals, from which dampness could not be excluded for $33,000,000, with no assurance that this amount would be sufficient, and with the certainty of great expenditures for maintenance and repairs, for tunnels must be pumped dry, ventilated, and perhaps thoroughly lighted. This is certainly not the kind of improvement that New York City is most in need [of], and it is not the kind of terminal railroad station which could meet the ever growing demands for greater convenience, safety, comfort and expeditious travelling.

Progress by 1882 on a Hudson River tunnel begun in 1874

> Imagine now, in a central part of New York City, within a stone's throw of its greatest avenue, a grand, imposing station, combined with every convenience and comfort of a first-class hotel, with numerous tracks and platforms, accommodating thirty trains at one time, arriving and departing, having all the elevated railroads running their trains directly into this station. Then imagine a massive stone viaduct and lofty columns supporting a six-track roadbed, through and over blocks of buildings to a magnificent bridge over the North river, leaping with a single span over its entire width, without a pier or other obstruction and with a clearance above highest tide of 140 ft., carrying six tracks. Then imagine the six tracks continued on a viaduct and gently descending to the level of the country in New Jersey to connections with all existing railroads and for future lines that will be built. No doubt such imagination may seem fantastic and profitless, though everybody will grant, were it possible to realize such a project, it would be a grand and eminently useful undertaking. But such a project can be realized. It is perfectly feasible and practicable to execute it at less cost than the proposed tunnels with corresponding terminals. The matter has been studied with the greatest possible care for a number of years, and all conditions have been weighed impartially and soberly. There cannot possibly be objections to a bridge spanning North river without a pier in the river and at such a height as to allow the largest steamers to pass under it freely. Bridge engineering has progressed so much that such a large bridge can be built with greater facility to-day than it was possible for the Brooklyn Bridge when it was proposed.

Later that year, Lindenthal prepared a four-page report on this solution of his to the problem, which he copyrighted in 1887 under his own name and had privately printed "not as a publication, but simply for convenience of the promoters of the project and for their exclusive use." His booklet was entitled *The Proposed New York City Terminal Railroad, Including North River Bridge and Grand Terminal Station, in New York City*, and the bridge was only one part of the integrated scheme. Six train tracks would be constructed on viaducts "high above the houses" of New York City between a huge bilevel Terminal Station, located "as close as convenient to the prin-

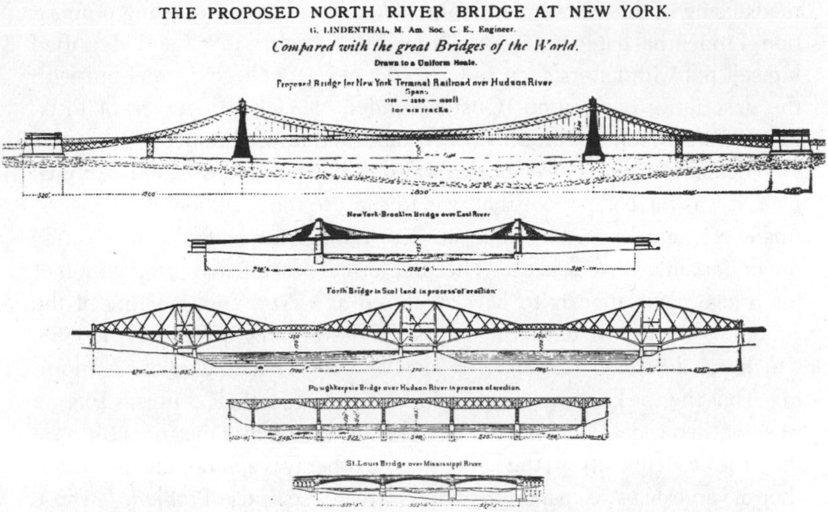

THE PROPOSED NORTH RIVER BRIDGE AT NEW YORK.
G. LINDENTHAL, M. Am. Soc. C. E., Engineer.
Compared with the great Bridges of the World.
Drawn to a Uniform Scale.

Lindenthal's proposed North River Bridge compared with the Brooklyn, Forth, Poughkeepsie, and Eads bridges, drawn to scale

cipal hotels," which then meant somewhere above Eighteenth Street and near Sixth Avenue, and the "great North River Bridge," also referred to as the Hudson River Bridge. Since at the time the Hudson could be considered the "most important water highway in the United States," any obstruction of it by bridge piers was out of the question. Thus Lindenthal proposed bridging the river "between established pier lines with a single span, 2,850 feet long and 145 feet above high tide."

As with all responsible engineering proposals, Lindenthal's report included an estimate of cost and a projection of revenue based on use. Since "surveys, plans and estimates for the entire project" had been made, and since, "except for its magnitude," the work was "as definite and free from experimental features as any other railroad or bridge project," Lindenthal must have been confident in his estimate of $23 million for the terminal station, viaducts, bridge, four miles of railroad, and a tunnel through Bergen Hill in New Jersey. When the cost of acquisition of the right of way was added, the total cost of the project was estimated to be $37 million. He projected that eight railroads, including the Pennsylvania, could run their trains directly to the Grand Terminal Station, collectively carrying about sixty thousand passengers per day plus freight. At ten cents each, those passengers alone would bring revenue of over $2 million annually. Because the expenses of operating the system were expected to be covered by the railroads using it, the overall plan looked like a sound moneymaking proposition. Lindenthal dated his report "New York, October, 1887," and identified himself not with letters such as C.E., denoting a college degree, but with the descriptive declaration "Gustav Lindenthal, Civil Engineer, of Pittsburgh, Pa.," which he certainly had established himself to be.

Whereas the North River Bridge was only one part of his Grand Terminal plan, it was the component that was to capture the attention of engineers, financiers, and laypersons alike, and to remain Lindenthal's unrelenting dream for almost five decades. The first formal professional presentation of his bridge plans appears to have occurred at a New York meeting of the American Society of Civil Engineers on the evening of January 4, 1888, which was described in a report in *The New York Times* the following morning. That the speaker was identified as "Prof. Lindenthal" confirms that he was not then widely known in New York, but the reporter may possibly have used the academic title in the belief that no other was appropriate for the author of "an exhaustive paper on 'The North River Bridge Problem' " whose reading "consumed over three and a half hours," even though the speaker "confined himself to the salient points of the general project." Nevertheless, Lindenthal, who may have done little to discourage the professorial image, apparently could not pass up an opportunity to criticize New York's Brooklyn Bridge, pointing out "enough defects in the East River Bridge to test the faith of any understanding mortal compelled to cross that iron thoroughfare in the course of his business." There seems little doubt that Lindenthal wanted to better the great achievement of Roebling and to build the greatest bridge in the world. Though he estimated then that it would cost no more than $15 million, he admitted in a report only three months later that the total railroad project might reach $50 million.

In his talk, Lindenthal also argued against a tunnel, which many engineers favored because of the great width of the river. Indeed, a tunnel was the greatest immediate threat to the realization of his dream, and he had concluded his report by citing the clear advantages of a bridge over a tunnel: "Utility, the greatest convenience, plenty of light and air, absence of smoke and noise shall be the leading features." Even though there had been some success with driving tunnels under water—Marc Brunel's tunnel under the Thames River in London having been completed over four decades earlier—there remained a general aversion to going underground and under a river in the dark for a mile or so, and bridges were the communication link of choice—if their costs could be afforded. However, a tunnel beneath the Hudson was already under construction, and the competition between tunnels and bridges would remain real and ever-present.

In the meantime, there was growing public interest in an interstate bridge. In late 1887, citizens of New Jersey had asked Congress to authorize and direct the president to appoint a commission of army engineers to look into the matter. This appeared to be the first "public move in a very ambitious project," according to *Engineering News*, which was sanguine in spite of the project's involving an "amount of money, for construction and real estate, that would have made a previous generation stand aghast at its mere mention." The journal that expressed on its masthead an interest in "all new engineering works or designs, large or small, of interest from their magnitude, novelty, or originality," believed in Lindenthal's dream, however, for the country then had "engineers capable of surmounting all the physical difficulties of the problem, and a people rich enough to pay for it, just as soon as the necessity is really felt for such a structure—and that time approaches." The necessity was already felt by the likes of Lindenthal, of course, but the time when enough others would feel it was to approach and recede for decades.

To complicate things further, rumor had it that some railroad men were becoming interested in developing plans for a bridge across the Hudson between Steven's Point, in Hoboken, New Jersey, and somewhere near 42nd Street, on Manhattan Island. Their scheme differed from Lindenthal's in several respects. For one, it was to carry "wagon-ways, foot-ways, and a cable road system," in addition to a good number of railroad tracks. For another, it was to be a cantilever bridge, with a maximum span of 780 feet and a headway of 165 feet above the water. The rumor had it that "no engineer has yet made plans, otherwise than to say it was feasible," which was certainly believable, since the cantilever bridge with multiple 548-foot spans was then under construction over the Hudson at Poughkeepsie and the 1,710-foot spans of the Firth of Forth bridge were nearing completion in Scotland.

Another group of investors was seeking approval for a bridge between Fort Lee, New Jersey, and the section on the New York side of the river known by its Dutch name, Spuyten Duyvil. They wished to place one or more piers in this relatively narrow part of the lower Hudson River, but steamboat operators were already complaining about the piers at Poughkeepsie, where the tides were not nearly so tricky as they were in the river around Spuyten Duyvil, past which tows of sixty to a hundred barges stretched out "anywhere from 200 or 300 feet to nearly a mile" (though the latter estimate was very possibly a zealot's hyperbole). Thus the stage was set for battles on several fronts, not only between the advocates of tunnels and those of bridges but also between proponents of cantilever and suspension designs, and, as always, between builders of bridges and operators of tugs and ferryboats, with all manner of variation in detail. These battles, not unfairly likened in emotion and intensity to those between the sheep- and cattle-herders of the Old West, would also rage in various forms and at various strategic locations for the next few decades.

True to its promise to give early publication to plans for new engineering works, *Engineering News* soon ran serially the details of Lindenthal's design, introducing them as the first item on the first page of the first issue of 1888 with assurances that the cost was "certainly not so formidable an obstacle for to-day as was that of the Brooklyn bridge for 1868," and that "there is probably no one on either side of the ocean who could be counted on more confidently to deal successfully with the intricate engineering problems involved than Mr. Lindenthal." His reputation—at least to editors of, and hence to readers of, *Engineering News*—seems to have been well served by his technical tracts and lectures of earlier years.

A profile diagram, with horizontal scale compressed five times more than that of the vertical, showed the bridge in context, complete with the proposed tunnel through New Jersey's Bergen Hill and the terminal with two track levels in New York City. An undistorted drawing of the bridge itself appeared above uniform scale drawings of the Brooklyn, Firth of Forth, Poughkeepsie, and Eads bridges. Unlike the chains used for his Seventh Street Bridge in Pittsburgh, Lindenthal proposed braced steel-wire cables enclosed in steel envelopes to "protect them absolutely against rain and weather." The stiffening trusses of the roadway proper were "principally designed to form the frame work for two large horizontal wind trusses [to] make the bridge safe against the most violent tornadoes," and the bridge was so designed that four additional railroad tracks could be added "at any time in the future, should it become necessary, making a double deck bridge." Actually, the first bridge to connect New York and New Jersey was still over forty years away, but it would share a remarkable number of features with Lindenthal's late-Victorian dream.

Lindenthal's plans, as published in *Engineering News*, showed him to have given considerable thought and effort to the great bridge. In addition to describing the technical details, his report kept returning to the *architecture* of the bridge, especially to the form of the towers, they being "the most prominent feature" of the structure. He acknowledged that the largest suspension bridges then built all had stone towers, but he cited the recent replacement of the cracked stone towers of the Niagara Gorge Suspension Bridge with metal ones, and explained that, "for bridge towers 500 ft. high, wrought iron or low steel is without question the most suitable material." The towers of his bridge would have columns shaped "for the double pur-

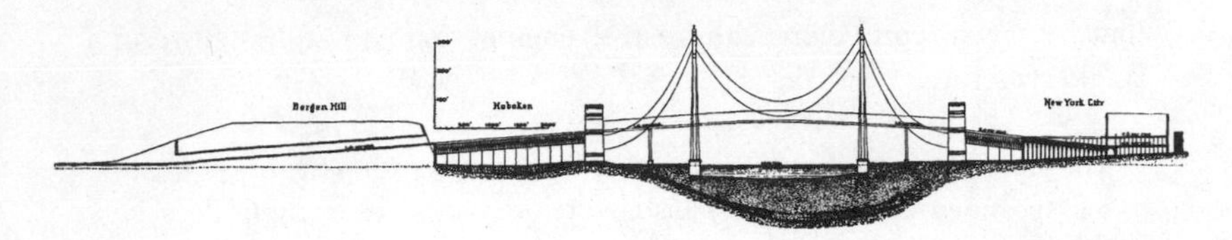

Lindenthal's New York City Terminal Railroad scheme, drawn with an exaggerated vertical scale and showing the proposed bridge and tunnel through Bergen Hill in New Jersey

pose of good appearance and to produce initial strains in the bracing between them, by which the rigidity of the towers is enhanced." The bracing itself was so arranged also to form "a grand and lofty portal" through which the train tracks would pass.

Engineering News was understandably proud to publish a "very liberal extract" from Lindenthal's paper, which it described as "the first definite description of a work which has at least a very fair chance of becoming the greatest of its kind on this continent, or in the world." This proponent of great schemes assured its readers that the fact that "some such structure will be built over the North River is as certain as any event still in the future can be," adding that its prospects were especially good because "it certainly has that solid basis which was so sadly lacking in the Panama canal scheme" that recently had been effectively abandoned by the French. *Engineering News* concluded its introduction to one of several extracts with optimism, for, "fortunately, engineering difficulties do not by any means vary in direct ratio with magnitude, as the cost does, and there seems to be little in the proposed design which previous experience does not indicate to be entirely practicable." Unfortunately, editor Wellington and engineer Lindenthal both seem to have underestimated the importance of nontechnical factors, which perhaps vary to an even greater extent with magnitude than does cost. The political and mercantile complications and competitions that accompanied such technically solid great projects as the Eads and Brooklyn bridges were evidently forgotten, at least by some, in the late 1880s in New York.

Lindenthal himself also seems to have worried less about general opposition to the plan than he did about attacks on the aesthetic integrity of his design. Considering the "architectural excellence of the bridge" to be of the "highest importance," he ridiculed the "hackneyed phrase" that "correctly designed structures have an innate architectural beauty, requiring no adornment, unless perhaps that of a well selected color of paint." Lindenthal pointed to various bridges (some recently completed) that he saw as embodying the best of engineering and architecture in a single structure:

> The graceful suspension in Buda Pesth (without question the finest existing specimen of this class of bridge-architecture), the early bridges in Paris, and a few over the River Rhine were built by engineer-architects, when the field of engineering did not yet justify exclusive devotion to one specialty, to the neglect of other branches of the science of building. But for the taste and stubborn persistence of the late Capt. Eads, the St. Louis bridge would have been built so as to be not the finest specimen of a metal arch-bridge architecture in this country, which it is, but on the plans of the old Omaha bridge, now worn out, and soon fit only for the scrap heap. . . .
>
> The standpoint of utility has, in our time, become with many almost the only professional point of view for judging of the merit of engineering work, so that the incentive for better things is wanting. A good deal of the blame is with the engineering schools. There is not one text book, to the author's knowledge, in the English language, on "Bridge Architecture," and no at-

An early version of Lindenthal's Hudson River Bridge design, with the Brooklyn Bridge in the background

tempt is made to teach the students even the rudiments of good designing. It is thought to be of more consequence to furnish an elegant graphical solution of the strains in a polygonal truss, or in the invention of a new formula for the very least weight of iron in a bridge, than to design with a decent regard for pleasing appearance, and for the feeling of fellow men and the opinion of posterity.

His apparent contempt for engineering schools may perhaps have stemmed partly from his own disappointment or private embarrassment at not having been more formally educated, and partly from the realization, based on his own achievements, that formal education was not a sine qua non for creating good bridge engineering and architecture. That Lindenthal had established himself so well by building significant bridges in Pittsburgh appears to have given him a self-confident, if not arrogant, belief that he was the pre-eminent American bridge engineer, and so entitled to serve as an arbiter of taste for another city's bridges. He went on to relate anecdotes of being told by a railroad manager how, "every time he hears of a new project for a larger bridge," he feared another "common hideous looking structure" would go up. When another gentleman spoke to him of "recklessly vulgar structures," Lindenthal defended engineers by arguing that they "cannot always do as they please and public sentiment must be educated" in appreciating "better things." He left little doubt that he was referring in particular to a "shameful conglomeration of iron structures as are found in New York and Brooklyn," which deserved better:

It is certainly true that if the New York Harbor, acknowledged to be the most beautiful in this country, should be defaced by a utility bridge of shabby appearance, it would be an unpardonable offense against the civilization of mankind. A pleasing architectural appearance of the bridge [proposed] was therefore held to be worthy of as much study as the engineering features, and the design aims to combine them in the best manner attainable out of a variety of designs made for the purpose.

Lindenthal expected his digression to be "pardoned in view of the importance of the project." Years later, his apparent inability to compromise on any front, functional or aesthetic, was blamed for the great dreamer's ineffectualness in getting the North River Bridge plans approved. But whereas Lindenthal was talking aesthetics, others were talking politics and economics, each of which *was* a sine qua non of great bridge building. And perhaps the biggest obstacle of all was the almost strictly technical decision as

Proposed cantilever bridge over the Hudson River

to whether a suspension bridge of the record span he proposed was indeed practicable. As the great bridge over the Firth of Forth neared completion, the cantilever type was day by day gaining support. In the end, it was Lindenthal himself who publicly raised the issue of a suspension versus a cantilever bridge.

3

The full title of the long paper Lindenthal read before the American Society of Civil Engineers was "The North River Bridge Problem, with a Discussion on Long Span Bridges." Many pages of *Engineering News* in the months of January and February 1888 were given over to Lindenthal's solution of the bridge problem, but many more for the month of March were devoted to his discussion. His definition of a "long span bridge" was one whose structural metal (concrete bridges were not even under consideration) was at least as heavy as the traffic it was designed to carry. There were four types of bridges most suitable for long spans, he asserted: the suspension bridge, which, after Eads, he termed a suspended arch; the erect arch, which was the familiar kind; the continuous girder, of which the Britannia tubular bridge was an example; and the cantilever.

First discussing the cantilever, Lindenthal pointed out that the type generally lacked rigidity under fast railroad trains, unless it was built with great height and depth, as over the Firth of Forth, at the sacrifice of headroom near the piers, which he thought unacceptable in the Hudson River. In addition to some more technical objections that he raised, he finally condemned cantilever bridges then built for "their general ugliness of appearance." Such aesthetic concerns, according to Lindenthal, "may not be of much consequence for a railroad bridge off in the woods, but even then it would not be more expensive to build them with an eye to better appearance, if for no other reason than to set a good example in imitative engineers." Among the things he found objectionable in contemporary cantilever bridges were the "indiscriminate use of eyebars and bulky compression sections in the same chord lines and the irregularity of truss frames," which he thought to be "as needless as it is ugly looking." He concluded his diatribe against cantilevers by alluding to two recently built ones in certain unnamed large cities. One of these bridges was the one in Philadelphia over Market Street, whose design was defended in a letter to the editor in a subsequent issue of *Engineering News*. The writer admitted that "cantilever fever" was prevalent at the time, but that Lindenthal was in fact "entirely ignorant of the special conditions of the case" which led to the bridge that he so criticized. Later in his own career, in fact, Lindenthal would argue precisely that some special conditions caused him to design an unusual bridge in an unusual location, but that continuous girder span was still then far in the future.

Lindenthal's discussion of arch bridges was quoted at considerable length from the 1868 report of Eads, including many of his illustrations, and his arguments were presented by Lindenthal as being as true in 1888 "as they were then, and as they always will be." The discussion of suspended arches, or suspension bridges, was the longest. Whereas Eads had, of course, found the erect arch superior to the suspended, Lindenthal concluded just the opposite—namely, that the suspension bridge presented "the most favorable conditions of stability and rigidity." Such a diametrically opposed conclusion from one engineer who had just agreed so much with the other with regard to the erect arch is not so much a contradiction as a demonstration of the complexity of the issue. In fact, there are in bridge design, as in all engineering problems, so many competing objectives and contrary constraints that in the final analysis the decision can be purely a question of personal preference and aesthetic taste, taking into account any special conditions at the bridge location. One engineer, because of his prejudices, might choose to design an arch rather than a suspension bridge for a particular site, whereas another might do just the opposite. Both bridge designs might be equally safe and reliable, but they might not have the same functional, aesthetic, and economic qualities.

In Lindenthal's case, he was so committed to the suspension concept for bridging the Hudson River that he turned the argument naturally and not unfairly to his use. Lindenthal admitted, for example, that it was "a popular assumption that suspension bridges cannot be well used for railroad purposes," and further conceded that throughout the world there was only one suspension bridge then carrying railroad tracks, Roebling's Niagara Gorge Bridge, completed in 1854, over which trains had to move slowly. However, rather than seeing this as scant evidence for his case, Lindenthal held up as a model the "greater moral courage and more abiding faith in the truth of constructive principles" that Roebling needed to build his bridge in the face of contemporary criticism by the "most eminent bridge engineers then living." In Lindenthal's time, three decades later, it was not merely a question of moral courage; "nowadays bridges are not built on faith," and there was "not another field of applied mechanics where results can be predicted with so much precision as in bridges of iron and steel." He did not promise such precision in the cost of his suspension bridge, however, and concluded his discussion by revealing an estimated cost of "approximately $15,000,000," which still might have seemed more dreamlike than the bridge itself.

In the meantime, the legislatures of New York and New Jersey had become involved again. In 1868, the New-York and New-Jersey Bridge Company had been chartered under the laws of New Jersey, which allowed one or two piers in the river as long as there was a thousand-foot clearance between them and a clear height of 130 feet in the center. With the renewed interest in a bridge, there was also renewed interest in getting the New York Legislature to pass a law so that cooperative progress could be made. There was, however, opposition on the New York side among supporters of a bridge farther up the Hudson River, nearer Albany. Early in 1888, a bill was introduced into the New York Legislature, but *Engineering News*, which by then had become an outspoken proponent of Lindenthal's plan, if not a downright mouthpiece of his, criticized the proposed legislation:

> It plainly contemplates the erection of a cantilever, and stipulates for the placing of one pier in the river channel, neither of which should be per-

> mitted unless found absolutely necessary, even if the cost be considerably increased. If there be one place in the world where a mere "utility structure" should not be permitted, but where dignity and beauty of form should be a controlling feature, it is over the North River at New York, and in that and other respects the suspension type seems to us to have great advantages for the location.

By midyear, federal legislation was also being proposed that would authorize a bridge company to build, within ten years of the approval of plans by the secretary of war, what was effectively a suspension bridge, for no piers were to be in the river. Among the promoters of the scheme were Lindenthal and Henry Flad, whose reputation, based on the Eads Bridge, was impeccable. As the wheels of Congress turned, there was considerable public discussion of the matter. With at least two bridge proposals competing for government approval, an editorial in *The New York Times* was optimistic that "we shall have a bridge across the Hudson into this city ere the century closes." Though the editorial did not mention Lindenthal by name, he was clearly being paid attention to: "An engineer of Pittsburgh, who makes bridges a specialty, has succeeded in gaining the ear of capitalists, and his calculations meet with respectful consideration from those who ought to know." The *Times* seemed to be alluding to the editors of *Engineering News*, but the newspaper itself had some reservations about his design: "The picture of this greatest of all wonders of bridge making offers much the same beauty of curve in the main span as the East River Bridge and more grace of outline in the towers, though the openwork steel construction of the latter compare unfavorably with the granite piers at Brooklyn." In fact, the open steel towers recalled unfavorably—to the editorial writer, at least—the Eiffel Tower, then under construction in Paris.

As *The American Architect and Building News* emphasized, the problem was not so much the length of Lindenthal's proposed bridge, for the "much-talked-of bridge over the English Channel would be 20 miles long." Furthermore, the longest bridge in the world was New York's elevated railroad, which consisted of a thirty-three mile-long "continuous bridge." What distinguished Lindenthal's proposal was the length of its main span, almost three thousand feet between two gigantic towers. Some readers of other publications were not so understanding. In the year following the public explication of the plans, London's *The Engineer* carried a critical appraisal by Max Am Emde. Lindenthal responded in a lengthy article in *Engineering News*, showing the more blunt and acerbic side of his personality, which included a tendency toward ad-hominem argument and sarcasm. Regarding the availability of information about the strength of steel

Late-nineteenth-century proposal for a railway bridge over the English Channel

wire for bridge cables, Lindenthal was critical of Am Emde's lack of knowledge: "Ignorance of it is inexcusable in an engineer, and unpardonable in a critic." And regarding the weight and number of trains that would be present at any given time, Lindenthal pointed out that "the bridge is not intended for use as a storage yard for loaded freight cars." This was an especially important point in designing bridges on the scale Lindenthal had proposed, and his argument that such bridges would be heavily loaded over their entire floor only during testing or with "special discipline" enabled him and subsequent American bridge engineers to design relatively light structures for their size, thus making them economically feasible, if potentially structurally unstable.

English engineers, on the other hand, still remembering the Tay and watching the Forth Bridge grow, remained sensitive to the consequences of too light a structure. That there was a rivalry in fact as well as in judgment between engineers on the two sides of the ocean was brought home by Lindenthal's closing his defense with the assertion that, in building bridges, American contractors had already established that they could hold their own with "contractors from England and other parts of the world." When the Forth Bridge opened in March 1890, *Engineering News* would also appeal to national pride: "If English and Scotch railways can afford to bridge the Firth of Forth . . . the great trunk lines of this country and the City of New York combined should surely afford to bridge the North River." The Brooklyn Bridge had for too short a time "stood unrivaled among bridge structures for its length of span," and the North River bridge project was an opportunity for American engineers once again to "eclipse the latest effort of the Old World's engineers."

Lindenthal's great bridge may have brought him some prominence among railroad executives and readers of *Engineering News*, but two years after his report he was still a newcomer to New Yorkers generally. This was evident, for example, in a front-page story in *The New York Times* that described the thirty-five-foot-long model being made to help convince representatives of the federal government to "take hold of the project and build a national bridge," since private capital seemed impossible to raise and both New York and New Jersey were balking at the price tag. In one place in the story, the "Pittsburgh man" had his first name misspelled "Gustave," and in another his surname was given as "Lilienthal." In spite of, or perhaps to overcome, such misnonymity, he seemed to take his case wherever there was an audience, which in 1889 included the meeting of the American Association for the Advancement of Science in Toronto. Lindenthal, furthermore, like Eads and Roebling and Baker before him, and like the builders of the greatest bridges after him, understood that different audiences had to be addressed in different ways. Politicians in Washington might best be swayed with a tangible model, but scientists meeting in Toronto would more likely listen to reason couched in terms of anthropology and natural science and of units of time approaching the scales used in geology.

"That facility and rapidity of communication is a primary cause of civilization is recognized as an axiomatic truth," Lindenthal began, and he proceeded to demonstrate that through practice his rhetoric was coming to be as sharp as his science: "The art of bridge building is ancient; the science of bridge building is modern." He traced the development of bridge types, concluding that the suspension bridge was "as old as mankind itself, perhaps even older," and he showed himself capable of scientific thinking in the Darwinian mode:

> Zoölogists tell us of the methods employed by apes in crossing streams. Failing to find a fallen tree to act as a beam or truss bridge, or failing to find meeting tree branches forming a sort of cantilever bridge, apes, we are told, form a chain by hanging together hands and tails, and suspended thus over a stream from tree to tree, the rest of the tribe climb along this living bridge from shore to shore. The strength of the chain with its weakest link, was, in this case, the weakest monkey; and there can be little doubt that occasional failures of such living bridges must have engrafted that bit of wisdom early on our anthropological ancestors. Modern bridge engineering, based on mathematical deductions, could not improve it.

A popular view of a possible stage in the evolution of bridges

The address was also full of technical details about strength and economy, demonstrating Lindenthal's rational approach to bridge design and sound judgment about it that would, decades later, influence favorably the best of those engineers who would work under and learn from him. Engineers who would not heed the lessons of the master, especially with regard to a sense of natural and artifactual history, would find themselves embarrassed.

Near the end of his address to the scientists, Lindenthal asked rhetorically of his dream, "How long will such a bridge last?" And he answered:

> If well maintained under the most competent engineering superintendence, there is no reason why it should not last as long as the Egyptian pyramids. They were built of more perishable material than steel and iron, provided iron and steel are kept well painted and free from rust. Rust and man are indeed the only destructive agencies for such a structure. No tornado could blow the structure over. No earthquake could shake it down, unless it were so great that the rock would cave and split, and swallow up the North River.

He also recognized, however, that there was a potentially more destructive force, and Lindenthal began to slip into a political mode that was out of character for this address but would become more and more a part of his rhetoric in his later years. Some of his observations were prophetic; in his desire for aesthetically pleasing bridges he found a silver lining even in the clouds of war.

> Man is more destructive to structures than decay and rust. The necessity of war may bring about the destruction of large bridges in the future as it has in the past. This may not always be an unmixed evil, if inferior structures are destroyed and rebuilt by grander ones. The taste and desire for architectural harmony is growing, though as yet the standpoint of utility, without regard to appearance, is too prominent in most of our bridge structures.

As Benjamin Baker, near the end of his lecture on the Forth Bridge, had invoked and quoted Thomas Pope, so Lindenthal closed his address by recalling that Pope, "an ingenious and ambitious shipwright," had eighty years earlier designed a gigantic wooden bridge, "partly cantilever, partly arch," that would have crossed the East River between New York and Brooklyn in one eighteen-hundred-foot span. At the time, Pope had exhibited a model of his bridge, which was never realized. But, rather than see in the story of

Thomas Pope's early-nineteenth-century proposal for a bridge across the East River between New York and Brooklyn

Pope a fatal paradigm for his own endeavors, Lindenthal saw hope. Pope had also proposed a single-span bridge to cross the North River, which he described, as he did other of his bridge plans, in "quaint verse":

> *Like half a rainbow rising on yon shore,*
> *While the twin partner spans the semi o'er,*
> *And makes a perfect whole that need not part*
> *Till time has furnished us a nobler art.*

Lindenthal took it upon himself to take over Pope's dream and to fulfill it in a more modern material and form. He believed that he was indeed a master of the "nobler art" that had evolved into an "exact science" by the end of the nineteenth century, and he was resolute. Determination alone, however, does not build bridges.

4

In the early spring of 1890, a bill was passed in the U.S. House of Representatives, and by early summer in the Senate, authorizing the North River Bridge Company to begin construction within three years, and requiring it to complete the structure within ten years after it was begun. With the approval of Washington, no further action by the notoriously contentious state legislatures was needed, nor was it sought, though it might have helped generate more solid local support for the bridge. In the meantime, the Consolidated New York & New Jersey Bridge Company had been formed out of the old 1868 charter issued by New Jersey and a more recent one issued by New York. The consolidated company did not have the federal authority needed to cross the interstate waterway, however, and so Lindenthal's North River company seemed to have the edge. *Engineering News*, comparing the two companies, noted that Consolidated had plans of which "the public has never seen even an outline," although there were newspaper reports that "the big bridge is started."

In fact, a ground-breaking did occur on Christmas Eve, 1891, but "the circumstances attending the ceremony of turning the first sod were somewhat inauspicious," for there was a pouring rainstorm, and the New York dignitaries and New Jersey delegates never did meet because of unclear directions to the site. However, even though some temporary trusswork was erected over an excavation for a tower on the Bergen County line, the company was believed to have had very little capital to proceed much further. It was reasonable to speculate, in fact, that Consolidated was hoping to have its charter bought out by the North River Bridge Company. *Engineering News* ridiculed the "spectacle" of "a few hackfuls of projectors trying in vain to find each other in the open country, turning over a single sod in a pouring rain storm," but the journal was dead serious when it commented that it would be "sorry, indeed, to see any bridge design carried out in which symmetry and dignity of appearance are ignored, or in which the river channel is needlessly obstructed." But two bridge companies continued to lay claim to charters for a Hudson River Bridge, and the similarities with the situation in St. Louis almost three decades earlier were not lost on close observers.

There is often a good deal of uncertainty as to exactly where a great bridge will be built, not the least for reasons connected with raising the capital. Among the most costly of items can be the purchase of land for the piers, anchorages, and approaches to the bridge, and if the location of these was fixed too early in the planning stage, real-estate speculation could increase the cost manyfold. Thus, as *Engineering News* pointed out in comparing the styles of the rival companies:

> The North River Bridge Co. is wise enough to see that before an enterprise involving the expenditure of from $60,000,000 to $80,000,000 can be honestly and successfully launched something besides printer's ink and wind is necessary. In enterprises of a great magnitude like this, everything depends, not upon the sale of isolated small blocks of stock to an uninformed public, but upon convincing great capitalists of the feasibility and future of the project. Before the latter can be done every detail of the plan must be worked out, every item of cost estimated upon sound data and

> traffic problems so carefully and exhaustively studied that the scheme can successfully stand the searching inquiry into its intrinsic merits that capitalists will surely inaugurate before putting money into it. It is worse than folly to invite general investment before this is done. And wide publicity of the exact location of proposed terminals and of other construction details is just the thing the business man will not seek until he has secured a considerable portion of the real estate necessary for his purposes. Nor will the promoter of a bona fide bridge take active steps in raising capital until he is fully informed as to the total cost of his scheme.

As an engineer, Lindenthal may have been cautious to a fault, for good engineering also involves decisiveness and an ability to fix on best estimates and go ahead with the business of raising money and turning sod. The strategy of detailed engineering analysis and uncertainty of location, as articulated in the journal that had virtually become Lindenthal's soapbox, was not exactly working. To give a sense of the obfuscation and the concomitant confusion that in some cases may have been deliberately introduced by both bridge companies during the years from 1886 to 1890, the location of the bridge was reported in various sources as terminating in Manhattan: "near Desbrosses St." . . . "somewhere between Seventieth and Eightieth streets" . . . "at about Sixtieth St." . . . "between 10th and 181st Sts." . . . "between Washington Heights and Spuyten Duyvil" . . . "at Fourteenth-Street" . . . "at Fort Washington" . . . "at any point in the city of New-York" . . . "near 13th St." . . . "about Forty-second St."

Financial conditions turned poor in 1893, and the prospects for any bridge across the Hudson looked bleak, especially as talk of tolls began to worry the railroad companies, among others. Cost of use was a real concern, for in early 1894 President Grover Cleveland had vetoed a bill that had passed Congress and that appeared to authorize the Consolidated New York & New Jersey Bridge Company to charge tolls on mail that passed over the proposed structure. The bill was vetoed also because it allowed for piers in the river, but they were still considered essential by those who did not believe a single span, suspended or not, to be possible.

With rival factions continuing to propose conflicting solutions, Cleveland appointed a commission of engineering experts "to recommend what length of span, not less than 2,000 ft., would be safe and practicable for a railway bridge across the Hudson River, between 59th and 69th Sts." The board comprised Louis Gustave F. Bouscaren, William H. Burr, Theodore Cooper, George S. Morison, and Colonel C. W. Raymond, "all well known to American engineers." Bouscaren had been born on the island of Guadeloupe in 1840 and was an 1863 graduate of France's Ecole Centrale des Arts et Manufactures. He had, among many other accomplishments, strengthened the cables of Roebling's suspension bridge across the Ohio River at Cincinnati, and had built a railroad bridge across the Ohio that was at one time the longest truss span in the world. Burr, born in Watertown, Connecticut, in 1851, was an 1872 C.E. graduate of Rensselaer Polytechnic Institute who had taught at Rensselaer, worked as assistant to the chief engineer of the Phoenix Bridge Company, and taught civil engineering at Harvard before joining the faculty at Columbia. Cooper was, of course, at the peak of his career. Morison, born in Bedford, Massachusetts, in 1842, was educated at Phillips Exeter Academy and Harvard, from which he was an 1863 arts graduate and an 1866 law graduate. Though admitted to the New York Bar in 1866, he never practiced the legal profession. He did go on to gain extensive experience in bridge designing and building, however, beginning in 1867 with work in Kansas City under Octave Chanute on the bridge over the Missouri River at Kansas City. On his own, Morison had been engaged in many bridge projects, including the Cairo Bridge over the Ohio River, which was among the longest bridges in the world in the 1880s. Charles Walker Raymond had been born in 1842 in Hartford, Connecticut, and graduated in 1860 from the newly formed Collegiate and Polytechnic Institute of Brooklyn, where his father was professor of English language and literature, and entered West Point the following year. He had a distinguished career with the Corps of Engineers, including charge of the Mississippi River levees and work on important harbor improvements, and would go on to play a key role in supervising the design and construction of the Pennsylvania Railroad Company's improvements in, under, and around New York City. It was Raymond (who had shown an early talent for mathematics) who was to draft the analytical discussion of the theory of suspension bridges contained in the report of the board over which he presided.

Within three months of its appointment, the board reported that it was "of the unanimous opinion that a cantilever span of 3,100 ft. in the clear could be built and would be a safe structure," but it would cost in excess of $50 million to cross the river thus without a pier. A pier in the center of the river would reduce the spans to two thousand feet and cut the cost of the bridge's superstructure in half, but the foundations would have to be dug to 260 feet below the water, which not only would be dangerous for the workers but also would add an uncertain amount of $10 million or so to the cost. A suspension bridge, with six tracks and a span of thirty-one hundred feet, could be built, it was thought, for about the same amount as the shorter-spanned cantilever. Furthermore, if a lighter bridge was acceptable, a safe but more flexible suspension bridge could be built for about $30 million.

On balance, taking into account the uncertainty associated with digging deep foundations, the board concluded in favor of a suspension bridge.

Another board of experts had been appointed earlier in 1894 by the secretary of war to look into questions relating to building bridges over navigable streams and, in particular, into the question of "the maximum length of span practicable for suspension bridges," and to look into matters of "strength of materials, loads, foundations, wind pressure, oscillations and bracing." The board comprised three members of the Corps of Engineers—then Major Raymond, and Captains William H. Bixby and Edward Burr—and its report acknowledged Lindenthal, Wilhelm Hildenbrand, and Leffert L. Buck, "for information and valuable suggestions." Appendixes were contributed by Lindenthal (on temperature strains in hinged arches) and Josef Melan (on the theory of the stiffening girder). Clearly, the board of army engineers had gone into considerable technical depth in the nine months it took to prepare its classic report, which treated in detail matters of oscillation and other causes of failure in suspension bridges, and thus provided "one of the most valuable and instructive engineering investigations of the day . . . in a field that has hitherto been practically unexplored," according to *Engineering News*. The conclusion of the board was that a six-track suspension bridge of thirty-two-hundred foot main span was practicable at an estimated cost of $23 million, and that traffic in 1894 warranted such a bridge, although it should be so constructed that its capacity could be increased in the future, as needed. In addition to addressing the Hudson River problem, the board had looked at the more general feasibility question, and concluded also that it was possible to build a suspension span as long as 4,335 feet.

Though the report of the army engineers removed technical objections to the suspension bridge, it did not fully dispose of financial objections. Indeed, even *Engineering News* admitted that, whereas it had been projected that there was rail traffic enough to cover the actual construction cost, it was not clear that the bridge could "attract a traffic sufficient to pay the interest on its cost." The Consolidated New York & New Jersey Bridge Company challenged the objections to a pier in the river, and also questioned whether the foundation for such a pier had to be dug so deep and therefore had to be so expensive as was feared, but the secretary of war continued to rule in favor of a suspension bridge. The argument for a cantilever did not end, however, in part because of the success of the Forth Bridge and in part because of the vulnerability of the suspension-bridge type to attack. Traffic on Brooklyn Bridge was an ongoing problem, aggravated in part by the structure's inability to carry heavy engines, thus requiring that cable cars be used on the bridge, and switching them about at the terminals presented an endless scheduling and capacity problem. To make matters worse, the bridge that had been held up as the counterexample to the persistent belief that suspension bridges could not carry railroad traffic, John Roebling's Niagara Gorge Bridge, was in the process of being replaced—and a cantilever was being proposed. The forty-year-old landmark bridge was showing signs of wear, and the weight of railroad trains had increased considerably since it was built. In reporting this development, the praise of *Engineering News* sounded faint indeed:

> To Mr. Roebling belongs all the credit for teaching engineers how to use wire in this form in a railway bridge; and that his connections were faulty in the light of modern practice, and that his stiffening truss was no such truss at all, does not detract from his boldness as an engineer and the services he performed in developing the manufacture of wire in this country.

The cantilever was well suited to the eight-hundred-foot span over the Niagara Gorge, and it would be the "cheaper, stiffer, and better structure," admitted *Engineering News*, but the suspension bridge was still the bridge of choice for spans on the order of three thousand feet.

The recent competition for a bridge over the Danube at Budapest was cited as an indication that "engineers are only now beginning to more carefully study the principles and details of suspension bridge construction." Indeed, first prize in that competition went to a thousand-foot-main-span wire-cable suspension bridge, but the design did not receive the bonus prize money that would have been awarded had the cost of the bridge not exceeded $1 million. In fact, the bridge—whose land spans looked like the side aisles of a Gothic cathedral, and whose tollbooths were built over the anchorages, also forming pedestals for equestrian statues—was estimated to cost almost twice that amount. Second and third places went to some very handsome cantilever designs, one of which looked like a suspension bridge in profile, and each of which was estimated to cost under $1 million. Of the remaining seventy-odd designs that were submitted from Europe and America, three additional ones were bought for possible use in Budapest. Among these was a chain suspension bridge, the only design of its kind submitted. This bridge and the top three winners were illustrated in *Engineering News* in 1894; the journal unfortunately used words alone to describe some others, "which seemed to be intended only to furnish amusement to the jurors in their arduous work":

> One design for a one-span bridge at Eskuter shows curiously curved trusses, freely resting on abutments. The widely separated chords of the

> trusses are stiffened by enormous rings, and are ornamented by a legion of saints' statues. The wagon traffic moves over a suspended roadway, while the foot passengers climb over the bold curve of the top chord. Another fantastic design consisted of an iron tube of 1,020 ft. span, made up of Mannesmann tubes placed parallel and connected with each other by iron bars, riveted in spirals to the tubes.

As *Engineering News* was to editorialize, on the occasion of the replacement of the Niagara Gorge Suspension Bridge with a stiffer steel structure, "there is no knowing to what flights over space the bridge of the future may attain." Money was admitted to be the limiting factor. In bridging the Hudson, the question was not of money alone, however, for the secretary of war would simply not approve a cantilever design with a pier in the river. The Consolidated New York & New Jersey Bridge Company thus asked Theodore Cooper, its consulting engineer, to prepare specifications for a suspension-bridge design. Since he had never designed such a bridge himself, his specifications only covered such things as the load the bridge was to carry, the foundation conditions, and materials of construction. Bidders were invited to select the geometrical outlines, and Cooper's firm leaned toward the design of the Union Bridge Company, which guaranteed to build for no more than $25 million a 3,110-foot span with "immense rigid trusses" supported by twelve cables. The design was that of Charles MacDonald, organizer and president of Union, who had been born in Ontario, Canada, in 1837. After working on surveys for the Grand Trunk railroads, he entered the United States in 1854, immediately began studies at Rensselaer Institute, and received a degree in civil engineering in 1857. Among much other railroad and bridge-building experience, MacDonald supervised the design of the great cantilever bridge across the Hudson at Poughkeepsie, but his suspension-bridge design for the New York crossing was an unharmonious concoction.

Lindenthal's North River Bridge Company was denying rumors that it was going to relinquish its charter, which was to expire in mid-1895, "unless something were done by that time showing the sincere purpose of the company to construct the work for which they have obtained powers." The company claimed that "work had quietly commenced some time ago upon the New Jersey anchorage," and that it had spent more money acquiring property and advancing the plans than its rival. In fact, according to a cornerstone reportedly snatched from the jaws of destruction almost a century later by Lindenthal's grandson, ground was indeed broken on June 8, 1895, and the first foundation masonry was laid at the site of the Hoboken anchorage, opposite Manhattan's 23rd Street. What was needed back then, however, was not ceremony but $21 million for the bridge proper and $15 million for property and accessories, which was admitted to be "an enormous sum of money, and the financiering of the bridge far exceeds in difficulty the engineering problems presented, unprecedented as these last are."

In the meantime, support was growing for completing construction of a tunnel under the Hudson, since the bridge companies continued to focus on elevated approaches, which was the costlier method of getting railroads into the city. Late in 1897, an editorial in *Engineering News* accused the bridge promoters of failing "to appreciate the fact that it is the suburban traffic, and practically that alone, on which their structures must depend for income" from tolls. The journal, which after the death of Wellington had no longer simply embraced Lindenthal's ideas, had become a voice of reason. In a letter challenging the editorial, Lindenthal simply reiterated his position on the bridge, which pretty much everyone by then knew, or was expected to know. But there were alternatives.

The Hudson Tunnel Railroad Company had been chartered in 1873, and ground was broken the following year. There were to be two tunnels, each containing a single track, but opposition lawsuits delayed the work until the end of the decade, and work was stopped in 1882, after over a million dollars had been spent but no more money could be raised in America. John Fowler and Benjamin Baker were approached in late 1887 and asked if they thought the tunnels could be completed for the amounts American contractors were estimating—namely, $900,000 for one tube and $1.2 million for the other. After consulting European tunnel engineers, and after a visit to the unique New York site by Baker, who inspected the books of the contractors to understand the cost of American labor, the designers of the incomplete but already famous Forth Bridge gave their support to the Hudson River tunnel scheme, which brought $1.5 million of British money into the project.

By the turn of the century, tunnel projects in and around New York were inextricably associated with the names of William Gibbs McAdoo, an entrepreneurial lawyer from Georgia, and Charles Matthias Jacobs, a Yorkshire-born, privately tutored, and apprentice-trained engineer who had come to America in 1889 and subsequently designed tunnels for rapid transit and gas lines under New York's East River. An exact contemporary of Lindenthal's, Jacobs had become involved with Hudson River tunnels in 1895. As Jacobs and McAdoo were demonstrating the feasibility of tunneling under the Hudson, electric-traction locomotives were being developed, obviating the objection that smoke would choke passengers in the tunnels. Thus the Pennsylvania Railroad decided to build its own rail tunnels under the river, thereby removing themselves as the most significant potential supporter of Lindenthal's bridge. In the meantime, over the past decade,

Lindenthal had become established and well known as a consulting engineer in New York. In 1902, he found himself appointed by reform mayor Seth Low as the city's commissioner of bridges; this necessarily redirected his attention from the North to the East River, which was contained wholly within the city of New York. But intracity and intrastate politics could complicate bridge design and construction at least as much as interstate issues.

5

Even before the Brooklyn Bridge was formally opened in 1883, there were calls for additional bridges between Manhattan Island and Long Island, on which the then separate city of Brooklyn was located. A new bridge was proposed to connect New York with Brooklyn's Williamsburg section. Another was proposed farther north; here the presence in the river of Blackwell's Island reduced the size of spans needed, while an approach convenient to Brooklyn's City Hall was still possible. A charter for a Williamsburg bridge was obtained in 1892 by Frederick Uhlmann, whose interest appears to have been in extending the Brooklyn elevated railways into New York; this would have been a lucrative endeavor, given the congestion on the nearby Brooklyn Bridge, which was being loaded to its limit. When an East River Bridge Commission was formed, it bought out Uhlmann's charter and appointed L. L. Buck as chief engineer to design a bridge capable of carrying an elevated railway as well as trolley cars.

Leffert Lefferts Buck was born in Canton, New York, in 1837 and received bachelor's and master's degrees from the local college, St. Lawrence, before attending Rensselaer Polytechnic Institute. He graduated in 1868, his studies having been interrupted by the Civil War, which he entered as a private in the Sixtieth New York Infantry. After a period as assistant engineer on New York's Croton Aqueduct project, he worked on railroad bridges and other engineering projects in Peru, Mexico, Aruba, and many locations around the United States. He oversaw the rebuilding of various parts of Roebling's Niagara Gorge Suspension Bridge during the period 1877–1886 and, later, its replacement with an arch, which had superseded the cantilever proposal. He became chief engineer of the Williamsburg Bridge project in 1895 and would continue in that capacity until the bridge was opened in 1903 as the largest suspension bridge in the world, with a central span of sixteen hundred feet—four feet six inches longer than the Brooklyn Bridge. Its approaches were to be so long that the entire length of the bridge would stretch for seventy-two hundred feet, between Clinton Street in Manhattan and Roebling Street in Brooklyn.

Leffert L. Buck, chief engineer of the Williamsburg Bridge

When plans for the Williamsburg Bridge were first published in *Engineering News*, in 1896, it was criticized "from an aesthetical point of view," and there appeared to be considerable visual discontinuity associated with the roadway at the towers, which were nothing like the monumental stone towers of the Brooklyn Bridge. Indeed, the shift at the towers from an above-deck truss to an under-deck truss made the truss itself look as if it had been severed by some angled guillotinelike device. In spite of this image, Buck's towers were remarkable in that they were all steel, and they were defended by *Engineering News*, which was "utterly opposed to false ornamentation in similar structures and to any attempt to disguise the real materials of construction or the chief lines of stress." The journal did admit, however, that, "in a monumental work of this character, in the center of a great city, good taste in design and proper ornamentation must be considered; and if a more pleasing effect can be secured . . . the effort should certainly be made and is worth the added cost."

In his report to the commissioners in September 1896, Buck asserted that the bridge could be completed by January 1, 1900, at a cost of $7 million, which compared very favorably with the $15-million final price tag of the Brooklyn Bridge. With regard to the cost, *Engineering News* criticized the commissioners for being overly frugal with the salary of the chief engineer, upon whose "judgment, skill and experience the safety and convenience of

Sketch of an early design detail for a tower and the roadway of the Williamsburg Bridge

the great numbers who will use the bridge for generations to come, depend solely." Buck's salary was $10,000 per year, whereas "the same commission pays out about $75,000 for two and one-half years' work of a legal counsellor." The editorial went on to anticipate what Lindenthal would say of Cooper a decade later—namely, that the low compensation granted engineers was a reflection of the "ignorance of the true value of the engineer" in such a large project, and it was "humiliating to the whole profession of engineers." The question of the compensation of engineers versus lawyers was especially keen at the time, because an injunction had been sought against the commissioners, who had limited bidding to those who could supply steel made by the "acid open-hearth process," as specified by Buck. The court refused to issue the injunction, and *Engineering News* praised the decision, concluding, "An engineer may not be infallible in his decision of engineering questions; but we shall make no gain by setting a lawyer to review his decision." An engineer like Buck was receptive to criticism, however, and the lines of the deck were much improved in a revised view of the bridge published later in the year. Though it retained the straight-cable profile on the land-based spans, because they were to be supported from below as girders and not suspended from the cables, the deck had achieved a continuity that Buck's earlier sketch had lacked. Whether the Williamsburg Bridge would be seen as a graceful swan or as an ugly duckling beside the Brooklyn Bridge would be largely a matter of taste.

Since a suspension bridge was required by the legislation authorizing the crossing, a cantilever could not be considered, even though it might have been more economical. However, economic considerations rather than aesthetic ones did strongly influence the appearance of the Williamsburg Bridge, especially with regard to its towers and cables, among the most costly components of any suspension bridge. Furthermore, economic and technical design factors are often intertwined. The decision, for example, to have the cables come straight down from the towers to the anchorages, and not support the land spans, meant shorter and lighter cables could be used, thus reducing their size and thereby their cost. Had stone or masonry towers been employed, they would have had to be very wide and heavy, in order to accommodate all the rails and roadways that would have had to pass through them. Employing lighter steel towers made smaller foundations (very costly components of any bridge) possible, and much time and cost were saved. "Roughly speaking, masonry towers would require foundations twice as large, would cost five times as much, and would take three times as long to build," according to a contemporary report. Moreover, using steel towers meant they could be built taller, thus allowing the cables to have a deeper curve; they did not have to be stretched so tight, and so could be smaller in diameter. Economic considerations also led to the choice of steel viaducts rather than masonry arches for the bridge approaches.

The construction of the Williamsburg Bridge was well under way when its administration was passed from a board of commissioners to the newly appointed commissioner of bridges, Gustav Lindenthal, on January 1, 1902, ending Buck's role as chief engineer. Though Lindenthal must have had severe reservations about the design and appearance of the Williamsburg Bridge, he avoided talking about them in his brief official address at the dedication ceremonies, in which he announced that the bridge was ready for traffic, on December 19, 1903. He simply described the monstrosity he

The Williamsburg Bridge, upon its dedication in December 1903

had inherited as "the heaviest suspension bridge in existence, and the largest bridge on this continent." In comparing the Williamsburg to the Brooklyn Bridge, he noted that the newer structure was twice as strong—something New Yorkers would have appreciated, since the limitations on the strength of the Brooklyn Bridge had curtailed the commuter traffic across it for some time. Nevertheless, it was the older bridge that was the architectural success: "The imposing and stately stone towers of the Brooklyn bridge give that structure the appearance of great strength, but in the steel towers of the new bridge, and in all its other elements, a greater power of resistance is hidden."

Rather than dwell on comparisons, however, Lindenthal spoke of the future of bridges. His words were prescient:

> So far as engineering science can foretell with confidence, this colossal structure, if protected against corrosion, its only deadly enemy, will stand hundreds of years in unimpaired strength. . . .
>
> Our city will be pre-eminently the city of great bridges, representing emphatically for centuries to come the civilization of our age, the age of iron and steel. A time must come, not many generations distant, perhaps not more distant than the crusades in the past, when the building of such colossal structures will cease because the principal material of which they are molded, that is, iron and steel, will not be [any] longer obtainable in

Gustav Lindenthal, as he appeared when he was commissioner of bridges for New York City, 1902–03

sufficient quantity and cheapness. When the iron age has gone, the great steel bridges of New York will be looked upon as even greater monuments than they are now.

For all of Lindenthal's grand projections into the future, the Williamsburg, along with many other New York bridges, would be in danger of collapsing long before the century was out. Forgetting his caveat about a bridge's "only deadly enemy," corrosion, New York and many other cities would during times of fiscal crisis neglect and defer maintenance of bridges like the Williamsburg to an alarming degree.

Even when the Williamsburg was young, there were problems with it. Within three years of its completion, headlines reported that, because it had "such a liking for the Borough," the bridge was "slipping to Brooklyn." Evidently, the bridge had been "out of place since it was built," but it was only then becoming known that "efforts to correct it had failed." According to *The New York Times*, "a piece of engineering computation of the utmost nicety" was taking into account every ounce of material in the structure to determine the needed adjustment, so that the heavy cars of the elevated railroad could be allowed to run over it. Studying, straightening, and strengthening the Williamsburg Bridge continued on and off for about a decade. Two additional supports were added under each of the (unsuspended) land spans, and additional steel was added to the deck so that it could carry the heavier subway cars that had been developed since the bridge was designed. In fact, something similar was going to happen to many of the bridges around the world, because of changing conditions and philosophy, as articulated in 1911 by one of the engineers involved with the Williamsburg Bridge strengthening project:

> Mr. Buck designed the bridge on the theory that traffic should adapt itself to the bridge; we are now proceeding on the theory that such a bridge should adapt itself to traffic and that it should be as good as any other for traffic purposes, and not be a weak link. Mr. Buck designed the bridge for small locomotives drawing trailers. To-day, in a six-car train, there are generally four motors, all heavier than any of the old locomotives. The trolley cars also have increased in weight. The bridge is perfectly able to carry its traffic to-day, but as it now stands it would be inadequate for the future. Ten-car steel trains will probably be run through the subway loop, for one thing, and for such conditions we must provide.

Buck had, however, designed a sound if unattractive bridge for the conditions he knew and under the conditions he worked. When some corrosion was discovered on the wrapping of the cables, they were uncovered in 1921 and found to be well preserved, and the steelwork generally to be in "perfect condition." At that time, *Engineering News-Record*, which had in 1917 been formed of a merger of *Engineering News* with the *Engineering Record*, noted that this condition of the cables "gave fair assurance that main parts of the great New York suspension bridges have an indefinite length of life." Indeed, the journal was tempted to say, they had an "unlimited life," if properly cared for:

> Such bridges, in other words, are not subject to perceptible decay, and so far as corrosion is concerned they may remain free from measurable deterioration if intelligent inspection and maintenance are applied. As we look at the ancient stone bridges of Europe we reflect with wonder and admiration on their endurance through the ages; yet it is not beyond the bounds of possibility that our great steel bridges may survive as long.

Intelligent inspection and maintenance are more readily called for than provided, however, as has been discovered in more recent years. Times and conditions change. Even the great stone monuments of Europe have been found to be susceptible to increasingly acidic environments. Inspection can uncover deterioration, but arresting or reversing it is another matter. Yet, in the early years of this century, when vehicle emissions were not even dreamed to pose the threat to stone and steel that we know them to today, bridges continued to be designed for the conditions of the time. And there were many bridges to design.

6

If the Williamsburg Bridge was well under way when Lindenthal became commissioner of bridges in 1902, two other East River crossings were not. Both the Blackwell's Island Bridge, farther north, and the Manhattan Bridge, to be constructed between the Brooklyn and the Williamsburg, were still on the drawing board. (Though the foundation for the Brooklyn tower of the Manhattan Bridge had actually been contracted for, this in no way meant that changes could not be made in the design of the towers themselves or the general superstructure.)

Shortly after a city ordinance authorizing a third bridge across the East River between Brooklyn and Manhattan was signed by the mayor early in 1900, bonds amounting to $1 million were issued, engineering work was begun, and bids for foundations were invited by early March. Since the Manhattan Bridge, as it was called from the beginning, was to be located wholly within the city of New York, the process was relatively efficient. The design had been largely completed when Lindenthal took over as bridge commissioner.

Among Lindenthal's early frustrations in the job were the delays accompanying the cable-making for the Williamsburg Bridge, which "dragged woefully" well into 1902. The cables were being spun by the John A. Roebling's Sons Company, and Lindenthal "was very much annoyed at the delays shown by the Roeblings in executing their contract." When the contract expired months before the cables were completed, Lindenthal deducted $1,000 a day from the payment to the company, which in the end amounted to a penalty of about $175,000. The Roebling firm, which had been excluded by New York politics from supplying the wire for John and Washington Roebling's own bridge, took the city to court, claiming that the bridge commissioner had not furnished them the space needed for their operations, and they were awarded the money that had been withheld. Whether it was the frustration accompanying the delay or the poor relations with Roebling's Sons, Lindenthal turned away from wire-cable suspension bridges and redesigned the Manhattan Bridge with eyebar chains, a system he had employed for the Seventh Street Bridge in Pittsburgh.

The original plans for the Manhattan Bridge were made under the supervision of chief engineer R. S. Buck, who, though no relation to L. L.

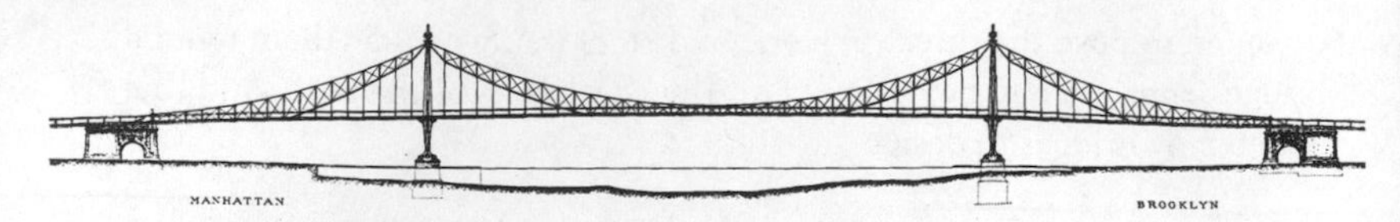

Lindenthal's design for the Manhattan Bridge, employing eyebar chains

Buck, had worked as assistant to him in calculating stresses and as resident engineer on the Niagara arch-bridge project. Shortly after the new bridge commissioner assumed office, R. S. Buck resigned, and Lindenthal assumed the engineering work on the East River structure. Commissioner Lindenthal's first semiannual report announced that changes in the plans for the Manhattan Bridge were prompted by the delays on the Williamsburg Bridge, but he also gave the positive reason of economy of construction and maintenance, an argument that was by no means universally accepted by bridge engineers. The new plans, published in *Engineering News* early in 1903, showed a radically different suspension bridge, with towers that were not rigidly fixed at their base, and with the stiffened eyebar cables that Lindenthal preferred. As with his still-unrealized North River Bridge, this meant that the stiffening system was incorporated into the cables rather than being part of the deck structure. The Manhattan design was said to be so stiff it could be thought of as an inverted arch.

Though Lindenthal argued that the design had architectural as well as engineering merit, the mayor submitted it to a board of five engineers that he had appointed: Lieutenant Colonel Charles W. Raymond, the army engineer; bridge engineers George S. Morison, C. C. Schneider, and Henry W. Hodge, of whom more shall have to be said later; and Professor Mansfield Merriman, an 1871 graduate of the Sheffield Scientific School at Yale and head of the Civil Engineering Department at Lehigh University since 1881. All of the board members were members of the American Society of Civil Engineers, which added credence to the appointments. Theodore Cooper—whose Quebec cantilever approach span was just under construction, but who had little experience with suspension bridges—would replace Raymond before the board reported.

While the board studied Lindenthal's plans, a debate raged in the pages of *Engineering News* over the relative merits of eyebar chains and wire cables for suspension bridges. Wilhelm Hildenbrand, who had drawn the earliest plans for the Brooklyn Bridge and who was now engineer of cable construction for the Roebling's Sons Company, pointed out that eyebars were not a novelty, having long been used in Europe almost exclusively and even in America for "small suspension bridges." Though the reasons given

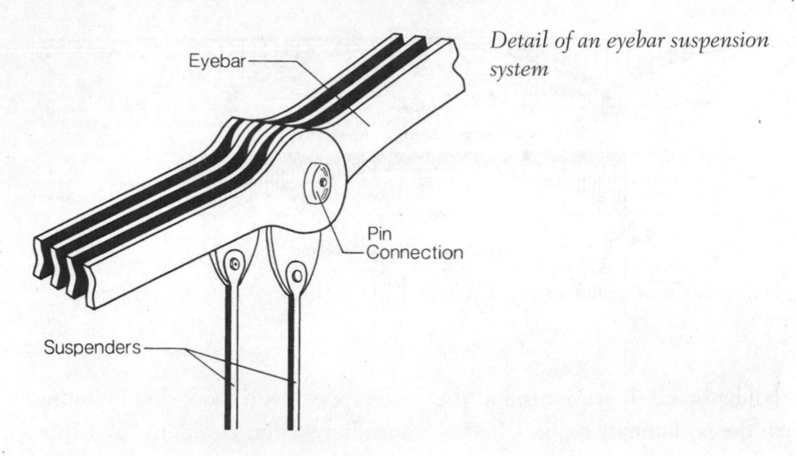

Detail of an eyebar suspension system

for the design change "might be accepted, unquestioned, by a private company from its chosen engineer, and in that case would probably remain unchallenged," Hildenbrand asserted that the commissioner of bridges of the city of New York could not prevail so easily, for the switch from cables to eyebars would add "two to three millions of dollars" to the cost of the Manhattan Bridge. By Hildenbrand's calculations, had the Williamsburg Bridge been built with eyebar chains rather than cables, its cost would have been increased by over $3 million. Although the editors of *Engineering News* might have been more supportive of Lindenthal in years past, they were not now, when he was being challenged by "one of the most experienced suspension bridge engineers in the country." Hildenbrand had been associated with New York engineering projects since 1867, but Lindenthal had yet to have a single design of his own realized in the politically charged city. If he had hoped to use the power of his office to redesign a New York bridge to his own prejudices, he was not going to have an easy time of it, although at first it appeared that he might get his way.

When Lindenthal was about to submit his plans for the Manhattan Bridge to the city's Art Commission, which offered opinions on the aesthetics of large structures, the board of engineers issued a preliminary report. According to the engineers, who believed that the chains had decided advantages with regard to erection and maintenance, "they are to be preferred to wire cables whenever the cost of the chains is not materially greater." The final report, which had awaited the results of tests of material and information on the availability and cost of eyebars, was issued in June, with a unanimous recommendation for the adoption and execution of Lindenthal's design, even though no firm cost comparisons were yet available.

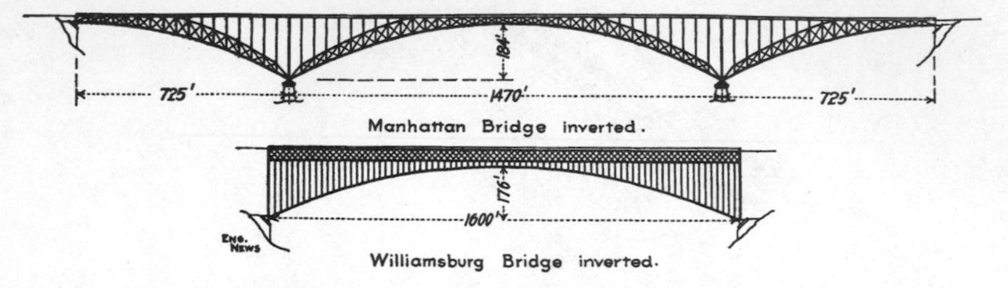

Diagrams showing how suspension bridges may be thought of as inverted arches

Hildenbrand, however, found the final report "even more disappointing than the preliminary report." It was "altogether so non-committal and spiritless that it is to be wondered whether the Mayor will consider it worth the money it has cost." The report was indeed confusingly terse and vague, and Hildenbrand was correct in observing, "That the experts are all engineers of high professional standing does not alter the facts and figures, nor does it make the weak points in their report stronger! It merely emphasizes the weakness." *Engineering News* and many other observers thought that a sensible way to resolve the issue would be to invite bids for both eyebar-chain and wire-cable designs and thus compare hard cost estimates, but the city appropriation mechanism did not allow such a commonsensical course, and so the debate continued.

Though the Municipal Art Commission approved Lindenthal's design, having reservations only about the decorations on the towers, there remained strong differences of opinion as to the comparative aesthetics of chains versus cables. George W. Colles, a Canadian engineer, wrote to the editor of *Engineering News* that "a chain-bridge is a very ugly thing—excusable only on grounds of engineering expediency." He thought that New York was "ugly enough already," and that it was "bad enough to have skeleton-bridge-towers," like those on the Williamsburg, "without the added eyesore of a chain-bridge." Colles decried the report of the experts as an example of a widespread "engineering impressionism," by which the opinions of experts were defended solely on the basis of their being declared experts. He observed that this "merely shifts the burden of the real decision from the engineer to the capitalist." Specifically addressing the claim of Lindenthal and the experts regarding the accessibility of chains for inspection, Colles correctly pointed to the "innumerable crevices and cracks which invite capillary absorption and subsequent corrosion," a problem that was to be the root cause of the collapse of the bridge over the Ohio River at Point Pleasant, Ohio, in 1967. That structure was nicknamed Silver Bridge because it was among the first to be painted with aluminum paint, but it was found not to be very easy to maintain or inspect the tight details where the eyebars were connected to one another, and where cracks could grow to dangerous proportions and lead to the bridge's sudden collapse.

The debate over the Manhattan Bridge continued, with Lindenthal himself responding to letters attacking his Pittsburgh chain bridge as the "ugliest of the three" over the Allegheny and pointing out its early foundation problems. In an editorial, *Engineering News* announced that some of its readers had gotten the impression that the preponderance of letters in support of wire cables reflected the editorial stance of the journal, and it invited more letters from supporters of eyebars. Perhaps in response to such criticism, which may very well have been coming from Lindenthal himself, the journal began to supply him with proof copies of letters, so that he might respond in the same issue. The commissioner's frustration began to show in his style, his longish letters seeming more and more frequently to end in sarcasm and condescension or worse. In response to a letter from Hildenbrand on the comparative strength of East River bridges, for example, the value of his opinions was estimated by Lindenthal to be "no greater than the value of the weather prophecies in a Farmer's Almanac." He dismissed another correspondent by saying that he had "yet much to learn in bridge engineering before essaying to discuss that intricate subject." By the end of 1903, an election year, all parties involved seemed to be out of patience, and *Engineering News* editorialized that "the best way for the new administration to decide whether it shall build an eye-bar bridge or a wire cable bridge across the East River at the Manhattan Bridge site will be not to build any bridge at all."

If *Engineering News* abandoned Lindenthal, *The New York Times* did not. The newly elected mayor appointed a new bridge commissioner, George E. Best, and he decided to throw out Lindenthal's design and return to the previous one. The *Times* in turn accused Best and his advisers of "personal spite against his predecessor, Commissioner Lindenthal, and a fixed purpose not to do anything that Mr. Lindenthal proposed." As late as 1906, the newspaper was still advocating a chain over a wire bridge and calling for competitive bids to settle the question. In the end, however, Lindenthal's tower designs, modified at their base to be more rigidly connected to the foundations, were all that survived of his ideas. The first strands of wire were run across the East River in 1908, after Mayor George McClellan announced that the bridge would be completed in December 1909 and that he would walk across it before his term of office expired. In late 1908, the

Views of the towers of the Manhattan Bridge, as redesigned in 1904 as a wire-cable structure

mayor himself pulled a lever sending the last of the wire strands across the East River; the Manhattan Bridge was formally opened on December 31, 1909, the last day of McClellan's administration.

When the Quebec Bridge collapsed during its construction in 1907, large bridges everywhere came under scrutiny. Rumors about the soundness of the Williamsburg Bridge began to surface, and concerns were also raised about the design of the Manhattan Bridge. The engineer appointed "to watch the construction of the new bridge" and check the plans was Ralph Modjeski, described in a contemporary report as "the leading engineering authority on bridges" in America, if not the world—in large part because of his membership on the board of engineers for the reconstruction of the

Quebec Bridge. Although he had begun his bridge-building career twenty years earlier with the design of a major double-deck railway-and-highway structure across the Mississippi at Rock Island, Modjeski was still often identified in the popular press as the son of the well-known actress Madame Modjeska, his name sometimes being misspelled with the feminine ending. In fact, the original spelling of their surname was much more complex than either had come to use in America.

Rudolphe Modrzejewski was born in Cracow, Poland, on January 27, 1861, the son of Gustav and Helena Modrzejewski, who, as Helena Modjeska, was to become known as "the première tragedienne of her time." According to his mother's memoirs, Rudolphe came to America with her for the first time in 1876, when they visited New York, Philadelphia, and the Centennial Exposition. As they crossed the Isthmus of Panama on the first transcontinental railroad, on their way to California, the young man declared that "someday he would build the Panama Canal." Although she remembers him as "even then determined to become a civil engineer," a career as a pianist was evidently also a possibility, for he had been well trained musically and was said to be a leading exponent of Chopin. Indeed, young Ralph, as he preferred to be called in America, was at one time a fellow student with Ignace Paderewski.

The engineer-to-be practiced piano a great deal while traveling with his mother; throughout his life, he was known to play the instrument almost every evening and for a couple of hours each Sunday. It may have been in his travels to America that Ralph Modjeski saw the needs and opportunities for artistry in a medium that used steel thicker than piano wire, but the young man decided to take up the study of civil engineering where it had first been taught, at the Ecole des Ponts et Chaussées, in Paris.

Writing three decades later, Madame Modjeska did not mention any difficulty her son had in getting into the prestigious school. However, decades later still, he himself would recall the experience, on the occasion of his receiving the Washington Award. This award was established by the Western Society of Engineers "as an honor conferred upon a brother engineer by his fellow engineers on account of accomplishments which preeminently promote the happiness, comfort, and well-being of humanity." The selection, made upon the recommendation of a commission representing the major national engineering societies and the Western Society of Engineers, was first bestowed in 1919 on Herbert Hoover, "for his pre-eminent services in behalf of the public welfare." Among the other eight engineers honored before Modjeski were Arthur Newell Talbot, founding professor of theoretical and applied mechanics at the University of Illinois, and Michael I. Pupin, for his work on long-distance telephoning and radio broadcasting. Ralph

Modjeski himself was recognized "for his contribution to transportation through superior skill and courage in bridge design and construction." At the awards ceremony, he remembered how he had become an engineer and why, and he spoke of his perseverance toward his goal:

> When I was four years old I got hold of a screwdriver. This gave me an idea. I immediately investigated what this screwdriver was for and practiced on a door lock of the drawing room of the house we lived in and took it all apart. I could not put it together again. And my father said, "You will be an engineer."
>
> I persisted in that until . . . I failed in the examination for entrance to the *Ecole des Ponts et Chaussées* where there were 25 places and 100 candidates. Then for about six months I practiced music six and eight hours a day. After six months I began to think, and at the end of nine months had thought out my problem and joined the preparatory school and three months later I passed the examination into the *Ecole des Ponts et Chaussées*.

In 1885, Modjeski graduated first in his class, and he returned to America to build bridges, beginning his career under George S. Morison, who has been described as the "father of bridge building in America." In 1893, the young engineer opened his own office in Chicago, as a senior member of Modjeski & Nickerson, one of the several engineering firms with which he would associate his name throughout his long career. He soon found himself engaged in enlarging the Rock Island Bridge over the Mississippi River. After that, Modjeski would work on or direct, often as chief engineer, the design and construction of a wide variety of bridges, in a variety of locations, including, in chronological order: Thebes, Illinois; Bismarck, North Dakota; Portland, Oregon; Peoria, Illinois; St. Louis, Missouri; Quebec, Canada; Toledo, Ohio; Memphis, Tennessee; Keokuk, Iowa; Metropolis, Illinois; New London, Connecticut; Poughkeepsie, New York; Cincinnati, Ohio; Omaha, Nebraska; Wenatchee, Washington; Clark's Ferry, Pennsylvania; Harrisburg, Pennsylvania; Tacony, Pennsylvania; Detroit, Michigan; Melville, Louisiana; Louisville, Kentucky; Evansville, Indiana; Washington, D.C.; Cairo, Illinois; Davenport, Iowa; and New York, New York. On the occasion of his receiving the Washington Award, Modjeski was said to have "to his credit more large bridges than any other man." But that alone was not why he would receive the Washington and other awards later in life. The reason, according to Ralph Budd, president of the Great Northern Railway, was simple:

> It is that Ralph Modjeski is inherently an artist. He has not chosen oil, or dry point, or marble, or even music, in which he doubtless would have ex-

Ralph Modjeski, perhaps in his late forties

celled, to express himself, but steel, and stone, and concrete. Using these as his chosen media, "by a pleasing simplicity of form and reliance upon the quiet dignity of the long spans whose members gracefully express function free from superfluities," he has made of bridge building a recognized art without in the least minimizing its importance as a science.

Many engineers who would receive such encomiums later in life often traveled a long and dispiriting road before reaching the podium among "prolonged applause," but few matched Modjeski's heightened sense of theater. Between 1905 and 1915, on the Oregon Trunk Railway, for example, Modjeski served as chief engineer for a series of bridges, including one over the Columbia River and a daring arch that spanned 340 feet at a height of 350 feet over a most dramatic construction site on the Crooked River. Overlapping with this responsibility was his much more visible presence as a member of the Government Board of Engineers for the second Quebec Bridge, from 1908 to its completion in 1917. His posing with other engineers astride one of the bridge's thirty-inch-diameter steel pins and standing inside one of the redesigned seven-by-ten-foot lower chord members, demonstrated his sense of showmanship.

Modjeski's theatrical personality would come even more to the fore when he worked as chairman of the board of engineers and chief engineer for the Delaware River Bridge between Philadelphia and Camden, New Jersey, now known as the Ben Franklin Bridge, which with a main span of 1,750

A sense of showmanship displayed by engineers on one of the thirty-inch-diameter pins for the Quebec Bridge (left to right: *G. F. Porter, engineer of construction, and G. H. Duggan, chief engineer, St. Lawrence Bridge Company; C. N. Monsarrat, chairman and chief engineer, and Ralph Modjeski, member, Government Board of Engineers*)

feet would be the longest suspension bridge in the world when completed in 1926. Modjeski's experience as an advance man for his mother and his gift for public relations would serve him well on that project, as it had in the earlier Quebec experience, for there survive photographs of him with various groups of engineers, directors, and city officials at seemingly each stage of construction. When the foot bridges for the Delaware River project were completed on August 8, 1924, chief engineer Modjeski would lead politicians from both states on the first official crossing from Philadelphia to Camden. It was to be a very warm day, and many in the party would shed their jackets during the steep climb to the top of the Philadelphia tower, but, as if to defy the sun itself, Modjeski would remove only his straw hat. After descending to the center of the main span, various members of the group, including Modjeski, would make speeches before a microphone that would be set up there by the Philadelphia radio station WLIT.

Public relations was not so effective for Modjeski or anyone else connected with the construction of the politically contested Manhattan Bridge, however. Aften ten months of study, Modjeski's detailed technical report on the bridge was issued in September 1909. Though it suggested that the foundation of the Manhattan anchorage might have had an improved design, overall it gave the structure "a clean bill of health." At 1,480 feet between towers, the Manhattan was shorter than the suspension bridges that flanked it, the Brooklyn and the Williamsburg, but it was distinguished for both its political and its technological significance.

The Manhattan Bridge was to have four trolley tracks and four elevated or subway tracks, but they were not yet installed, nor was all the pedestrian decking, when the bridge was dedicated on the final day of 1909, in part because a new city organization had taken rapid-transit arrangements out of the control of the bridge commissioner. When the Manhattan Bridge carried heavy rail traffic over the ensuing years, it would be found that, because of the arrangement of the rails, considerable twisting of the deck would take its toll on the structure. It would turn out that this was due in large part to the bridge's having been designed by employing the analytical tools of the then new deflection theory devised by the Austrian Josef Melan to take into account the coupled action of the deck and suspension cables. Leon Moisseiff, an engineer with the Bridge Department, introduced the method of calculation into American bridge building through his work on the Manhattan Bridge. (Lindenthal actually raised some questions about the new calculation scheme, but his concerns were dismissed.) Moisseiff would go on to employ the method as engineer of design under Modjeski on the Delaware River Bridge, and virtually all other large American suspension bridges were to be designed in the same way—until the Tacoma Narrows Bridge failed the very year it was completed, in 1940. But this is getting ahead of the story.

7

When the Manhattan Bridge opened, a fourth crossing of the East River was also under construction, one that in years past had actively been referred to as the second East River crossing. It was to connect the boroughs of Queens and Manhattan in the vicinity of Blackwell's Island, now called Roosevelt Island, which provided dry land for midriver piers. However, even though great suspension spans would not be required, the realization of a Blackwell's Island Bridge was also a long and arduous process. A suspension bridge was actually proposed as early as 1838, and in 1867 a structure was authorized, but nothing much happened until 1872, when the "scheme was rescued from impending oblivion" by the formation of a bridge company with William Steinway, the piano manufacturer, as president. Finan-

An 1838 proposal for a bridge at Blackwell's Island

cial hard times derailed the project for a while, and it was not resumed in earnest until the late 1870s, when Steinway gave up the presidency of the New York and Long Island Bridge Company to Dr. Thomas Rainey, a native of North Carolina who had become successful operating steamships between New York and South America, and whose title was believed to be self-proclaimed. As was often the case with such projects, the business, manufacturing, and real-estate interests would ultimately form partnerships with the engineering interests to bring the bridge to fruition.

Steinway and other Long Island property owners had a clear economic interest in supporting a bridge between New York and Long Island, but they obviously did not have the experience to oversee its design and construction. Though Rainey was willing to devote his time and energy to such an endeavor, he did not have engineering experience or judgment, and so early structural plans were not very satisfactory. The bridge was projected to have piers on the northern end of Blackwell's Island and thus enter Manhattan at about 77th Street. The total length of the bridge was to be about two miles, and access where it crossed high above the New York avenues was to be "by enormous passenger elevators like those in use in first-class hotels." Not having to construct long suspended spans would keep the cost of the bridge under $5 million, and the great number of tolls collected would give a handsome return on investment, which for the Long Island investors would be in addition to the increased value of their real estate. The legality of running railroad tracks across the bridge was

An 1881 proposal for a "second bridge over the East River," at Blackwell's Island

challenged; by 1893, amid financial hard times, only one pier had been constructed, and the project once more lay dormant, ultimately to be abandoned.

In 1894, the company proposed a new cantilever bridge with major spans over the two river channels and over Blackwell's Island itself. The new location was to be more southerly, so that Manhattan approaches would be between 62nd and 63rd streets, and chief engineer Charles M. Jacobs expected it could be ready for traffic in 1897. Though the bridge's 846-foot channel spans would be modest by Firth of Forth standards, the proposed bridge would have the second-longest cantilever spans in the world, and twice as many tracks as the Forth Bridge as well as two driveways. A contract was awarded early in 1895, but later that year the Supreme Court ruled against allowing railroad traffic across the bridge, and so the project was again thwarted. The idea of a Blackwell's Island bridge was revitalized in 1898, after the creation of a consolidated New York City, by its first mayor, Robert C. Van Wyck.

In his first report as bridge commissioner in 1902, Lindenthal announced that the plans for a Blackwell's Island bridge had been changed to give a narrower roadway and to provide for access to Blackwell's Island itself, on which the city Departments of Charities and of Corrections both maintained institutions, and hence the later name of Welfare Island. Lindenthal's new plans, revealed a year later, called for two large cantilever spans, of 1,182 feet and 984 feet, which, unlike the Firth of Forth spans, would not incorporate suspended portions. Though this created some complications in design calculations, the outline of the bridge had a more continuous top

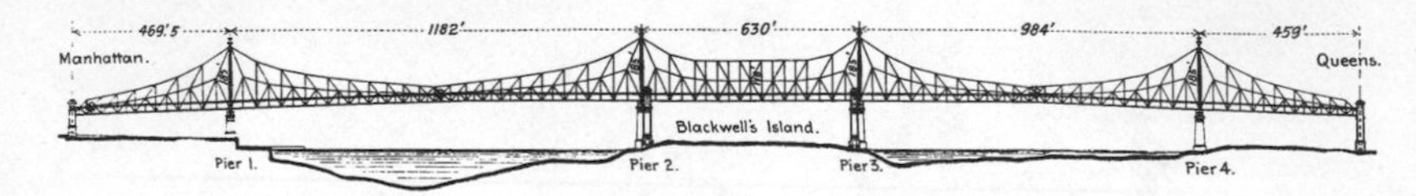

Blackwell's Island Bridge, 1903 design

curve, and was defined below by a flat roadway, thus giving the structure somewhat the appearance of a suspension bridge—the distinctly different genre with which it would often be confused. Lindenthal's specifications also called for the use of nickel steel in the eyebars and pins in the upper chord, to assure ductile rather than brittle behavior; this would be the first bridge to use so much of that material.

Unlike the Manhattan Bridge, Lindenthal's Blackwell's Island structure was built essentially as it had been designed when he left office. A steel strike in 1905 did delay the beginning of construction, but the "largest cantilever bridge in the United States" was begun in earnest in 1906, with a projected cost of $18 million, 50 percent higher than originally estimated. Later that year, lightning struck a section under construction "and so weakened it that two or three heavy gusts of wind brought the whole piece smashing to the ground," but that was nothing compared with what happened in Canada in 1907.

After the collapse of the contemporary cantilever under construction at Quebec, there was naturally some concern for the stability of the one over Blackwell's Island. Work gangs at the New York construction site were pitted against each other in competition to close overhanging sections of the bridge before the March gales began. It was not only the forces of nature that were feared, however; on one occasion, dynamite was found where an explosion would have brought down the incomplete central span, and union opposition to the open-shop project was suspected. In spite of all this, in March 1908, the last link in the superstructure was completed, and what had "seemed to be defying the law of gravity" was then reported to look "perfectly safe."

Regardless of how safe the bridge looked to reporters, *Scientific American* raised the concern that changes had been made to Lindenthal's plans and so perhaps introduced weaknesses not unlike those that had brought down the Quebec Bridge. Two independent consultants, Professor William H. Burr of Columbia University, one of the experts who had been appointed to consider a suspension bridge across the Hudson, and the New York engineering firm of Boller & Hodge, were called upon "to examine and report upon the design and structure" of the bridge at Blackwell's Island. Though they raised some caveats, involving the weight of steel in the bridge and the load it should be allowed to carry, the consultants found no reason to think that the bridge was in imminent danger of collapse. Burr did recommend full-scale tests of compression members, and Lindenthal concurred, saying that the bridge "must not be opened for public use until the strength of its compression members has been proved by actual test." The Quebec failure had changed the focus entirely from concerns over eyebar tension members to built-up compression members, and engineers knew that there had never been conclusive tests of the theories they employed. Burr also urged removal of some of what he considered excess steel in the bridge before letting it carry elevated railroad tracks. In effect, he argued, the bridge was straining more to hold itself up than it was supposed to under the weight of regulated rapid-transit cars.

In the meantime, construction crews continued to work on the bridge, connecting it up with the approaches in August, and fearless pedestrians immediately began to enjoy the mile walk between Queens and Manhattan. When it came to the attention of a reporter that ducks, pigeons, and swallows roosted nightly on the structure, and in such large flocks that they could chase away threatening cats, newspaper headlines declared that the birds were vouching for the strength of the bridge. The authorities cited in support of this idea ranged from Rudyard Kipling, who had written about bridge builders, to ornithologists, who were reported to have said that "birds in large flocks will not settle on a weak structure." While criticism continued in the technical press, the popular press was reassuring the public that "architects and structural engineers are among the few persons who will always do their work just as well as they can possibly do it, irrespective of whether they make any money out of the job or not."

Sometimes the name of a bridge and its formal opening generate more interest than its safety. Real-estate interests in Queens objected to the name Blackwell's Island Bridge because they considered it a misnomer, and one that was "unpleasantly suggestive of a penal institution and a poorhouse," which occupied the island. Amid some protest that there was a plot among Irish Americans to obliterate "from the map of the United States all names of places of English origin," the name Queensboro Bridge was in the end accepted. With that settled, attention could focus on planning the week-long festivities that would open the bridge in June 1909. Planning began well in advance, but not everything went according to plan.

Early in the year, it was announced that the "Bridge Queen-to-Be" had disappeared. Miss Elinor Dolbert was a French-born eighteen-year-old clerk in the Bloomingdale Brothers department store, who had been discovered to be a "fine vocalist" after she sang at a concert given by the

Bloomingdale Mutual Aid Society. She astonished the large audience, and "was encored so many times that she fainted from the stress." Since the young singer became famous throughout the store, she was asked to sing for "the firm," who "decided to have her voice tested." When several voice teachers confirmed that she did indeed have a fine voice, lacking only in "voice culture," the Bloomingdale company "resolved to defray the expense of the cultivation of Miss Dolbert's voice."

All the while, the Queensboro Celebration Committee, with which Samuel J. Bloomingdale himself may have been influential, was looking for someone to sing the song written expressly for the big day. After an audition, Miss Dolbert was selected to be "Queen of the Bridge," but only from the Manhattan side; the rivalry across the river was presumably too great to expect agreement in matters of voice and beauty. An entertainment was scheduled by the Bloomingdale Mutual Aid Society to raise money for a wardrobe and pin money for Miss Dolbert, but it was called off when she disappeared. As far as was known, she "had no love affairs," and it was speculated that the young lady, "to whom fortune was being kind," had simply "run away from good fortune." In the hope that someone might recognize her, a description was printed in the newspaper: "She is about 5 feet 6 inches in height, and has a mass of golden hair, and large black eyes. She wore a black cheviot tailor-made suit, the collar and cuffs of the jacket trimmed with black velvet, a jaunty black hat, with a single ostrich plume, and a bow of black velvet ribbons, a black lynx fur boa, tan gloves and tan shoes." Although it is difficult to imagine her lost in a crowd, Miss Dolbert seems never to have been found; the bridge show, however, had to go on.

If Miss Dolbert could not sing on the bridge, perhaps Wilbur Wright could fly around it. Not to be outdone by the Hudson-Fulton memorial-exhibition planners, who announced a $10,000 prize for an airship flight between New York and Albany, the bridge-celebration committee announced similar prizes for airplane and dirigible-balloon contests in conjunction with their festival in Long Island City. While they were waiting to hear from Wright, the committee also let it be known that they were negotiating with Roy Knabenshue, who three years earlier had circled the Times Building in his airship, to scatter leaflets entitling the finders to participate in a drawing for 250 lots in Queens.

The "veracious press agent employed by the Queensboro Bridge Celebration Committee" also let the newspapers know that 235 people had applied for permission to jump off the bridge on the day of its opening. The applications had been analyzed, and they were reported to be classified as follows:

Professional high divers	168
Freaks, employing inventions to break their fall	34
Would-be suicides	9
Unemployed	24
Total	235

The would-be suicides were identified as young women, who gave "unrequited love, unhappy matrimonial experiences, and a struggle for existence" as reasons they wanted to jump off the bridge. The unemployed were all men, hoping to land a job. One of them reasoned that if he survived he would get a good position, and if he did not survive he would not need work, and it could be "given to some other unfortunate." In the final analysis, no applications were approved, and no bridge jumping was to be allowed at the celebration.

Although the formal opening ceremonies were to be held the week of June 12, the mayor, bridge commissioner, and bridge engineers drove across the structure in late March and opened it to public traffic. Questions of safety had been relegated to closing paragraphs in newspaper stories about the planned festivities, and the thirty-seven survivors of the Committee of Forty, prominent businessmen who had promoted the bridge, made their plans to cross the structure as a group. They decided to engage a fife-and-drum corps to lead their march, after which they would have dinner at Strack's Casino, in Astoria, where they had met 133 times during their now successful campaign to have the bridge built.

A more clandestine first crossing took place one morning in May, when Dr. Rainey, "shorn of his wealth and enfeebled by age and ill-health," left his home on Lexington Avenue and all alone walked across the bridge. His own bridge, which he had spent twenty-five years and almost a million dollars promoting, was never to be. According to the report in *The New York Times*, headed, "Sees His Dream Bridge,"

> Dr. Rainey is nearly 85 years old, and made a pathetic figure as he shuffled along, his steps feeble and uncertain, his former towering frame shrunk and bent. He wore a pair of house slippers, a soft cap, and a sack-coat, and had no overcoat. . . .
>
> "This is my bridge," said the doctor as he wiped away the tears that trickled down his withered cheeks. "At least it is the child of my thought, of my long years of arduous toil and sacrifice. Just over there," pointing to a ruined heap of stone along the river front, "are the old towers of my

> bridge, which I began to build many years ago. I spent all I owned on the project, and then New York, with all its great wealth and power, came in and took away my possessions, and now in my old age I am left in ill health and alone to eke out my remaining days.
>
> "It is a grand bridge," he said, "much greater than the one I had in mind. It will be of great service to thousands in the years to come, when Dr. Rainey and his bridge projects will long have been gathered into the archives of the past."

A tablet was planned for the new bridge, to commemorate Rainey's efforts and dream. Among those on the committee arranging for it were Charles H. Steinway and Samuel J. Bloomingdale. The official bridge-opening ceremonies took place in June, as planned. About a quarter of a million onlookers strained to hear speeches in the early afternoon and would gawk to see the "new bridge ablaze with red fire and electricity in the evening." Before that, however, and just before the parade approached the reviewing stand, Dr. Rainey was introduced as "the father of the Queensboro Bridge idea" and received a rousing cheer.

8

Gustav Lindenthal had no such prominent acknowledgment at the Queensboro Bridge ceremonies, if he was there at all; his thoughts were again with railroads, not with pedestrians. Since 1904, he had been engaged as a consultant for a project to connect the tracks of the Pennsylvania Railroad, which included those of the Long Island Railroad, with the New York, New Haven & Hartford Railroad, thus enabling continuous rail traffic to flow from Long Island and New England into Manhattan and from there westward through the Pennsylvania Railroad's tunnels under the Hudson River, which had obviated the need for Lindenthal's North River Bridge, at least for that line. The new project involved a three-mile-long steel viaduct that would cross the East River about three miles north of the Queensboro Bridge in a sinuous curve that passed from Long Island over a treacherous channel known as Hell Gate, over Ward's Island, across Little Hell Gate, over Randall's Island, and finally across the Bronx Kills into the northernmost borough of New York City. The centerpiece of Lindenthal's plan was to be a steel arch on the order of one thousand feet between abutments, the largest arch bridge in the world, and it would carry four railroad tracks. By 1907, he had progressed far enough with the design that it could be submitted to the Art Commission.

As with all great projects, the chief engineer required assistance for its detailed design and supervision. Among those Lindenthal enlisted to help him were Othmar Ammann, a young Swiss-born engineer who had worked for the Pennsylvania Steel Company, which built the Queensboro Bridge, and who had participated in the investigation of the collapse of the Quebec Bridge. Ammann was to serve as principal assistant engineer on the Hell Gate project; his story will be told more fully in the next chapter. Another of Lindenthal's assistants on the Hell Gate was to be New York–born David Steinman, who was very nearly exactly Ammann's contemporary, and whose story also requires a chapter of its own. In addition to engineering help, Lindenthal early on sought the assistance of Henry F. Hornbostel, then a young consulting architect, who had studied at Columbia and the Ecole des Beaux-Arts, and who would go on to design the campuses of the Carnegie Institute of Technology, in Pittsburgh, and Emory University, in Atlanta. Though Hornbostel had been retained by Commissioner Lindenthal to help with the Manhattan Bridge, he had been dismissed by the succeeding administration when he refused to submit new plans unless he received additional compensation for his services. Hornbostel's main contribution to Lindenthal's Hell Gate Bridge was to be the addition of "a pair of immense pylons" of granite topped by concrete that framed the arch and its entrances.

The concept of a railroad link with a bridge across Hell Gate as its single most significant structure was created in 1892 by Oliver W. Barnes, an engineer with extensive railroad experience and Pennsylvania Railroad connections, and Lindenthal, who then saw the plan as part of a greater scheme that included his North River Bridge. The New York Connecting Railroad Company was incorporated that same year to construct a steam railroad about ten miles in length with termini "in Westchester County, east of the Bronx River, and in the City of Brooklyn." Among those involved in the incorporation with Barnes was Alfred P. Boller, an 1861 graduate of Rensselaer Polytechnic Institute who had most recently worked on various projects in and around New York. It was Boller who had worked out the first plans for a bridge at Hell Gate—a cantilever design—in 1900. At that time, the Pennsylvania Railroad was leaning toward a scheme that had Lindenthal's North River Bridge bringing rail traffic from New Jersey into Manhattan, which would from there connect to the Long Island Railroad through the Steinway Tunnel, named after the president of the tunnel company for which Barnes served as chief engineer, and ultimately linking up with New England through the Hell Gate connection. Pennsylvania Railroad President Alexander J. Cassatt, brother of the painter Mary Cassatt, subscribed to the scheme, but with more affordable and exclusive tunnels under the Hudson River instead of a great common bridge over it. Vice-President Samuel Rae was to be made president of the New York Connecting Railroad, which by then the Pennsylvania had acquired. When Lindenthal's term as bridge commissioner ended, Rae appointed him to direct the Hell Gate project as consulting engineer and bridge architect, a title he must have relished.

Under Lindenthal's direction, three comparative designs for a bridge with an 850-foot main span were considered: a stiffened suspension bridge with eyebar chains, a smaller version of his Manhattan Bridge, which in turn was a smaller version of his North River Bridge; a three-span continuous truss of unremarkable profile; and a three-span cantilever, more graceful than Boller's design and bearing some resemblance to Lindenthal's plans for the Queensboro Bridge. The bridge architect would no doubt have preferred the suspension design, for the truss would have the appearance of "a utilitarian structure," a fault also common to cantilevers, which provide "no opportunity for monumental towers or abutments at the ends, because the absence of a large horizontal thrust or pull does not justify a large mass of masonry at those points, as in the case of an arch or a suspension bridge." Although the words are Ammann's, in his definitive report on the Hell Gate Bridge, they can be assumed to have been approved of if not inspired by Lindenthal. An arch design would also have satisfied the requirements of having no piers in the water and of being erectable without falsework to obstruct the waterway, in the manner of the Eads Bridge, but the arch was not considered at first, because there were no natural abutments to take the thrust. The suspension bridge would not normally be economically competitive with the other designs for an 850-foot span; according to Ammann, however, "whatever differences in cost may be found by comparative designs are largely due to the individual judgment of the designer in the selection of the truss system, material, permissible unit stresses, foundations, and architectural features."

Ammann's list of factors to be weighed by a designer did not include the one that ultimately had the most effect on the bridge type chosen. Among the existing structures on Ward's Island were state-hospital buildings, and in 1905 the line of the railroad had to be moved farther north to increase its distance from them. In order to fit the approaches to a suspension or cantilever design onto the island, a tight curve in the railroad would have been necessary, and this was undesirable in conjunction with the heavy grade that was needed to provide the proper clearance for ships to pass under the bridge. An arch bridge was considered, and it was found that one could be built with less steel than the alternative designs required; even with its more costly foundations, it was a competitive choice. In the final analysis, the favorable appearance of the arch led to its adoption.

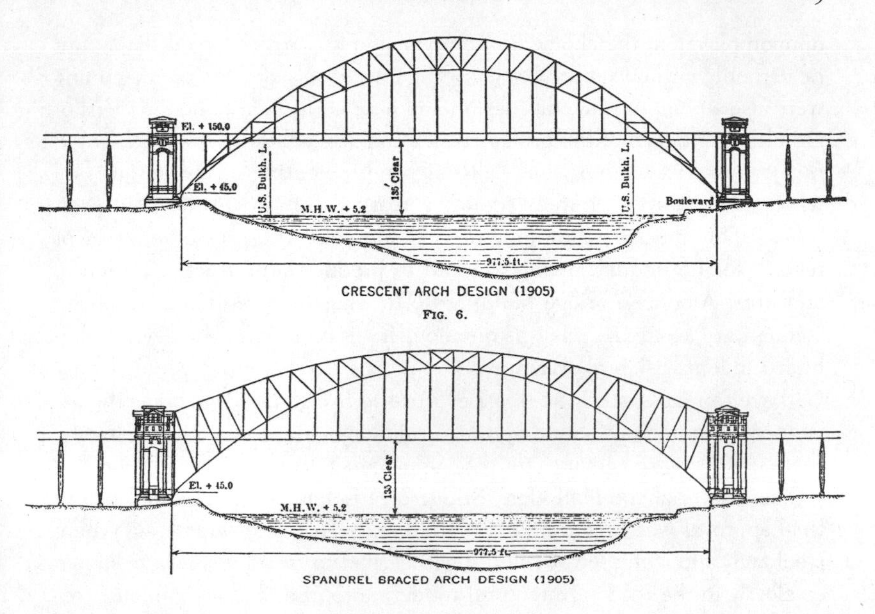

Two arch designs for the Hell Gate Bridge

Lindenthal's first arch designs were modeled after Eiffel's Garabit Viaduct, a crescent-type arch over the Truyère River in France, as well as some German spandrel-arch bridges over the Rhine. The latter arch type was selected, in part because it was "more expressive of rigidity than the crescent arch, the ends of which appear to be unnaturally slim in comparison with the great height at the center." David Billington, the premier structural critic of the later twentieth century, has interpreted this as indicative of Lindenthal's predilection for "massiveness over lightness: the German over the French." Before detailed design work began, however, the top chord of the arch was given "a slight reversal of curve toward the ends," partly to provide some wind bracing, but also to "improve the silhouette of the arch." This structural fillip would ultimately contribute much to the characteristic profile of the Hell Gate Bridge, as well as to the feature of the design that has been most questioned.

Whereas Ammann wrote that "the artistic outlines of the steel superstructure are the result of the proper interpretation of the economic and engineering requirements of the structure," Billington sees the stone towers above their base as structurally unnecessary to take the loads from the steel, and thus he sees the towers as nonfunctional and "rather a massive frill." It is not clear how much of a role, if any, the consulting architect Hornbostel's opinion played in the choice of arch type and its final recurved shape, but he certainly influenced the design of the towers, which from the beginning were also a point of some discussion and have continued to be the focus of structural criticism of the bridge. When the original design was presented in 1907, the Art Commission, "although not objecting to the design as a whole, disapproved of the decorative features of the towers and their bases." This must have disappointed Lindenthal, the bridge architect of record, for he unquestionably wanted to produce an attractive structure and bring American bridge building up to what he considered European standards of aesthetics. He was not alone in his concerns. The never-to-be-built Lindenthal-Hornbostel towers dominated the frontispiece of Henry G. Tyrrell's 1912 "systematic treatise," *Artistic Bridge Design*, and as the existence of numerous contemporary municipal art commissions attests, there was rising sensitivity about the appearance of large urban structures.

The towers of the Hell Gate Bridge clearly had to be modified before final approval was sought and construction begun, and one detail, where steel arch and stone tower came together, had to be addressed. As shown so clearly in the 1906 architectural rendering reproduced in Ammann's report, the original tower design left a gap of about fifteen feet between the masonry and the steel, an arrangement that might have precluded Billington's criticism of the final design three-quarters of a century later. As described in a recent account, Lindenthal was aware that this represented the "correct engineering solution," clearly showing all the thrust transmitted through the lower chord to the abutment:

> Lindenthal feared, however, that the public, supposing that the towers supported the bridge, might think he had forgotten something. To deal with this possible psychological hazard, he grafted stubs of girders to the ends of the upper chords and placed T-shaped forms of concrete within the side openings of the towers. At a distance, they appear to be connected, but in fact a space of about six inches separates them.

Whether he really feared the public's perception or the art commission's, the towers themselves were modified, but there appears to be no official record of approval of their final design. As he did in Pittsburgh in concealing the slender end posts of the Smithfield Street Bridge with ornate portals, so Lindenthal appears also to have employed an architectural treatment to conceal the potentially confusing structural detail of the recurved top chord of the Hell Gate arch. Perhaps he did not really want to or know how to end his masterpieces. There seems little doubt, however, that appearance was important to Lindenthal, who also envisaged extra-

A 1906 design detail for the Hell Gate Bridge tower and arch

technical and extra-utilitarian functions for the Hell Gate Bridge. According to Ammann:

> Mr. Lindenthal conceived the bridge as a monumental portal for the steamers which enter New York Harbor from Long Island Sound. He also realized that this bridge, forming a conspicuous object which can be seen from both shores of the river and from almost every elevated point of the

Completed Hell Gate Bridge, showing the steelwork of the upper chord carried into the towers and the long curving viaduct over Ward's Island

city, and will be observed daily by thousands of passengers, should be an impressive structure. The arch, flanked by massive masonry towers, was most favorably adapted to that purpose.

The visual appearance of his bridges was thus of considerable importance to Lindenthal, and now that he was in charge of a privately financed project rather than a municipal one, with its many constituencies, the bridge architect not only could but had an obligation to consider the important factor of aesthetics. As Ammann explained his mentor's method, or perhaps echoed him:

> A great work of art evolves from an idea in the mind of its creator. It is brought on paper or into a more contemplative form and then changed and remodeled. Not until the plans have passed through changes and corrections, and have been submitted to an almost endless series of finishing touches, does the great work attain its perfection.
>
> A great bridge in a great city, although primarily utilitarian in its purpose, should nevertheless be a work of art to which Science lends its aid. An elaborate stress sheet, worked out on a purely economic and scientific basis, does not make a great bridge. It is only with a broad sense for beauty and harmony, coupled with wide experience in the scientific and technical field, that a monumental bridge can be created. Fortunately, the Hell Gate Bridge was evolved under such conditions, and therefore may well be said to be one of the finest creations of engineering art of great size which this century has produced.

Questions of aesthetics and symbolism aside, a great engineering project still needs a great engineering staff, and there were many more details than towers and recurved chords to be considered and calculated. How would the individual steel members, some to be twice as heavy as the largest previously used in construction, be made and joined? How would the weight of locomotives and railroad cars be borne by the various parts of the bridge, individually and acting in concert? To answer such questions required detailed thought and calculation, of such depth and magnitude that they were beyond the mental or physical capacity of one engineer. As chief engineer and bridge architect, Lindenthal directed his staff to explore various options and to consider and compare alternatives. Though he could indeed direct that this or that tower design be chosen, someone else would be expected to calculate the volume of masonry or concrete it would require, to estimate the time needed to construct it, and to prepare whatever detailed drawings were needed to make sure the abutments were located and aligned to meet and match the steelwork that someone else was thinking about in equally precise detail. Other engineers would later have the responsibility for overseeing and inspecting the construction to make sure that the plans were being followed so that things did meet as they were designed to. During the construction phase, Lindenthal was assisted by an engineering staff of ninety-five, and Ammann, as assistant chief engineer, "had general charge of the office, field, and inspection work."

In early 1914, *Engineering News* reported that the Hell Gate Bridge was then actively under construction, "with minor architectural changes in the terminal towers," but other details began to attract the attention of some ever-critical readers. In a letter to the editor, "an admirer of the central span" wondered why the drawings showed a steel-viaduct approach to the bridge, and why the Art Commission did not object to it. The reader knew "the deteriorating and nerve-racking noise which is likely to accrue from trains passing over such a structure." He suggested that concrete arches would have been just as economical. Furthermore,

> The arches could be built with fine architectural effect, and in such a way that screen walls could be carried up on both sides to, say, 15 ft. above base of rail, with the result that noise of traffic would be almost entirely eliminated. Such a structure would be almost as silent and picturesque as one of the beautiful old aqueducts still to be seen in the older countries.

Lindenthal responded almost by return mail, explaining the choice of steel over concrete viaduct spans. The weight of a large masonry viaduct could not be supported easily or economically by the ground conditions on Ward's or Long Island. He pointed out that with development and the introduction of sewer lines, the ground would be drained, and the piers would settle. Steel girders could be adjusted under such circumstances, but masonry arches would develop unsightly and possibly threatening cracks. Whether this was part of the design logic or a subsequent rationalization, Lindenthal had taken all such criticism seriously and responded accordingly. With regard to noise, he pointed out that the rails would be embedded in broken-stone ballast fourteen inches deep, carried in troughs of reinforced concrete, thus deadening much of the sound.

He also took the opportunity to explain that the towers "were necessary parts of the structure, and not mere ornamental parts." For the thrust of the arch to be properly resisted, the towers had to provide a certain weight to the foundation, and Lindenthal chose to accomplish this by building tall rather than squat towers, which "would have been unsightly." His solution, in other words, was much like erecting the buttresses of a Gothic cathedral higher than appearances demand in order to add weight and maintain slenderness. Since the weight was needed in the towers for structural reasons, Lindenthal chose to provide it without sacrificing proportion. In the case of the viaducts, the additional costs would have been prohibitive.

In fact, there eventually was a change in the design of the viaduct from the original to the revised drawings, published seven years apart, in *Engineering News*. In 1907, the viaduct over Ward's Island was shown to be steel girders resting on steel piers, but in 1914 sketches, though the steel girders remained, the piers were shown as concrete. The decidedly social rather than technical reason for the change was mentioned in passing by Lindenthal during the discussion appended to Ammann's paper. According to Lindenthal, an "objection was made by the authorities of Ward's and Randall's Islands to the steel columns, because they feared that inmates of the municipal institutions on those islands would climb them and make their escape. It was insisted that the design adopted should prevent this." Ward's Island held the state mental hospital, of course, and Randall's Island, over which the viaduct also passed, was the location of a correctional institution. Presumably, a technical answer was found to the prior objections to heavy concrete piers, or more money was simply spent on them.

By the end of the year, when the foundations were complete and the steel had begun to be erected, a small item headed "Hell Gate Arch Bridge Not a New Thing" appeared in *Engineering News*. In spite of its headline, the item signaled that all was once again well between the bridge builder and the journal. It proudly quoted this passage from Carlyle's *Sartor Resartus*: "Never perhaps since our first Bridge-Builders, Sin and Death, built that stupendous Arch from Hell-gate to the Earth, did any Pontifex, or Pontiff, undertake such a task." *Engineering News* thereupon christened Gustav Lindenthal "pontifex" of the modern Hell Gate arch.

Construction progressed; the two trajectories of steel met over the unobstructed water in the fall of 1915, and the arch bridge and viaducts were completed a year later. The first passenger train to cross the bridge was the Federal

Express, the previously established night train between Boston and Washington, D.C. The Federal Express route had for a long time included a fourteen-mile car-ferry transit through the crowded waters of the East River, which in winter was subject to delays caused by ice. When the car ferry was discontinued in 1912, the Express was operated over the Poughkeepsie Bridge until early 1916; at that point, regular service was discontinued. A weekly ferried Express was reinstituted briefly during the summer of 1916, so that travelers could bypass New York City during an infantile-paralysis epidemic. The completion of Hell Gate Bridge and the New York Connecting Railroad allowed the restoration of regular Federal Express service in 1917.

9

Though the Hell Gate Bridge would be mentioned in Lindenthal's obituary as his "chief memorial," even with its completion his career was far from over. Before the Hell Gate was finished, Lindenthal became consulting and chief engineer for a railroad bridge on the road between Sciotoville, Ohio, and Fullerton, Kentucky, across the Ohio River, about 120 miles above Cincinnati. The bridge was to carry heavy freight traffic, mostly coal trains, on a new branch of the Chesapeake & Ohio Railroad. The Sciotoville was the first large continuous truss bridge, which means that it consisted of truss elements rigidly connected across piers rather than of separate elements between them, and the longest and heaviest fully riveted one then erected in America. With two 775-foot river spans, it has been called "perhaps the boldest continuous bridge in existence" and "the ultimate expression of mass and power among American truss bridges." J. E. Greiner, a Baltimore consulting engineer, in a written discussion of Lindenthal's paper on it, called the completed bridge a "daring and handsome structure, decidedly 'Lindenthalic' in all its features," and declared it to be another of the master's structures evidencing, in Lindenthal's own words, the "genius that originates as distinguished from routine which merely imitates." Another discussion was submitted by Charles Evan Fowler, the New York consulting engineer who had published in 1914 plans for a cantilever bridge between San Francisco and Oakland that would have surpassed the Quebec in span. One structural critic called Fowler's "the boldest bridge plan ever made" but did not think the tolls from wagon and automobile traffic using it would pay for the upkeep of the roadway. Nevertheless, Fowler seemed to be more interested in size than suitability, in giantism than genius, and his discussion of Lindenthal's paper revealed, however subtly, that a massive cantilever across San Fran-

Sciotoville Bridge

cisco Bay would bring the record for that kind of span below the Canadian border to the United States, the home of all other great spans:

> The Sciotoville Bridge is a striking example as to what may be accomplished by the use of continuous bridges. It is the longest of that type ever constructed and now gives to America the proud distinction of having the longest spans for every type of bridge construction, namely, the Sciotoville Continuous Bridge, the Hell Gate Arch, the Quebec Cantilever, the Williamsburg Suspension, the Metropolis [Illinois] Simple Truss Span, and the Willamette River Draw Bridge.

Lindenthal seems to have been drawn not so much to sheer size as to monumentality, however, and his definitive professional paper on the Sciotoville Bridge was in a sense a monument to the achievement of its engineers. His principal assistant engineer on the Sciotoville project was, as on the Hell Gate, Othmar Ammann, who might have been expected to write up and present a description of the project for the archival transactions of the American Society of Civil Engineers had he not been called to military service in his native Switzerland. As in the Hell Gate project, Ammann was succeeded by David Steinman, but it was Lindenthal himself who wrote up the Sciotoville Bridge, which in his own words was a "detailed, although somewhat belated description." However, unlike Ammann's paper on the Hell Gate, which was read within months of the completion of the bridge, Lindenthal's Sciotoville paper did not appear until five full years after that bridge was completed. Nevertheless, the paper was awarded the same Rowland Prize that he had won thirty-nine years earlier for his description of the Monongahela bridge. Lindenthal closed his paper on the design and construction of the Sciotoville Bridge with acknowledgments of his assistants "in this unusual work, bristling with new problems and difficulties." First to be mentioned was Ammann, and second Steinman, but understanding why the seventy-two-year-old chief engineer prepared the paper, rather than assigning it to his chief assistant, who by then had returned from Switzerland, remains for a subsequent chapter. Whatever Ammann's disposition, however, Lindenthal was busy with many projects, including writing endeavors, and it is understandable that his report on Sciotoville was not contemporary with the bridge. He seemed more inclined to write about future projects, like a North River Bridge, than completed ones, like Sciotoville, no matter how inspired or gigantic they might be.

A decade after the Sciotoville Bridge was completed, Lindenthal was asked to be responsible for the design and construction of three bridges across the Willamette River in Portland, Oregon, following a political scandal there regarding the awarding of municipal bridge contracts. The bridges—the Burnside, Sellwood, and Ross Island—were completed in the mid-1920s in that city, the largest bridges on the West Coast until the great structures at San Francisco were built in the next decade.

Lindenthal's group of Portland bridges, like his New York spans, stands today as testimony to what was often said of him, that "he never built two bridges alike." The memoir of him in the *Transactions of the American Society of Civil Engineers* expanded on this truth to speak of "his habit of looking on each bridge problem as new and unique, a problem whose proper solution could hardly be the same as that of any prior bridge problem." Furthermore, "he took up each bridge project broadly, seeking first a conception of general form that would offer the best solution and going on to stresses and details only as the last step." It takes nothing away from Lindenthal to say that this is a habit shared by all great bridge designers.

Not all bridges are of the scale of the great ones Lindenthal designed, however, and not all engineers always agree on what is the "best solution" for a given problem, as the Manhattan Bridge debate so clearly demonstrated. And nothing seems to provide so good an opportunity to discuss such differences of opinion as the publication of a major book and the reviews it may elicit. While the Hell Gate and Sciotoville bridges were still under construction, a two-volume illustrated treatise of well over two thousand pages was published by John Wiley & Sons and sold for the remarkable price of ten dollars. The treatise was titled simply *Bridge Engineering*, and it was written by "one of the masters of the art," J. A. L. Waddell.

John Alexander Low Waddell, a contemporary of Lindenthal's, was born in Port Hope, Ontario, Canada, in 1854. Waddell graduated from Rensselaer Polytechnic Institute with the degree of C.E. in 1875 and worked in Canada as a draftsman and engineer on field work before serving as an assistant professor of rational and technical mechanics at Rensselaer. He then took up studies again, this time at McGill University, receiving both a bachelor's and a master's degree in 1882, after which he went to Tokyo to become professor of civil engineering at the Imperial University of Japan. He returned to the United States four years later to join the Phoenix Bridge Company, and soon opened an office to serve as an agent for the company and as a consulting engineer in his own right in Kansas City, where he spent much of his early American bridge-building career. It would develop into a distinguished one.

While in Japan, Waddell published two books, with the pedestrian titles *The Designing of Ordinary Iron Highway Bridges* and *A System of Iron Railroad Bridges for Japan*. In 1898, he published a small "pocket-book," which he titled in the tradition of Latin treatises, *De Pontibus*, and for which he was much better known. The 1916 *Bridge Engineering* was a much-expanded form of the pocket book, and in an editorial *Engineering News* explained the difficulty "of finding someone to prepare a critical review . . . who will be, at least, a peer of the author in reputation." Thus the journal was proud to announce that it believed it had "rendered a notable service to the profession in securing the consent of Gustav Lindenthal, the Nestor of American bridge engineers, to review Mr. Waddell's great work." *Engineering News* also no doubt wanted to give its readers some explanation for Lindenthal's lengthy review, which was "very much more than a book review." It was, in fact, much more of a scathing challenge to the authority of Waddell's treatise than the editors may have expected, and the editorial closed with an acknowledgment that, though Waddell's book was "destined for many years to come to rank as an authority in its field," it was also of great value to have "a critical study made of its recommendations, so that the engineer may know in what parts of the book there is a disagreement among doctors as to the soundness of the principles which are there stated."

Lindenthal's "illuminating review" is a model of the form, and he discussed not only the content but also the style of the work. He found the latter to lack uniformity, which he speculated was due to "the fact that parts of the book were prepared by different assistants, to whose helpful labor the author gives proper credit in the preface." Perhaps Lindenthal, who has been described as being throughout his life "too active to find the leisure necessary for writing books," did not know or made no allowance for the fact that Waddell kept his staff employed with work on the book during a period of increasing war and decreasing bridge building. Lindenthal also criticized the generally "breezy and often gossipy narrative form" that Waddell apparently preferred because he intended the book to be somewhat autobiographical. Among the mannerisms that Lindenthal singled out for criticism was "an affected, though innocuous, punctiliousness in attaching to names inconsequential titles, as Esquire, C.E., member of, etc., as if marking some for social or professional distinction while others, not less distinguished, go without it." Perhaps Lindenthal was a bit oversensitive to this topic because of his own uncertain background, and it was easier for him to attack Waddell than to correct him and the record.

As could be expected in a book comprising eighty chapters, there were some inconsistencies of style and substance, but the sixty-page index was excellent. Thus Lindenthal could easily look up references to himself on seven pages, and what he found on some of those pages must have galled him. Waddell's first mention of the Hell Gate did not associate it with Lindenthal, but in a discussion of it six hundred pages later its designer, "the noted bridge engineer, Gustav Lindenthal, Esq.," is acknowledged as the source of data and a picture of the bridge, which is described as being "certainly of aesthetic appearance" and reflecting "great credit upon the artistic ability of its designer." Elsewhere, the engineer is described as "Gustav Lindenthal, Esq., C.E.," but the unearned degree must have been less galling than the unwelcome criticism.

In his chapter on cantilever bridges, Waddell began his treatment of Lindenthal's Blackwell's Island structure by describing how a refiguring of the stresses in the completed bridge found them to be "so great (due to both ambiguity of stress distribution and overrun of dead load) that some of the roadways had to be omitted." After beginning with such sharp criticism, Waddell continued his discussion of the design with ridicule that is suggestive of the common modern characterization of chaos theory, in which a single flap of the wings of a butterfly in Australia is said to be able to affect the weather in Philadelphia:

> A New York engineer connected with the bridge once remarked that the structure is so complicated that, if a man were to stand at the first panel point of the farthest span and were to spit into the river, his doing so would affect the stress in every main truss member of every span in the entire structure—and the statement is actually correct. The layout of this bridge is a constructive lie. The top chords of the long spans were made into a continuous curve to resemble the cables of a suspension bridge, the object being aesthetics; but the attempt thus to beautify the structure was a failure, and the damage done to the bridge by the omission of the suspended span is measured by millions of dollars.

Waddell was referring to the indeterminate nature of the structure from a calculational point of view. By omitting a true suspended span, Lindenthal had indeed made the stresses in the structure so interdependent with its deflections that a small movement or a change in the load at one point on the bridge did affect it everywhere else. As for the faux-suspension criticism, it may be that Lindenthal brought that on himself by so liking eyebar suspension bridges that he consciously or unconsciously mimicked the form in a continuous cantilever.

Perhaps the aspect of *Bridge Engineering* that most irritated Lindenthal was what must have seemed to be Waddell's slighting of him. Whereas no engineer might have minded having his name omitted from criticism such as that leveled against the Blackwell's Island Bridge, reference to another of one's bridges without one's name attached might have been a different matter. Furthermore, in a discussion of impact loads on bridges, a 1912 paper of Lindenthal's on the subject was described to have "much valuable information; but the formula proposed is far too complicated, being based on many theoretical assumptions," and some of its statements and deductions were criticized as being "not in accord with the latest experiments on impact." Though Waddell allowed that Lindenthal was "one of the most prominent" bridge engineers, they nevertheless disagreed on the subject of continuous truss spans. The author of *Bridge Engineering* believed that the Sciotoville Bridge, in which Lindenthal "resurrected" the subdivided triangular truss form, worked only because foundation conditions were "exceedingly favorable" at the site. But perhaps the single most difficult part of Waddell's book for Lindenthal to take was the treatment of the suspension bridge, his form of choice. In discussing proposals to bridge the North River, Waddell mentioned three and offered his opinion on their likelihood of being realized:

> Messrs. Geo. S. Morison, Gustav Lindenthal, and Henry W. Hodge have made designs for that crossing; and it is not at all unlikely that the last-mentioned engineer and his financial associates in the not very distant future will succeed in consummating the enterprise. For the sake of the engineering profession as well as for other good reasons, it is to be hoped that they will be successful. The building of such a structure as the one they contemplate would be a fitting climax to an already brilliant professional career.

Hodge's plan alone is illustrated, and it is described by Waddell in some detail compared with those of Morison and Lindenthal. Lindenthal's is given especially short shrift, being in fact merely referred to as being "made in the late eighties" rather than described. Indeed, as we shall see, Waddell seems to have made the proper assessment at the time, and Hodge and his

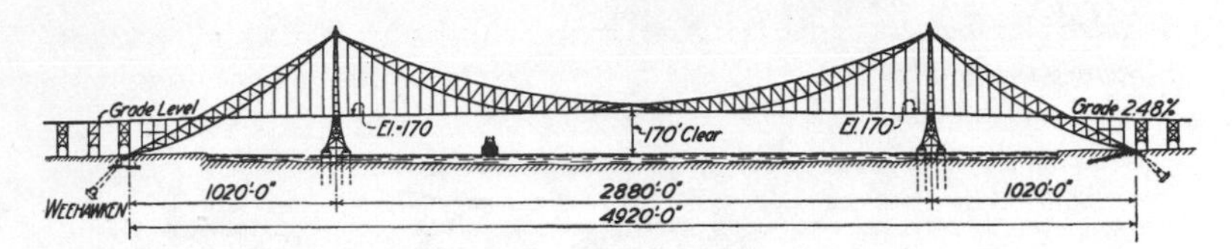

Henry Hodge's proposed Hudson River Bridge

associates might very well have been the ones that first bridged the North River had Hodge, fifteen years Lindenthal's junior, not become ill and died before bridge building was revitalized after the war.

Though Lindenthal may have had a personal ax to grind in writing his devastating review of Waddell's treatise, the sweeping contents of the two large volumes did leave room for disagreement. Waddell had established his reputation on, among other things, the Halsted Street Lift-Bridge in Chicago, which, when completed in 1895, certainly solved the technical dilemma of spanning the Chicago River at street level while making provisions for water traffic to pass. However, the Halsted Street Bridge did so in an arguably ugly way. A swing bridge would have worked at the location, of course, but with a midspan pivot on an undesirable pier in the river channel or with a land-based pivot about which the span swung into such a position as to obstruct valuable riverside property that otherwise could be used for wharves or piers. Bascule or leaf drawbridges, like the contemporary Tower Bridge in London, were another possibility, but they presented different mechanical problems. Indeed, all bridge designs with movable spans presented major aesthetic problems, and they have been among the most criticized for their appearance. Though the tall structural towers of Waddell's Halsted Street Bridge did allow the 130-foot span to be raised over 140 feet in one minute, they were an eyesore whether the span was up or down, and the bridge had an ungainly look. To those who drove the streets of Chicago or plied the waters of its river, however, the function may have excused the form, and it was to such clients, or, rather, their elected representatives, that the bridges had to be sold.

Catalogues of bridges previously designed and constructed, or heavily illustrated reports of major projects, were important to consulting engineers like Waddell and bridge-building companies alike, for it was through such catalogues that they often made their initial contact with prospective clients. Most catalogues showed a sense of design themselves and presented their bridges from the most attractive perspectives. Since consulting engineers often evolved their associations and partnerships as new and different design challenges, conditions, and opportunities arose, the same

Waddell's lift bridge across the South Branch of the Chicago River at Halsted Street

bridge may often have appeared in the catalogues of seemingly different firms, creating a confusion of attribution. J. A. L. Waddell, for example, after his return from Tokyo, practiced under his own name in Kansas City until 1899, when he began a series of partnerships: with Ira G. Hedrick, as Waddell & Hedrick (1899–1907); with John Lyle Harrington, as Waddell & Harrington (1907–17); with N. Everett Waddell, as Waddell & Son (1917–19); alone, as J. A. L. Waddell, after his son's death, until 1927, during which period Waddell moved from Kansas City to New York; and with Shortridge Hardesty, formerly his principal assistant engineer, as Waddell & Hardesty (1927–45). Though Waddell's name remained associated with the firm for some years after his death in 1938, it was dropped in 1945, when Clinton D. Hanover, Jr., joined Hardesty to form Hardesty & Hanover, which now identifies itself as one of the oldest consulting-engineering firms in the United States, tracing its origin back to Waddell. As late as the early 1990s, Hardesty & Hanover's current brochure listed major and recent projects of the firm dating back to 1890, and included an illustrated entry for the Halsted Street Bridge. In contrast to the handsome layout and attractive photographs in Hardesty & Hanover's catalogue, including some movable bridges that can be described as being pleasing to the eye, a Waddell & Son catalogue dating from about 1917 does not at all present their bridges in the most attractive context. That the difference in catalogues is not just a matter of different graphical standards is demonstrated by a catalogue contemporary with that of Waddell & Son. In contrast to the weedy and littered foregrounds in some of the Waddell pictures, a Strauss Bascule Bridge Company catalogue from about 1920 presents bridges, at least some of which are just as unsightly, in thoughtfully cropped photographs that show the structures in a much more favorable light.

Waddell seems to have paid considerably more attention to photographs of himself than to those of his bridges. The frontispiece of *Bridge Engineering*, for example, is a photographic portrait of the author. Waddell's forelocks appear to have been deliberately curled, his long mustache combed and waxed, and his lapel pinned with two medals, one probably that designated him Knight Commander, Order of the Rising Sun, presented to him in Japan in 1888. The other medal is most likely the one designating him Knight First Class, Order of Société de Bienfaisance of Grand Duchess Olga of Russia, presented to him in 1909 for his services as principal engineer of the Trans-Alaskan-Siberian Railroad project. In the three-quarter-length portrait of a standing Waddell that appeared as the frontispiece of his *Memoirs and Addresses of Two Decades*, published in 1928, his head and facial hair appear unchanged, his dress is even more formal, and he sports three new medals. These probably were for the Order of Sacred Treasure, Japan (1921); Order of Chia Ho, China (1922); and Cavaliere of the Crown of Italy (1923). Waddell must have presented an imposing figure, and in his obituary the London journal *Engineering* conferred upon him the title "Pontifex Maximus," which it noted that the English poet Robert Southey had bestowed on his friend Thomas Telford. Waddell, like Telford, the journal reasoned, had been in "possession of a constitution apparently indifferent to the rigours of field work in all weathers."

Lindenthal, whose own career was then in its twilight, was to be swayed neither by a portrait nor by the myth of Waddell the prolific author, who in a biographical sketch that may very well have been autobiographical was said to write out "in longhand his accurate and well-finished papers and discussions in his office or during numerous long railroad trips back and forth across the continent." No matter how dedicated to his writing Waddell might have been, Lindenthal expected it to stand up to engineering scrutiny. His basic technical criticism of the 1916 book was nicely summarized in the lengthy review:

J. A. L. Waddell, from the frontispiece of Memoirs and Addresses

> The book . . . appears to be valuable and authoritative only in so far as it deals with the engineering of bridges of ordinary span and type, that have already become more or less standardized. When the author ventures outside of this field into the domain of long-span and indeterminate structures, his insufficiency of knowledge and experience are betrayed [on the specific pages cited], and his lack of grasp of the big questions in this field becomes evident. This may sound like harsh criticism, but it appears to be justified in view of some of the author's pretentious but erroneous judgments on higher-class structures.

Lindenthal takes Waddell's discussions of several topics as an opportunity to set the record straight, as on the characteristics of cantilever bridges, on the safety of suspension bridges, and on aesthetics. On Waddell's "witticisms" about the Queensboro Bridge, for example, Lindenthal remarks that "there are few structures, even of those designed by the author, about which some amusing things could not be written, but such do not furnish instruction to engineers," and he goes on to discuss the political system in which "engineers are as often abettors as victims." With regard to aesthetics, Lindenthal wonders if Waddell's "taste will always be shared by other designers." When he elaborates, Lindenthal appears again to have something of his own experience on his mind:

> The author's repeated reference to aesthetic appearances based on nothing more than curves in the top or bottom chords will appear to others as rather naïve. This chapter on architecture is well meant as an earnest plea to bridge engineers to show themselves in their work as cultured men. The author, however, in undertaking to furnish instruction and guidance to seekers of the aesthetic in bridge construction, has set himself a task that is evidently beyond his scope. For any bridge structure requiring architectural consideration the bridge engineer will do well to consult a competent architect; and experience has shown that not every architect is competent here.

10

The question of bridge aesthetics and the role of the engineer versus the architect in bridge design was one that was to grow well beyond a difference in point of view between Waddell and Lindenthal, and some of Lindenthal's views may inadvertently have threatened the position of engineers generally. In 1919, a Delaware River Bridge Joint Commission was created by the legislatures of Pennsylvania and New Jersey, and among its first orders of business was the appointment of a board of engineers to study specific sites and types of bridges. In the meantime, the Pennsylvania State Art Commission wrote to the governor calling for an architect to be put in charge, stating that the commission members were "convinced that the question of 'where' and 'what' are of greater importance, and more difficult to answer, than 'how' to build it." Indeed, to their mind, the "how" was "after all but a detail." The insinuations and arrogance of the art commissioners would have been enough to incite the engineering community, but the final straw was contained in a statement that made patently false claims about the history of bridge building in America: "The great bridges of New York have all been planned by architects, though, of course, built by engineers. They are beautiful because they fulfill their purpose, and are fittingly designed, with due consideration to the 'where' rather than only the 'how.' " Perhaps Bridge Commissioner Lindenthal had retained architect Hornbostel, and perhaps engineer Lindenthal's ego had driven him to demand the title of architect as well as engineer on the Hell Gate project, but it was a gross misstatement to say that architects had decided where and what bridges were built across the East

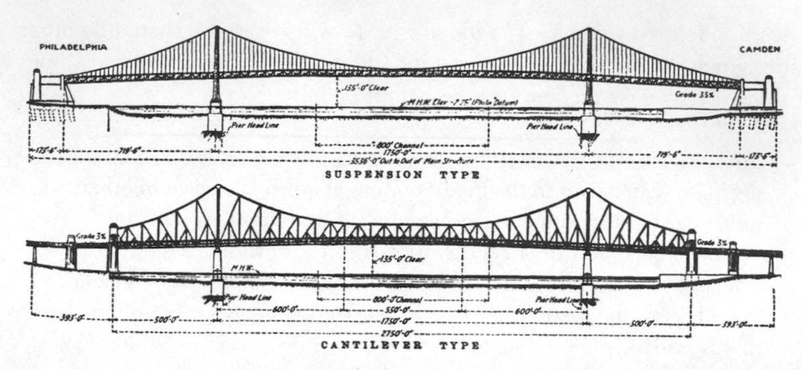

Suspension- and cantilever-type bridge designs proposed to cross the Delaware River between Philadelphia and Camden, New Jersey

River. Indeed, as the stories of the bridges reveal, they were conceived, located, relocated, and designed by the (sometimes conflicting) recommendations of engineers, and there were long and continuing disagreements as to whether any of the bridges was even beautiful or fulfilled its purpose.

Various engineering groups responded with "resolutions of remonstrance," demonstrating their own firm grasp of the history of the New York bridges. Philadelphia's Society of Municipal Engineers pointed out that "engineers are as keenly aware as any class of citizens, of the need for taste and beauty in structures erected in public view," and they pointed out a difference between the conditions under which engineers and architects sometimes work. In designing a bridge or some other structure, the engineer is often directed by the client to provide the most economical, no-frills structure. The architect, on the other hand, is "oftenest called in by clients who wish to pay the necessary price for taste and beauty," and so does not work under the handicap of strict economy in design. The engineer-architect issue was a threatening one, but it became moot, at least in Philadelphia, when Ralph Modjeski was chosen as chief of the board of engineers charged with making recommendations as to site and type of bridge.

After the location of a suspension-bridge design was settled upon and approved, construction on the Delaware River Bridge was begun early in 1922. The final decision was probably made more expeditiously than it might have had there not been a strong desire to have the bridge ready for the Sesquicentennial of the Declaration of Independence, on July 4, 1926, and an experienced engineer was most likely to be able to do that. Among the critics of the design was Lindenthal, perhaps remembering Modjeski unkindly as the engineer who gave the stamp of approval to the Manhattan Bridge, whose final plans had, of course, been altered and modified from

Perspective drawing of Delaware River Bridge

Lindenthal's own changed design. Thus he wrote that "the engineer who thinks merely of stresses must combine with the architect, who deals with artistic forms." In focusing on the towers, the "most prominent feature" of suspension bridges, he asserted that "from the aesthetic point of view, metal towers, no matter how finely designed, will never equal stone towers." At this time, as we shall see, his own design for the North River Bridge was evolving toward stone from its original steel towers, whose curve had resembled the Eiffel Tower. That was the structure that he still considered "the finest example of an artistic metal tower," even though "as an architectural creation it does not impress the beholder with that feeling of dignity and majesty which he experiences at the sight of any of the great spires in famous cathedrals." The Williamsburg Bridge towers, for which "no architect was consulted," were "bandy-legged," and the Delaware River Bridge towers were "too much on the utilitarian principle of braced telegraph poles or derricks, holding up ropes." However, Lindenthal also knew as an engineer that metal towers were "lighter and cheaper," thus requiring not only less costly and time-consuming foundations but also less capital investment. In a situation where a bridge is desired to be ready for a sesquicentennial, for example, such considerations are naturally persuasive, though seldom sufficiently so to all. In this case, one of the most effective critics was an artist.

Joseph Pennell, who was born in Philadelphia in 1857, attended evening classes at the Philadelphia School of Industrial Art and then the Pennsylvania Academy of the Fine Arts on and off around 1880, his talents as an illustrator blossoming. He then worked mostly in Europe, traveling back to the United States to record American engineering projects, from which developed the "Wonder of Work" theme of his sketches of projects under construction, such as the Panama Canal and Hell Gate Bridge. World War I caused him to return to the United States more or less permanently, and in 1924 he sketched the Delaware River Bridge under construction. The title of his etching, *The Ugliest Bridge in the World*, was unkind to the incomplete structure, which was not yet a fully formed bridge. However, the relatively wide towers, necessarily so to accommodate the eight lanes of traffic, were rather squat-looking for their height. It is one of these towers, which Lindenthal found wanting in their design, that is the focus of Pennell's drawing.

Like the story of great bridges generally, the history of the Delaware River Bridge was long and tortuous. Plans were proposed to cross the river between Philadelphia and Camden, New Jersey, as early as 1818. In 1843, a model suspension bridge was exhibited at the Franklin Institute Fair, and in 1851 a suspension bridge with four-thousand-foot spans was proposed by

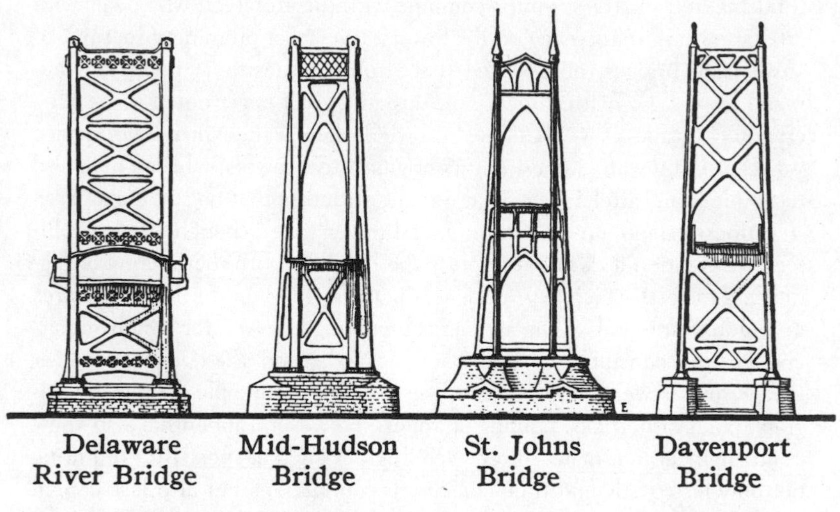

Comparison of steel-tower designs of several contemporary suspension bridges

John C. Trautwine, Sr., but neither attracted much serious interest. New Jersey and Pennsylvania bridge commissions were in place in 1918, when the firm of Waddell & Son was retained to make a consulting-engineering study. The report concluded that a bridge would be preferable to a tunnel and presented a suspension design with helical incline approaches, no doubt prompted by the high cost of land for the conventional means. The same approach scheme had also been employed in a Waddell design for a bridge across the entrance to the harbor at Havana, Cuba, and had been used as an illustration in the chapter on aesthetics in *Bridge Engineering*. A consulting architect, Warren P. Laird, was subsequently engaged to advise

A photograph of the Delaware River Bridge under construction, and Joseph Pennell's etching of "The Ugliest Bridge in the World."

on the Delaware River Bridge's location, and the brouhaha over whether engineers or architects should take the lead in such projects resulted.

A Delaware River Bridge Joint Commission, created by the two states in 1919, the following year appointed the board of engineers with Ralph Modjeski as chairman and with the distinguished Philadelphia engineers George S. Webster and Lawrence A. Ball as the other members. Suspension and cantilever designs were considered, the former winning out on economic grounds. Modjeski effectively became chief engineer of the project, and he selected Leon Moisseiff engineer of design and Clement E. Chase as principal assistant engineer. Paul P. Cret served as architect to the project, but under the engineer Modjeski, whose personality and predilection for the dramatic dominated throughout construction. Physical construction began with a ceremony early in 1922, but, "in lieu of the traditional digging of the first spadeful of earth, a plank was torn loose" from a pier that the bridge would replace.

Before the bridge was finished, a seemingly unresolvable difference of opinion arose between the states of New Jersey and Pennsylvania as to whether or not tolls would be charged. The New Jersey commissioners voted to halt the awarding of contracts—including one to wrap the cables before they began to rust—until it was agreed that tolls would provide funds for interest on and amortization of the bonds that were issued for construction. Philadelphia residents, on the other side of the river and the issue, preferred a free bridge paid for by taxes. *Engineering News-Record* reminded its readers that, "under the pressure of post-war costs and overdue public needs," taxation had "largely ceased to be an attractive means of financing large public improvements." After the Supreme Court granted Pennsylvania permission to sue New Jersey over the issue, the journal observed that, "if the outcome of Pennsylvania's attempt to outwit the New Jersey taxpayer succeeds, future public toll-bridge proposals will not have an easy time of it." In the end, it appears to have been public sentiment in Pennsylvania in favor of a toll that swayed that state's legislators to retreat from their position. Construction resumed, and the bridge was officially opened to traffic on July 4, 1926. *Engineering News-Record* reported then that "it probably ranks as the largest public toll enterprise ever carried out," but that distinction would not hold for long, although the bridge itself was expected by some to last indefinitely. At the twenty-fifty-anniversary ceremonies, for example, it was written: "There will be many more anniversaries, for no man can place a limit upon the time this bridge, magnificently designed, honestly constructed and scrupulously maintained, shall endure as the link between the states." Such conditions of maintenance can continue practically, however, only as long as toll revenue or some other source of funds provides the resources.

Sketch of Charles Evan Fowler's proposal for suspension bridges of 3,500-to-4,000-foot main span to cross the Hudson River at three locations (59th Street, 83rd Street, and 178th Street) for a total cost of about $100 million, essentially the same design he proposed for a bridge between Detroit and Windsor, Canada

By the mid-1920s, great suspension bridges were under construction or being considered in large cities across America, including at Detroit. The Ambassador Bridge, completed only three years after the Delaware River Bridge, would best its 1,750-foot main span by one hundred feet and thus hold the world's record for a short while. But the record was to be almost doubled in New York in 1931, and increased by another 20 percent in San Francisco in 1937. Though Lindenthal may have conceived and submitted designs for some of these projects, it was the bridge across New York's Hudson River that he really wanted to build, and for which he still held out hope. His latest design had steel towers enclosed in masonry, following the aesthetic he had recently espoused, and the 825-foot-tall, thirty-five-thousand-ton towers would be higher and more massive than the Woolworth Building, designed by the architect Cass Gilbert and opened in 1913. However, Lin-

Lindenthal's 1921 design for a Hudson River bridge at 57th Street

denthal seems not to have completely lost his affection for the bare metal towers of his early North River Bridge design; it, and not its masonry-clad descendant, was illustrated in a long reflective article entitled "Bridge Engineering," prepared by Lindenthal for the fiftieth-anniversary number of *Engineering News-Record*, in 1924.

II

Approaching his seventy-fifth birthday, Lindenthal was both pragmatic and philosophical. Regarding financial constraints, on which he had a considerable opinion, he wrote: "Engineers are sometimes under the authority of laymen with whom financial considerations may seem more important than

An illustration of Lindenthal's proposed bridge, demonstrating how long and heavy its main span would be

A 1921 comparison of a tower of Lindenthal's bridge with the Woolworth Building, then the tallest skyscraper in the world

safety. If the pressure for cheapness comes from them, then the engineer should decline responsibility for the work." After looking back over the previous half-century of railroad-bridge building, he began to shift into prognostication: "Large bridges costing millions of dollars were comparatively few and will probably become less frequent in the next fifty years." He did not want to leave the wrong impression, however, for he continued to point out that the "size of bridges never was limited by question of what we could fabricate, but rather by financial considerations." Perhaps Lindenthal was beginning to resign himself to the fact that his greatest dream would never be realized, but, then again, he knew that it was not because of the limitations of engineers or engineering.

Looking further into the future, Lindenthal thought it "probable that the zenith of large bridge construction" would be reached in the late twentieth or early twenty-first century, because of the "increasing cost of iron and coal" to make steel. He thought the "iron beds would become exhausted long before the coal mines," and "the production and use of portland cement (which requires coal for calcination) will also cease," and that stone bridges would "again be the only practicable lasting kind." He continued, in a rambling historical mode:

> Bridge construction and bridge architecture will be to posterity in a certain sense a surer index of the progress of our present day civilization than houses, temples or cathedrals appear to us of past ages. This will be so because the economizing of iron, when it becomes costly, will probably begin in bridge and structural construction before it begins in other kinds of construction. The large sources of energy in nature, coal, water power, wind, tides, heat of the sun, etc., can none of them be utilized without large masses of iron for the tools, machinery and power plants necessary for the conversion of these energies into power for the use of man—all of them more necessary than iron bridges, which may be then structures *de luxe*. Surely they would have appeared as such to the armored knights only one thousand years ago, when an iron armor was worth nearly half its weight in silver. Iron bridges, iron ships and railroads will then be curiosities. The colossal consumption of iron will have come to an end. In a span of time much shorter than that from Tut-ankh-amen to the present, steel bridges will probably have disappeared from the face of the earth through corrosion and neglect. Iron is a more perishable material, particularly in northern climates, than stone of which were built the Pyramids and the Greek temples and the wonderful Roman arch aqueducts—all in frostless, benign climates. These could be built again, but not iron bridges.

Lindenthal underestimated the world's store of iron and did not foresee in 1924 the enormous amount of steel that the automobile would consume, nor did he seem to foresee the tremendous amount of pollution it and the steel mills in its service would create. Such developments would invalidate his climatic argument and threaten the Pyramids as much as iron bridges. But, in spite of his flight of iron fancy, the old man was still not willing to give up his dream completely; his last paragraph held out a glimmer of hope in civilization's retreat from the iron age:

> The creative genius of mankind, which now in this glorious age of iron is ascending to hyperbolic eminence in every branch of technical science, will leave few if any durable iron monuments to distant prosperity. Among them should surely be, if possible, substantial large iron bridges that could be made to last several thousand years, if properly cared for in countries with stable, high civilizations. But who can look that far into the future?

Lindenthal may have been able on paper to wrap his steel towers in masonry in the hopes of preserving them for thousands of years, but he would have had to make many other pragmatic modifications to his great bridge if he had hoped to see it begun in his lifetime.

After the Hell Gate and Sciotoville spans were complete, Lindenthal had returned in earnest to the related issues of the North River Bridge and the "port problem of New York." In 1918, during an "exceptionally severe winter," in which New York was cut off from coal and food supplies by a frozen Hudson River, he issued a privately printed pamphlet reiterating much of what he had written three decades earlier, but he was no longer advocating a railroad bridge only—he was now predicting that six million cars per year would also pass over such a bridge. With an uncharacteristic dropping of the definite article, he anticipated "the slowing up and congestion of vehicles at ends of bridge for the purpose of paying toll," and suggested making the bridge free for highway traffic to avoid the problem. He had come to recognize the motor vehicle as something to be taken into account, but did not see it as a source of revenue. This was especially curious since he repeatedly pointed out that finances, not engineering, were the impediment to progress with his bridge plans. He felt the times were still not right financially, however, and "it would be folly even to think of starting construction and diverting capital to it until a year or two after the war—whenever that may be." The great toll bridges, like the one across the Delaware at Philadelphia, were still a few years away.

The 1918 tract was signed "Dr. Engr. Gustav Lindenthal," reflecting his possession of the honorary doctor-of-engineering degree conferred upon him in Dresden in 1911, but also lending credence to his lack of earlier degrees—why else would he not have appended them to his name before, or made an issue of the matter of degrees in his review of Waddell's book? He also signed another pamphlet, published the following winter, "Dr. Eng. Gustav Lindenthal," this time using another form of the abbreviation, suggesting that he was experimenting with the title he had only recently come to use. This study, addressed to the New York, New Jersey Port and Harbor Development Commission, was a comprehensive plan for railroad-terminal plans that included a double-deck bridge with the following capacity:

Lower Deck:

4 Railroad tracks for freight (all needed from the start).
4 Railroad tracks for passenger trains from 7 railroad systems, with together 24 tracks to the Union Station (all 4 tracks needed from the start).
2 Tracks for moving (or conveyor) platform from under 57th Street.

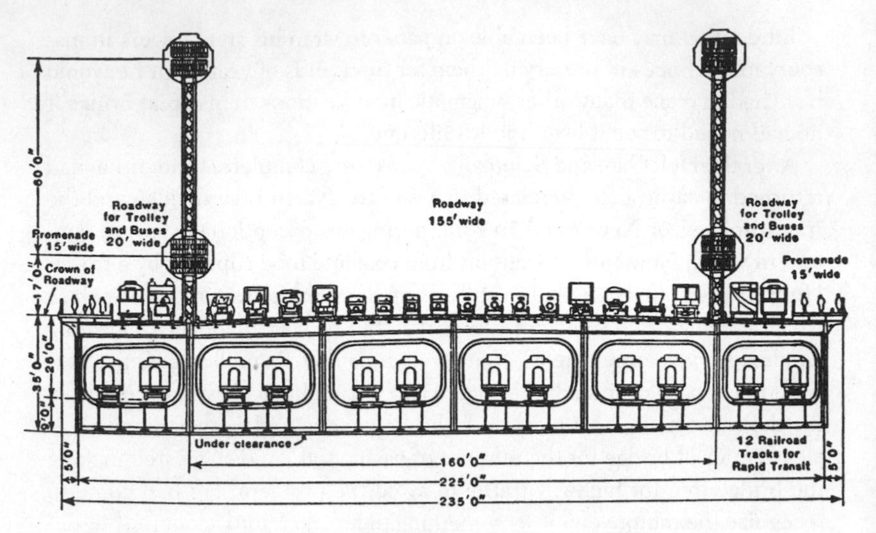

Roadway configuration for Lindenthal's Hudson River Bridge, 1923 version

Upper Deck:

2 Tracks for rapid transit trains to 9th Avenue Elevated.
2 Trolley tracks for surface cars.
6 Lines of vehicular traffic.
2 Sidewalks.

Lindenthal summarized the history of attempts to bridge the North River, discussed tunnels versus bridges, and referred to his earlier pamphlet discussing the advantages of having private capital build the bridge. Finance and political economy, "the most backward and the least understood" of the "departments of knowledge," were also the focus of a treatise written by Lindenthal and first published in 1922. As revised in 1933, *A Sound Scientific Money System, as a Cure for Unemployment*, was the engineer's Depression-era attempt to apply scientific-engineering principles to "deduce and predict the safety and stability of a money system." None of his efforts or writings was to bring sufficient investors to his bridge project, however.

As Lindenthal grew older, each birthday was noted by the press. On the occasion of his eightieth, in 1930, for example, he was reported to have planned to spend part of the day at the office of the North River Bridge Company, in Jersey City, and the rest of the day at his home in Metuchen, New Jersey. He later admitted to being a bit miffed that his associate on the bridge project, the consulting engineer Francis Lee Stuart, called him to come into Manhattan for an important business lunch at the Engineers Club. It proved to be a surprise birthday luncheon, perhaps marred only by his being asked if he was not cheered that the war was over and that the War Department was going to reconsider his proposed North River Bridge. "He waved the inquiry aside," however, saying "he would rather not discuss the bridge on his birthday."

On his eighty-first birthday, in the year when the George Washington Bridge was to be opened at 179th Street, *The New York Times* reported that the War Department had still not ruled on Lindenthal's bridge 120 blocks to the south. Nevertheless, he was now confident that it would, as he spent the day hard at work in his office from eight-thirty in the morning till five

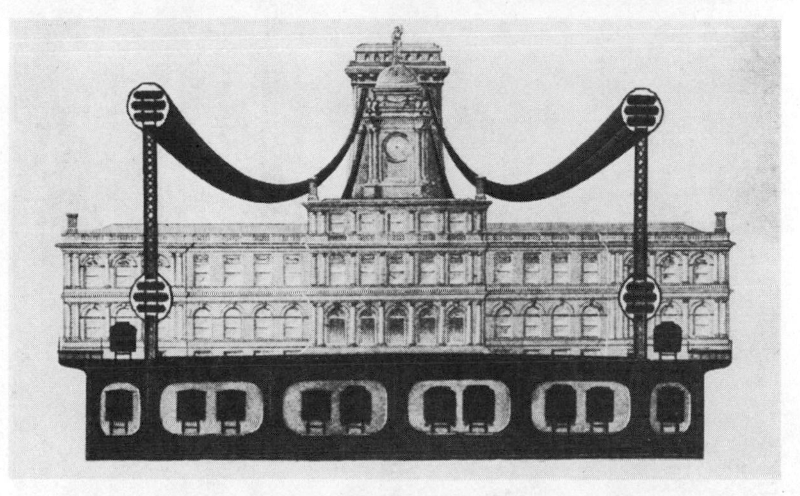

An illustration showing how New York's City Hall could fit across the roadway of Lindenthal's bridge

in the evening. The application for a permit was eventually "pigeonholed for eight or nine years," however, and approval was never to come. Honors, not bridges, came to Lindenthal in his old age. Late in 1932, for example, he was hailed at a dinner given by the Architecture League as the "grand old man of engineering." In response to his introduction as one of the guests of honor, Lindenthal spoke of his career and "of his difficulties with politicians particularly in the building of the Manhattan, Williamsburg, and Queensborough bridges and of his feeling of satisfaction when unhampered by political interference he built the great Hell Gate arch." He evidently never

Another view of Lindenthal's never-realized Hudson River Bridge

did learn how to or want to deal with politicians, and this, more than any other factor, kept him from realizing his dream of a North River Bridge.

For all the hailing and reminiscing, there must also have been a certain tension at the dinner, for two other engineers were honored along with Lindenthal. One was Ralph Modjeski, who had already built the Delaware River Bridge, and was currently chairman of the board of consulting engi-

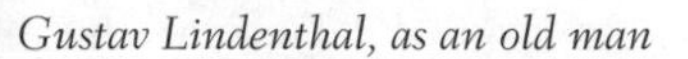
Gustav Lindenthal, as an old man

neers for the San Francisco–Oakland Bay Bridge, then under construction. Though soon to be overshadowed by the longer central span of the Golden Gate Bridge, whose construction was just about to begin, the Bay Bridge was in fact a greater project in total length. Comprising back-to-back suspension spans, each with 2,310-foot main spans, a massive tunnel through Yerba Buena Island, and a fourteen-hundred-foot cantilever span across the East Bay, the overall bridge was to dwarf all others. The third honoree at the architects' league dinner was Othmar Ammann, once Lindenthal's assistant, but now being honored as the designer and builder of the George Washington Bridge, which, with a thirty-five-hundred-foot main span, was not only by far the longest suspension bridge then in existence, but also the first to cross the Hudson at New York, albeit much farther north than the bridge of Lindenthal's dreams.

Among the dinner speakers was Cass Gilbert, who had served as consulting architect to the George Washington Bridge. His speech "emphasized the necessity of cooperation between the engineer and the architect." Francis Lee Stuart, then a consulting engineer for the city, also spoke, citing "the long span bridge as the most outstanding advance in the science of engineering during the last century." It will take another chapter to untangle all the past battles, victories, and defeats their remarks must have evoked in the minds of the engineers and architects present at the dinner.

Lindenthal did not go to his office on his eighty-fifth birthday. He spent it at his New Jersey home, which he had named The Lindens, recuperating after a long illness. He was never to regain his health, and he died two months later. Until his final illness, he had remained active as president and chief engineer of the North River Bridge Company, working on "his dream of forty years." The funeral was held at the family home. Of the two younger engineers most closely associated with Lindenthal during the construction of the Hell Gate Bridge, only David Steinman was reported to be in attendance.

AMMANN

Engineers can dream alone, but they can seldom bring their dreams to fruition by solitary effort. The roots of large-project engineering, such as bridge building, lie in military operations, which necessarily involve generals and soldiers, chieftains and Indians. Though the chief engineer may be the one who holds the grand plan in his head, it may ever remain his dream alone unless he can command a staff of engineers to direct still others to commit the concept to paper, to carry out the voluminous calculations that flesh out the dream and put a price on it, and to make the numerous drawings of details that allow it all to come together in the field. As generals begin as soldiers, and chieftains as Indians, so engineers of record begin as anonymous draftsmen and calculators. The best remember their professional roots.

Othmar Hermann Ammann was born in 1879 in the Swiss city of Schaffhausen and grew up just across the Rhine River in the small town of Feuerthalen, in the canton of Zurich. The river takes a sharp bend there, just above the Rhine Falls, and bridges played an important role in the communication between the people living on the opposite banks. Othmar's father, Emanuel Christian Ammann, the descendant of a long line of physicians, clergymen, lawyers, merchants, and government figures, was a well-to-do hat manufacturer, and his mother, born Emilie Rosa Labhardt, was the daughter of the noted landscape painter and lithographer Emanuel Labhardt. Thus it was not thought unusual that young Othmar often carried a sketching pad and pencil to the riverbank, where he drew images of a four-hundred-foot-long wooden bridge that had been built there in the previous century by the town carpenter, Hans Ulrich Grubenmann, and that was still remembered throughout the world as the largest of its kind. Two of Othmar's brothers would grow up to be painters, but he seems to have known early on that he would be either an architect or an engineer.

Othmar continued to sketch throughout his teenage years, and he enrolled in the state college in Zurich intending to study architecture. However, after he found himself excelling in mathematics, becoming the top student in the subject, and being drawn more and more to scientific subjects, he entered the Swiss Federal Institute of Technology to study engineering. At this prestigious school, known around the world by the initials ETH, which stand for Eidgenössische Technische Hochschule, Ammann studied under such influential and active engineers as Wilhelm Ritter, who had spent three months of 1893 in America, attending the World's Fair in Chicago and visiting many bridge sites. He incorporated his experiences into his lectures, and a written record of bridge building as he found it in the United States was published in 1895 as *Der Brückenbau in den Vereinigten Staaten Amerikas*. The final span illustrated in Ritter's survey was Lindenthal's proposed Hudson River crossing, and there can be little doubt that Ammann in Switzerland was aware of this.

While still a student, Ammann had worked one summer at a bridge-fabricating plant; by the time he graduated with a civil-engineering degree in 1902, he knew what kind of structures he wanted to design and build. He began his career as a structural draftsman with the firm of Wartmann & Valette, in Brugg, Switzerland, where he gained some experience surveying for a mountain railway and designing a modest stone arch. After a year or so, he moved to Frankfurt, Germany, where he joined the firm of Buchheim & Heister as an assistant engineer working on reinforced-concrete designs. Apparently looking for a more satisfying experience, Ammann took the advice of another of his polytechnic professors, Karl Emil Hilgard, who had worked in America for three years as a bridge engineer for the Northern Pacific Railroad, and sailed for New York. In America, Ammann thought, he could gain a few years' experience in an environment where some of the greatest bridges in the world were being built. Furthermore, according to Professor Hilgard, there was more opportunity for a young engineer in America, where "the engineer has greater freedom in applying individual ideas." He showed his students pictures of the Brooklyn Bridge and of bridges over the Ohio and Mississippi rivers, and told them that, in the United States, he had "seen youngsters in charge of work which, in Europe, only graybeards would be allowed to perform." Whereas some of young

Othmar's classmates had accused Hilgard of "being more American than the Americans," Ammann took him at his word.

I

A 1904 photograph shows a young Ammann looking dreamily out a window, his cheek resting on a closed hand and his elbow resting on a table full of books and plans. An engineer's scale and slide rule are in the foreground, at the ready for the designing of great bridges—as was the young engineer, though only in his dreams at the time:

> My first serious interest in the problem of bridging the Hudson was awakened shortly after my arrival in New York on a visit to the top of the Palisades Cliffs from where I obtained a splendid view of the majestic river. For the first time I could envisage the bold undertaking, the spanning of the broad waterway with a single leap of 3000 feet from shore to shore, nearly twice the longest span in existence. This visit came at that time as near to a dream to see the ambitious effort materialized. Nevertheless, for a young engineer it was a thrill to contemplate its possibility, and from that moment as my interest in great bridges grew I followed all developments with respect to the bridging of the Hudson River with keenest interest.

By this time, Ammann no doubt knew all about the proposals of Lindenthal and others to span the Hudson, and he realized that, for all the opportunity America presented to youth, the matter of credibility still had to be addressed, as well as confidence. As he would reflect and advise a quarter-century later, when he was actually building his dream:

> Get all the experience you can. . . . Learn from those who mastered your trade or profession before you. I have known many ambitious young men to fret themselves and waste their energies early in life because they could not achieve at once great things, for which, as a matter of fact, they were not prepared.
>
> It is true of other careers as it is of the engineer's—the first thing a man must decide is whether or not he has the ability to follow the calling he has chosen. Once convinced of this, it is a matter of hard work and experience; if the experience you need isn't thrown in your way, you must move heaven and earth to get it. . . .
>
> Let me put it another way: Study the career of any man of real achievement, and you will almost certainly find that this is true: from the very start he was not only willing but *eager* to profit by the experience of oth-

> ers. How true it is that there is nothing absolutely new under the sun! However great a man's achievement may be, it rests, in the final analysis, not upon radical departures from the experience of those who went before him, but upon the way in which he adapts their experience to his own purpose.

Ammann, in short, had a vision and a plan, and he assiduously followed the advice he would someday give to others. He knew in 1904, fresh off the boat from Switzerland, that he was not yet in fact ready to build great bridges, but he began immediately to plan to, and to move as much of heaven and earth to his advantage as he could. With letters of introduction from Professor Hilgard, who had advised him to keep his eyes and ears open and his mouth shut while he was gaining experience, Ammann promptly found a position as assistant in the office of Joseph Mayer, on lower Broadway, "the first door he knocked on." Mayer was a consulting engineer in New York who had been chief engineer of the Union Bridge Company, Lindenthal's rival for spanning the Hudson, and had produced the monstrous cantilever design for the 70th Street location. No doubt Mayer saw many advantages in hiring this well-trained and talented young immigrant, not least of which were his multilingual abilities, a substantial asset in polyglot New York. It is easy to imagine that Ammann also saw real advantages in associating himself with an engineer so close to large and important projects, albeit still dreams.

The relationship between Ammann and Mayer was brief, lasting only from the spring to late fall of 1904, during which time the young engineer "designed twenty-five or thirty railroad bridges." Othmar wrote to his parents in early December that his "boss was simply a skinflint and thought he could save a few dollars, in that he could cut my salary in half during the time that he had no work." The later image of a shy and retreating Ammann is belied by his report that he did not accept such treatment and extracted from a "morally obliged" Mayer a good recommendation that helped him get a new position with the Pennsylvania Steel Company, located in Steelton, just south of Harrisburg, near where the Pennsylvania Turnpike now crosses the Susquehanna River. Ammann wrote to his parents from Harrisburg, where he lived, that he was working for the second-largest bridge-construction firm in America.

He was enthusiastic about his new position, in which he was immediately assigned the design of a bridge almost five hundred feet long. He described the office, in which about one hundred engineers and technicians worked, as "very modern and practically organized" and located close to the bridge workshops, which he visited in his free time and in which he gained

Othmar Ammann in 1904

still more experience. Ammann reported to his parents that his salary was "about $70 per month" plus overtime, but he asked them, without giving a reason, not to share this information with anyone. He wrote that "the more I learn the more ambitious I become and the more I enjoy my work," and he assured his mother that he was eating well and not drinking so much as the young people back home. To his father he explained that the only photographs he had taken in New York were technical; he promised to take some different ones to send home. In the summer of 1905, Ammann returned to Switzerland, where he married his school sweetheart, Lilly Selma Wehrli, the sister of the Wehrli Brothers, who were well-known photographers. Lilly and Othmar returned to Pennsylvania, and he continued to work in Steelton.

Among the projects in Ammann's division at the Pennsylvania Steel Company was the fourth bridge across New York's East River, at Blackwell's Island, to be known as the Queensboro Bridge. This was, of course, to be the great cantilever connecting Manhattan with the borough of Queens, and Ammann worked on it under chief engineer Frederic C. Kunz, who was in charge of construction. Ammann no doubt saw this as a rare opportunity to learn about the practical aspects of bringing to completion a bridge project almost as massive as that needed to span the Hudson. Like much of the work of engineers in subordinate positions, Ammann's under Kunz was largely anonymous. That does not mean it was insignificant, however; when Kunz's book, *Design of Steel Bridges*, was published in 1915, he would acknowledge Ammann in the preface, along with two other engineers, "for their able assistance in the preparation" of the volume. Among the many plates in that book is one showing the elevations of notable cantilever bridges, and the Queensboro, on which Ammann worked, was clearly among the most notable. Drawn to scale between the Forth Bridge and the second Quebec Bridge, then under construction, the Queensboro was clearly a distinct and significant span, regardless of what some critics would say.

It was while the Queensboro Bridge was still under construction that the first Quebec Bridge collapsed, and Kunz, Ammann, and every other engineer in Steelton, just seventy-five miles from the Phoenix Bridge Company's design office at Phoenixville, felt the shock. It was immediately clear that, as with all major structural failures, there would be an investigation. When C. C. Schneider, formerly with the American Bridge Company, was named to lead the investigation, Ammann, presumably through Kunz, offered his assistance. Ammann could clearly have seen this as another opportunity to be involved with one of the most significant current bridge problems, but there was also a clear advantage to his employer, the Pennsylvania Steel Company, in being as close to the investigation as possible, so that lessons learned might be applied to their own great cantilever project. Though the report that appeared under Schneider's name owed much to Ammann, he was still an assistant, and thus did not receive the explicit formal recognition he may have deserved. He is said by hagiographers to have become "the actual boss of the study, and to this very day his report is considered as a model of thorough investigation," but the truth of that may hinge on one's definition of "boss." Whatever the respective roles of Schneider, Kunz, and Ammann, after the Quebec Bridge report was published in 1908 and the Queensboro Bridge opened in 1909, they all three found themselves together in Philadelphia, in the newly constituted engineering firm of Schneider & Kunz. Ammann, who became principal assistant engineer with the firm, was clearly still a junior and not yet a named partner in engineering endeavors and enterprises. He was in a position to have assisted Kunz in designing a new Quebec Bridge in 1909, but Ammann must have been disappointed that Kunz's span had none of the grace of the Queensboro Bridge and was not in the end selected to be built. After eight years with Kunz, Ammann sought a change, and his superiors recommended him to Gustav Lindenthal.

Whereas Kunz had been chief engineer of construction of the Queensboro, Lindenthal, of course, was chief engineer of its design. Ammann must have admired the technical ambition of Lindenthal, whose two-decades-old proposal for a Hudson River crossing was then still the grandest dream of all, the kind of dream that had actually lured Ammann to America. Thus, when he had the opportunity to take a position with Lindenthal's firm, he accepted it eagerly. In a characteristically unemotional understatement of the event, he wrote in his diary in the third person on July 1, 1912, "OHA started position with G.L.," and a little ways down the page added, "Mr. L. stated: I estimate an Engineer ⅓ by his character, ⅓ by his ability and ⅓ by his experience." Lindenthal must have estimated Ammann high in all three thirds, for before the end of three months on the job Ammann could add to his diary, on September 24, "I am appointed Assistant Chief Engineer of East River Bridge Division, New York Connecting Railroad, by Mr. G. Lindenthal." In his new position, Ammann was to be in general charge of the office, field, and inspection work for the $20-million project, whose centerpiece was the Hell Gate Bridge across the East River.

Ammann was Lindenthal's chief assistant among the staff of ninety-five engineers, which included the special assistant engineer David Steinman, who would later become Ammann's main rival. War in Europe overshadowed such nascent competitions, however. After only two years of work on the Hell Gate project, and just before erection of the arch proper was begun, Ammann left for Switzerland to help in the possible fight against Germany. Although still a Swiss citizen and reserve officer, the thirty-five-year-old lieutenant's return to his homeland after a decade in America turned out to be less than the glorious military campaign he may have anticipated on his eastward sailing. He managed a discharge after only eighty-one days of active service, and that mostly in a supervisory capacity building fortifications on the St. Gotthard Mountains in the Swiss Alps. Barely four months after he had left Lindenthal's office, Ammann reported back to resume work on the Hell Gate Bridge—and thereby to displace Steinman.

Over the next year, Ammann supervised the completion of the great arch; the last rivet was driven into the bridge in September 1916. By this time, he held the position of deputy chief engineer, and he was the logical choice to draft a report on the completed project. Although national meetings of the American Society of Civil Engineers were omitted in the years 1917 and 1918 because of the war, local meetings continued to be held in metropolitan areas like New York, and on November 21, 1917, Ammann presented a full and most authoritative account of the planning and construction of the Hell Gate Bridge. It was published in the *Transactions of the American Society of Civil Engineers* for 1918 and awarded the Rowland Prize that year.

Whereas Ammann played an anonymous role of uncertain extent in the actual authorship of Schneider's report on the Quebec Bridge and Kunz's book on the design of steel bridges, there can be little doubt as to who wrote the 150-page paper on the Hell Gate Bridge. Ammann does acknowledge his "obligation, for permission to present this paper and for valuable information, to Gustav Lindenthal," but the paper carries only a single author's name: O. H. Ammann. To Lindenthal's credit, he did not pull rank to have his name displace Ammann's or be added to it, nor did he squelch Ammann's opportunity to get full credit at last for his ability to plan and execute engineering reports of uncommon clarity and style. This talent would in later years often be remarked about in popular profiles of the engineer, but it was not lost on the engineering audience either. Indeed, one member of the society, Henry Quimby of Philadelphia, closed his discussion of Ammann's paper with some extraordinary remarks:

> The paper is an unusually satisfying one, both in the fact that it appears while the public and the professional interest in the remarkable feat is still fresh, and in that it discusses so freely the reasons for the various features of the design. The oral presentation of the subject by the author was also exceptionally felicitous, summarizing and supplementing the paper rather than repeating it by reading word for word, as is too often done with preprinted papers.

Writing, not to mention speaking ability, is an often overlooked talent of successful engineers. There can be little doubt that John Roebling's ability to put pen to paper made it immensely easier for him to gain political and financial support for his milestone Niagara and Brooklyn bridge projects. Eads and Cooper wrote voluminously, as did Lindenthal, though his apparent inability to keep his pen from drifting from the main objective of his words into diatribe must have taken away from the sound and otherwise convincing arguments that he advanced. Ammann, on the other hand, seems to have approached his engineering reports with all the circumspection and rationality that he did design projects, without having to sacrifice aesthetics or style in either. Later in life, he would speak often to reporters about his writing, confessing that reports were no easier to design than bridges, and that he usually had to take his manuscripts home "and work on them until two in the morning." Among Margot Ammann's earliest recollections of her father was of him "bent over his desk, writing a report." He wrote "on a block of yellow lined paper" with a pen that had a thick nib, judging from the documents that survive. Her recollection also speaks to his discipline with regard to correctness and revision: "He frequently consulted the dictionary that was always by his elbow and [made] revisions with much slashing, writing in the margins and changing of sequence by cutting paragraphs with scissors and then pasting them elsewhere in the report." A New Jersey neighbor who was often awake to attend to a sick mother throughout the night corroborated Ammann's work habits: "Whenever I looked over to the Ammann house, at one o'clock, three o'clock, there was always a light burning in Mr. Ammann's study and I knew he was working."

Ammann would have plenty of opportunity to hone his writing skills, for he was to produce over one hundred full-length reports during his career, suggesting the large number of projects on which he worked. In the time between the completion of the Hell Gate project and the presentation of his paper on it, Ammann was principal assistant engineer to consulting and chief engineer Lindenthal for the steel superstructure, erected by the McClintic-Marshall Company of Pittsburgh, for the Sciotoville Bridge over the Ohio River. This bridge too had added considerably to Lindenthal's reputation as among the greatest bridge builders of his age, of course, and he himself wrote the paper reporting on it. However, as opposed to the timeliness of Ammann's report on the Hell Gate, Lindenthal's paper came five years after the bridge was completed. Indeed, the paper's opening sentence acknowledges that the "peculiar construction" of the bridge had "been the subject of frequent inquiries," and offered the "detailed, although somewhat belated, description" as the "permanent record" of the project. In contrast to Ammann's fluid and inclusive style, Lindenthal's, in his forty-five-page paper, is jerky and contentious, if not curt at times, with statements of fact and opinion intermixed. In a section on the history of continuous truss bridges, for example, after describing Robert Stephenson's classic Britannia Bridge, Lindenthal remarks that "too much credit cannot be given to that galaxy of early English bridge engineers," which included Stephenson, and Lindenthal goes on to express his clear approval of their ways: "They did their own thinking; they did not wait for precedents, but created them." Lindenthal clearly must have thought of himself, and his Hudson River Bridge, in the tradition of these engineers. He seems to have gained resolve from such a reading of history, much as Eads had found in Telford's ideas a precedent for a great arch bridge at St. Louis.

As was customary, Lindenthal acknowledged those who had helped with the project. In contrast with an engineer like Waddell, whose paper on the Halsted Street Lift-Bridge recognized first the politicians who made the project possible, Lindenthal made no mention of the commissioners of the bridge, the Chesapeake & Ohio Northern Railway, other than matter-of-factly in the paper's title, and then essentially only to locate precisely the artifact itself, rather than to flatter its owners. Only the engineers and constructors were acknowledged for their assistance "in this unusual work, bristling with new problems and difficulties." Those singled out included David Steinman, for "computations of superstructure," but Ammann was mentioned first and foremost as "Principal Assistant Engineer in general charge."

Unlike the abrupt closing of Lindenthal's paper, Ammann's on the Hell Gate summarized in systematic list-like form "some broader engineering questions," or lessons learned from the project. His remarks are clearly laudatory toward Lindenthal, but the reader cannot help thinking that assistants like Ammann gain in stature by their association with the chief engineer and his projects:

> A great engineering work cannot be spontaneously created in its final, perfect form, but has to grow and develop gradually, in its entirety as well as in its constitutent parts. Although the layman can only judge such a work in the light of an accomplished fact, the engineer must ever be conscious that it is only through extensive and laborious preliminary studies, and untiring efforts to improve, that he can hope to achieve a perfect work.
>
> In the execution of a great and complex engineering or scientific undertaking, collaboration of experts in various fields is essential, but a great structure of monumental character must be the product of an individual creative and directive mind.
>
> A great structure cannot be the result of a set of rules and specifications, nor of elaborate mathematical computations. Such a work requires wide experience and sound judgment, and therefore, should be entrusted only to engineers of high professional attainments and reputation.

Lindenthal's plan for a Hudson River crossing indeed fell into the category of "a great structure of monumental character," but as the great engineer approached his seventieth year he had become less and less flexible about how the project might evolve with the needs of an evolving metropolitan area. The war had slowed bridge construction generally, and it was a time of inactivity for engineering firms like Gustav Lindenthal's. This might have been an opportunity to do speculative work on the Hudson River plan, making it more economical and therefore more attractive to potential supporters, but Lindenthal apparently chose not to do that. After the Hell Gate and Sciotoville projects, there was little to do in the office even for Ammann, and Lindenthal suggested that he try to get work elsewhere until there was something for which to call him back. Nothing else was available, however, and Ammann was thinking about entering war service when Lindenthal offered him a position as manager of a clay mine jointly owned by Lindenthal himself and a New Jersey judge, later governor, George S. Silzer. Ammann subsequently admitted that "the position was not attractive," but he "accepted it [so] as to be on hand in case Mr. Lindenthal needed my assistance."

Othmar Ammann spent the next few years effectively exiled in the central–New Jersey county of Middlesex, managing the obscure mine of the Such Clay Pottery Company rather than building grand steel bridges. When he took over the operation of the mine, it was unprofitable, and so his compensation was in jeopardy. However, he turned the situation around, thereby demonstrating a sound managerial sense and business acumen. His performance could hardly have been lost on Lindenthal and Silzer; the latter especially would no doubt recall it years hence.

Standard biographical sketches of Ammann do not mention his time working at the New Jersey mine. Rather, all the years between 1912 and 1923 in his career are accounted for as being spent working for Lindenthal, as he technically was. In addition to work on the major Hell Gate and Sciotoville projects, studies for the 57th Street Bridge did in fact continue throughout that period, though in limited form. In 1920, when the war and the recession were over, questions of a Hudson River crossing again became paramount in New York and New Jersey, and Ammann was appointed assistant chief engineer of the North River Bridge Company. Lindenthal's dream sprang from nineteenth-century assumptions about the importance of an over-water railroad link between the states, and his plans had grown to accommodate twenty vehicle lanes as well as twelve railway tracks. There were also to be terminal facilities and a moving platform for pedestrians. The estimated cost for the bridge alone, of over $200 million, was prohibitively high, and solid backing continued to be elusive. In 1921, the formation of the North River Bridge Corporation was announced, with over $250 million in capital stock, but the general feeling in banking circles at the time was that such a large capital undertaking would not be very viable. Soon, an organization known as the Hudson River Bridge and Terminal Association was incorporated, its purpose being "to obtain public support for the undertaking, projected by Gustav Lindenthal, an eminent bridge engineer, to build a great double-deck highway and railroad bridge from Manhattan to Weehawken."

2

Among the complicating factors in the early 1920s in winning approval for bridging the Hudson was the alternative of vehicular tunnels, which was gaining support. Railroad tunnels had by then long operated successfully under the Hudson, and subway tunnels under the East River had become almost unremarkable feats. There was only one principal remaining doubt as to the efficacy of going under rather than over the river: it remained to be seen whether the exhaust gases of automobile and truck traffic could be effectively removed from a subaqueous tunnel a mile or two in length. In spite of this, as early as 1918 a meeting between the New York State Bridge and Tunnel Commission and the New Jersey Interstate Bridge and Tunnel Commission resulted in a joint commission organized to promote the crossing of the Hudson River, and the preference was for a tunnel.

The commission debated Hudson River crossings during the winter of 1918, a particularly severe one. New York Harbor was icebound, and the inability of delivery trucks to get across the river on ferries resulted in a "coal famine." Since the joint commission had recently asked a consulting engineer to look at proposed tunnel plans and report critically on them, the time was propitious for an underwater crossing to gain support. The fact that the consulting engineer was George Washington Goethals ensured that the issue received much publicity.

George Goethals was born in Brooklyn in 1858, but his family moved to Manhattan when the quiet and somewhat shy, studious, and yet well-liked boy was eleven. His education in New York City public schools, the City College of New York, and West Point, from which he was to graduate in 1880, would later in life give him a sense of obligation to public works, at which he would excel. Six feet tall, with blue eyes and a ruddy complexion, young Goethals was an impressive figure, and his early inclination was to follow the profession of either law or medicine. However, mathematics attracted him in school, and he became increasingly interested in engineering, leaving City College before completing his degree to take advantage of a vacancy in the Cadet Corps that had opened up at West Point. He remained at the military academy for a year after graduation, teaching astronomy, but soon afterward was assigned to work on the Columbia River and later the Ohio River, to assist in making improvements for navigation. He returned to the military academy in the late 1880s to teach civil and military engineering, and then was assigned to work on improvements on the Cumberland and Tennessee rivers, including work on the Muscle Shoals Canal, near Chattanooga, and the Colbert Shoals Lock. He served at army headquarters in Washington, D.C., for a period, in the Spanish-American War, and on various river-and-harbor improvement projects in Rhode Island and southern Massachusetts. As if this broad experience were not enough, his service on the General Staff from 1903 to 1907 gave him considerable exposure in Washington, and by then he was a natural choice to take on leadership of the Panama Canal project. No doubt sensitive to the debate as to whether the canal should be a private or a military project, Lieutenant Colonel Goethals never wore his uniform in Panama. When the canal, which had been under discussion for centuries and under construction for decades, was finally opened in 1914, Goethals was a hero, if not a legend, for his ability to complete what so many before him had started. He was promoted to major general in 1915 and retired from the army in 1916, whereupon he moved to New York to work as a consulting engineer.

In that capacity, he was extremely influential in the growing debates over the nature of transportation in the New York area. The governor of New Jersey asked Goethals to lay out a new highway system for that state, and he was involved with questions of moving military goods in and through the New York and New Jersey area during World War I. Thus he was an almost unassailable authority on how to forestall a future coal famine, which in 1918 "was due almost entirely to the city's inability because of the ice-choked river to transport thousands of tons of coal that were literally in sight on the other side of the river, and yet as unattainable as if they were still in the mines."

Goethals estimated that the 1913 vehicular-tunnel proposal of the firm of Jacobs & Davies, with some of his own modifications, could be built in three years and could be paid for with tolls that were less than the ferries charged. Furthermore, if it was lined with concrete blocks instead of iron, the tunnel construction could proceed without interfering with the war effort. The joint commission, which had originally been charged with considering an interstate bridge at either 59th, 110th, or 179th Street, was thus now leaning toward a tunnel entering Manhattan at Canal Street, where the terrain was favorable. Goethals had convinced them, in the midst of the coal famine, that such a solution was the quickest and cheapest one and, by replacing the need for ferry slips, it would free up valuable waterfront space for commerce. Furthermore, since war conditions were making it imperative that vast amounts of materiel be moved through New York Harbor, the government might share the cost of a tunnel. Though this seems not to have happened, by early 1919 Goethals's plan had been fleshed out to comprise a single tunnel with two levels, each to accommodate three lanes of traffic in a roadway twenty-four and a half feet wide. The cost of $12 million was to be shared equally by New York and New Jersey; toll revenue was expected to pay for the tunnel in twenty years while at the same time establishing a maintenance fund.

In June 1919, with the necessary state legislation finally passed, the joint commission appointed as chief engineer Clifford M. Holland, because, although "the youngest chief tunnel engineer in the United States and probably in all the world," he had extensive experience in building subways and tunnels in New York. Holland was born in Somerset, Massachusetts, in 1883, and he graduated from Harvard in 1906 with both bachelor of arts and civil-engineering degrees. He went to New York and became an assistant engineer with the Rapid Transit Commission, which was then building the city's subways. It would be said of Holland that "he spent more time underground, particularly in compressed air, than any other civil engineer on similar work." In accepting responsibility for tunneling under the Hudson River, Holland insisted that he be given free rein in selecting his engineer-

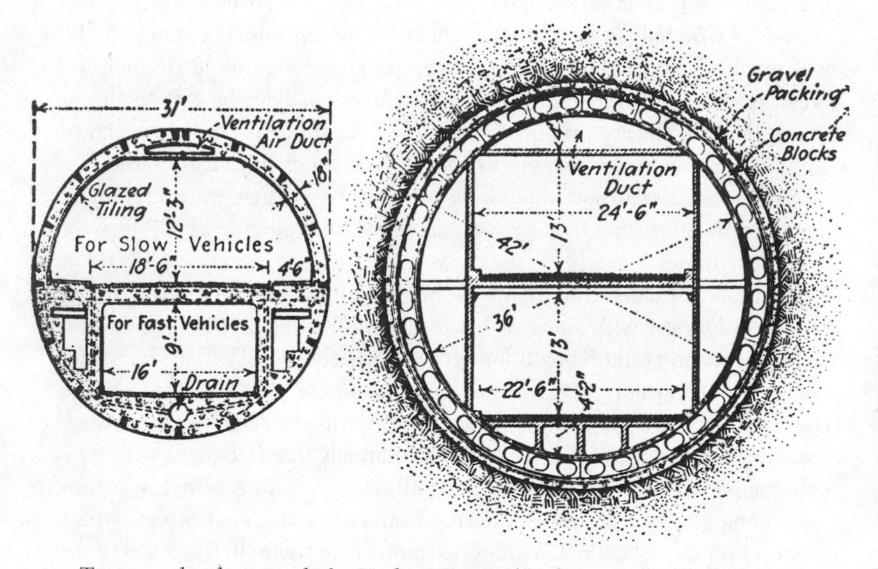

Two unrealized proposals for Hudson River vehicular tunnels, by the firm of Jacobs & Davies in 1910 (left) *and by O'Rourke and Goethals in 1919*

ing staff, and his strong sense of conviction and determination would be essential to his prevailing when it came to the nature of the tunnel that would be designed. The report of his appointment made clear that he would indeed have broad responsibilities and discretion:

> The duties of the chief engineer are to organize an office and field staff sufficient to gather data concerning physical conditions of site, to make the necessary surveys, to prepare estimates of cost and to decide upon the type, size and location of the proposed tunnel. After this work is completed the work of drawing plans and preparing specifications in detail so that contractors may bid will be taken up.

Goethals had prepared a conceptual plan and made a gross determination of feasibility and an estimate of cost, but it was now time to look carefully and critically at all aspects of such designs, consider alternatives, and work out details so as to ensure that the tunnel was buildable and workable. For his responsible charge of the novel undertaking, Holland was to be paid a salary of $10,000 per year. Each member of the board of consulting engineers was also to be paid the same sum, and they were required to meet biweekly or oftener until the type of tunnel was agreed upon. The board consisted of J. Vipond Davies, a partner in the firm of Jacobs & Davies, who before proposing a vehicular tunnel had built the twin tunnels for the Pennsylvania Railroad and the two pairs of the so-called McAdoo (later, Hudson) tubes that carried the Hudson & Manhattan Railroad; Henry W. Hodge, whose extensive experience with bridge design, including a proposed Hudson River crossing, had made him familiar with conditions on the river bottom; William H. Burr, professor emeritus of civil engineering at Columbia University, who had had wide experience in bridge and harbor engineering and had been appointed by President Theodore Roosevelt to the international board created to settle the question of what kind of canal to build in Panama; and Colonel William J. Wilgus and Major John H. Bensel, representing military experience and interests.

As with all large engineering projects, the chief engineer was to be assisted by many others. Several key appointments were approved at a meeting of the joint commission on July 1, 1919: Jesse B. Snow was appointed principal assistant engineer, at $5,400 per annum; Milton H. Freeman, resident engineer, $4,200; Ole Singstad, designing engineer, $4,200; and Ronald M. Beck, assistant engineer, $3,000. These were only some of the engineering expenses to be associated with the project, of course; the total cost of engineering services would be on the order of 6 percent of the total project cost, which through the end of 1919 continued to be taken at Goethals's estimate of $12 million. Thus, when Holland's report was made public in early 1920, it was full of surprises. Not only was the cost of the project put at nearly twice what had been thought, but Goethals's design was severely criticized and rejected by Holland. According to his report:

> After a very careful investigation of this plan your engineer found that: First, the capacity of the tunnel is greatly overestimated, as the width of roadway is not sufficient for three lines of ordinary and usual traffic, but is sufficient only for two lines of traffic and a three-foot walk; second, the lining of concrete blocks is not of sufficient strength to withstand the external load, and is not suited to the Hudson River conditions; third, the difficulties of construction would be enormously increased and the meth-

> ods supposed to overcome them are undeveloped and do not insure the safe prosecution of the work, so that its successful completion is a matter of conjecture; fourth, the estimated cost of construction is entirely too low; fifth, the time for construction, fixed at three years, is very much less than would be required.

The report was signed by only four consulting engineers, for Henry Hodge had died in the meantime, and they gave their unanimous support to Holland's plan for twin cast-iron lined tubes each twenty-nine feet in diameter over Goethals's proposal for a single forty-two-foot-diameter concrete lined tube. By backing the more conservative plan, they were keeping within the state of the art and demonstrated practice. Although a total of

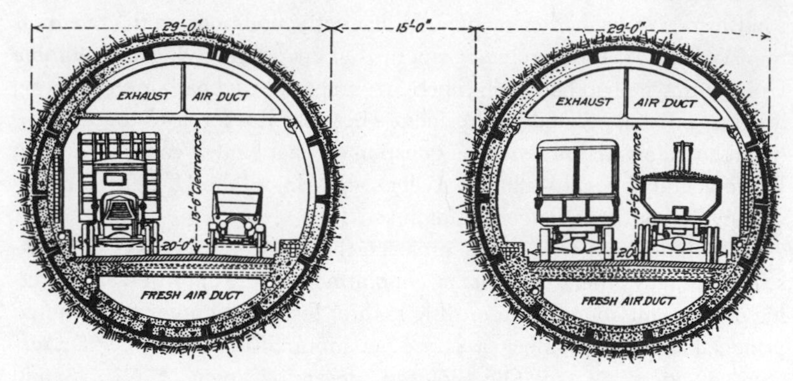

Clifford Holland's design for a Hudson River vehicular tunnel made up of twin tubes

eleven plans were considered, Goethals's had gotten the most publicity, and it was his that caught the most visible criticism, no doubt because it had theretofore appeared to be the design of choice.

Within days of the release of Holland's report, Goethals wrote to the joint commission requesting data and analysis substantiating the conclusions against his design. Holland suggested that Goethals, like any citizen, could come to the office to examine what was a matter of public record, but the engineer's staff could not afford to spend the three weeks it would take to assemble the material for him. In the meantime, Edward A. Byrne, chief engineer of the Department of Plant and Structure for New York City and a member of the consulting board serving without special compensation, let it be known that he did not support the report's conclusion. New York had, of course, a long history of corruption and influence-peddling in the construction of public works, especially in the context of the political machine known as Tammany Hall, and it appears that the situation was once again rife with abuses. Byrne, sitting on a board as part of his job while his colleagues were drawing annual retainers of $10,000 for their participation, might have been especially vulnerable to bribes and kickbacks.

It turned out that Goethals's tunnel plan would have employed a process patented by John F. O'Rourke, who had stood to gain a healthy royalty. O'Rourke came to America with his family two years after his birth in 1854 in Tipperary, Ireland, attended Cooper Union, had taught there for some time, had been construction engineer on the Poughkeepsie railroad bridge, had extensive experience with tunnel projects, and held numerous patents for construction methods and equipment. In 1920, hoping to salvage something of the Canal Street tunnel project, he proposed to release his patents in exchange for payments of 25 percent of the estimated $9 million savings that the use of concrete blocks would provide over cast-iron. In other words, O'Rourke would realize in excess of $2 million, which he might be willing to share with those who might help him get it.

With an arrogance that must have been fueled by fame, Goethals continued to promote his tunnel scheme, and the young Holland persisted in rebuffing it. He explained that calculations showed that the large-diameter tunnel would actually have so much buoyancy in the notorious Hudson silt, as the fluidlike material of the river bottom was called, that the tube would float. Furthermore, hardly any of Goethals's assumptions or assertions withstood the close scrutiny of Holland and his engineering staff; no one had the experience to serve as a guide, because no tunnel of the kind Goethals proposed had ever been driven under any river. In early March, Holland summarized the situation to a meeting of the New Jersey commissioners as follows: "The proposal to build a tunnel of unprecedented diameter of untried materials abandons all that experience in tunnel building has taught in the gradual development to the present state of knowledge and enters upon a new field of uncertainty." A week later, the joint commissioners passed a resolution effectively directing Holland to devote no more time to considering the Goethals-O'Rourke scheme and to proceed with work on the twin cast-iron tubes.

Public confusion and debate about the tunnel design continued for over a year, and disagreements arose between the New York and New Jersey commissions. After the board of consulting engineers finally rejected the Goethals-O'Rourke scheme, the New Jersey Commission dismissed the current board and ceased compensating them. Two new engineering-board members were appointed by the New Jersey Commission, but the New York Commission refused to recognize them. The New York chapter of the

American Association of Engineers, which would evolve into a group that promoted registration for engineers and concerned itself with the status and employment of engineers, issued a report seeking "answers to questions whether the action of the New Jersey Commission did not reflect unfavorably on the members of the board, whether it did not tend to injure the whole engineering profession and was contrary to the public interests, and whether it would not seem undignified and unprofessional for other engineers to accept appointment to the places thus made vacant." The issue of engineering professionalism and ethics was a growing one, and the report on the dismissal was an occasion to articulate it:

> There can be no question that if anybody, private individual, corporation, or public commission, is dissatisfied with the services of any engineer in his or their employ, be he the most eminent consulting engineer or a student just out of college, he has the right to dismiss such engineer and seek other assistance. It is, however, a well established, even though an unwritten rule of professional ethics, that should an employer discharge an engineer, and especially an engineer in responsible charge of work or one employed as a consultant, without reasonable cause, that it is unethical for another engineer to take over the work or the position.
>
> The board of consulting engineers which was appointed and which, in conjunction with the chief engineer and his staff, has brought the enterprise to the point where its successful completion can be practically visualized, is not only a body of eminent engineers but also a body of eminent, well known and public-spirited citizens, and such a board, in connection with an enterprise of this character, should not be, and as a matter of fact cannot be, discharged just as one might discharge an office boy, and especially is it true at such a critical period of the work. Whatever the technical legal status of the matter may be, public sentiment, where properly informed, will not permit it.

Such was the climate in which large public-works projects were conducted. As matters of professional engineering continued to be debated in trade journals like *Engineering News-Record*, so did the practical matter of awarding contracts to sink shafts for the tunnel continue with the commissions. New Jersey commissioners favored awarding the contract to the Shaft Construction Company, which was incorporated on the very day that the bids were received, and whose low bid was even below the engineers' estimate. The commissioner of public safety in Bayonne, New Jersey, was reported to be glad that his brother, the principal in the Shaft Construction Company, was "successful enough to underbid one of the largest construction firms in New York." However, the New York commissioners agreed with Holland, who reported that "in the opinion of your engineer . . . the fact that a new organization, inexperienced in this work, has submitted such a very low figure raises the question as to whether this company has a proper appreciation of the character and quality of the work to be carried out." Since each commission was independent, and since there was no formal way to compel agreement between the two state bodies, the matter had to be resolved through public debate and efforts to influence public opinion. In the meantime, the joint commission did agree to send Holland to Europe to inspect vehicle tunnels in such cities as London, Glasgow, Paris, and Hamburg, in order to collect firsthand information for making final design decisions on the Hudson River tunnels, for which invitations to bid on final construction contracts had yet to be advertised.

The New York commissioners continued to oppose the Shaft bid, but that was not the only obstacle to beginning work on the New Jersey side. The land needed for the entrance-and-exit plaza was owned by the Erie Railroad, which was unwilling to make concessions regarding the perpetual retention of some of its tracks alongside a cold-storage plant in which T. Albeus Adams, one of the New Jersey commissioners, had an interest. When an agreement was finally reached, it involved combining the contract for driving the New Jersey access shaft with that for driving the tunnel tubes proper. The delays were estimated to have cost the two states about a half-million dollars, out of a total estimate that had in the meantime grown to almost $29 million. However, the investment was thought to be well worth it, for the tunnel would have a capacity of fifteen million vehicles per year—about twice the number that were crossing the Hudson on ferryboats in 1921.

Even as final plans were being made, there were still some serious doubts as to whether the toxic exhaust of automobiles and trucks could be effectively removed from a tunnel. This was such a critical question that the Bureau of Mines had constructed in Pittsburgh an experimental circular tunnel four hundred feet long through which automobiles could be driven to test a ventilation system. A group of New York and New Jersey commissioners and engineers went to Pittsburgh to ride around this tunnel in cars forty feet apart for almost an hour, having their pulse, blood pressure, and blood samples taken and analyzed for effects of carbon monoxide. After the automobile tests were successfully completed, a smoke bomb was set off in the tunnel to demonstrate how effectively the ventilation system could exhaust the dense and visible gases. The New Jersey commissioners, at least, were not satisfied with such perks of their position, however. With the light at the beginning of the tunnel finally turning green for construction, it was proposed that a twenty-foot granite shaft be erected in the center of the New Jersey plaza to commemorate the commissioners themselves. When presented with this proposal, the New York commissioners declined the honor, but not before one of them suggested that the name and face of the originator of the idea be done in brass rather than bronze.

With the prime contract awarded to the low bidder, the experienced tunnel-driving firm of Booth & Flinn, work could begin under the general superintendence of Michael L. Quinn. A ceremonial ground-breaking took place on March 31, 1922, at which, according to a newspaper report, "Mr. Holland took a pick from the hands of John Bazone, a laborer, and drove it into the earth. Mr. Quinn thrust a shovel under the loosened earth and threw it to one side." Though the workmen wanted their tools back, neither Holland nor Quinn would give his up, for each was to be plated with silver and displayed in the respective ceremonial wielder's office.

Ground-breaking in New Jersey did not take place until May 31, and then clandestinely, according to a newspaper account. Crossing the river from New York, Holland and several representatives of the Erie Railroad Company, including chief engineer R. C. Falconer, and of the contractor, Booth & Flinn—perhaps about a dozen in all—stole across ferry property toward the Erie Railroad yard, in order to avoid being seen on the streets. After convincing several policemen that the group was not up to anything improper, they reached the railroad yard, where "Mr. Falconer, armed with a pinch bar, removed a section of track, George H. Flinn drove a pick in the ground, Mr. Holland dug up a shovelful of dirt, and the breaking of ground was over." Although it was a private affair, there seems to have been at least

Clandestine ground-breaking in New Jersey for the Canal Street Tunnel by (left to right)*: C. M. Holland, chief engineer, for whom the tunnel would be named posthumously; G. H. Flinn, of the contractor Booth & Flinn; and R. C. Falconer, chief engineer of the Erie Railroad, in whose yard the ceremony was taking place*

one reporter/photographer at the scene: a picture of the ceremonial event and another of a large group of attending engineers standing as if in a lineup appeared in the next week's issue of *Engineering News-Record*. The ceremony was no doubt brief, and the general concern was less with capturing tools to silverplate than with getting on with the construction job. Holland had reasoned that once ground was broken, it would be more difficult to stop work. He had been increasingly concerned that the New Jersey commissioners would hold to their threat that work would not be permitted to begin until concessions were made regarding street improvements, not to mention the threat of a gala Fourth of July ground-breaking that not only would have delayed the start of work but might also have upped the stakes in the fight for concessions once plans for the gala were set.

3

Even with the Canal Street tunnel under way, the relative merits of bridges and tunnels continued to be much discussed, but it was the low cost of a tunnel compared with grandiose bridge schemes like Lindenthal's that began to tip the balance. Furthermore, it was argued, if several tunnels were built at various locations along the river, traffic between New York and New Jersey could be diffused. This addressed directly one of the continuing concerns over Lindenthal's 57th Street bridge: the concentrated volume of traffic it would bring to already congested midtown streets. Private capital began to see the possibilities of the alternative, and the Interstate Vehicular Tunnel Company made known its plans for three new tubes under the Hudson, though one of the company's incorporators, Darwin R. James, would not reveal the locations being considered, so as not to inflate real-estate prices. Even though the question of ventilation had yet to be proved on the full scale, James asserted that for the "purely commercial" project "the engineering problems had all been solved."

In early March 1923, plans were announced for a tunnel to connect Manhattan in the vicinity of 125th Street to Bergen County across the river. The projected growth of vehicle traffic and subsequent development of New Jersey made the estimated cost of $75 million appear to be a sound investment. Before long, a Harlem and Bergen Tunnel organization was seeking incorporation for the purpose of "the construction of one or more tunnels for vehicular and pedestrian traffic between a point in Harlem, between West 120th and West 140th Streets, and a point in Bergen County, N.J." At the time, another company was also seeking incorporation, in order to construct a tunnel near 42nd Street.

The possible proliferation of uncoordinated river crossings, combined with the problems of two independent commissions, was no doubt a factor in the growing talk about bringing all tunnels, including the one already begun, under the jurisdiction of the Port of New York Authority, which had been established in 1921 to develop and regulate movement about the port shared by the neighboring states. New York's Governor Al Smith proposed early in 1923 that, with the authority to issue bonds, such a body could finance public works whose toll revenue not only would pay off the bonds but also would provide the funds necessary for ongoing maintenance and operation, without the need of expending tax revenues. In late May, Smith vetoed two tunnel bills before him and let it be known that he opposed private control of such facilities, thus helping to complete the groundwork for a new era in Hudson River crossings, whose need was growing.

In the meantime, Holland's tunnel was progressing, but not without troubles for its engineer. Sandhogs, as the tunnel workers were known, went on strike for "more wages, fewer hours, and lower air pressure," to reduce the risk of contracting the bends. Costs were also rising, in part because of increased prices of materials and labor, but also because of design changes such as those involving the approaches. Toward the end of 1923, when the tunnel was about 60 percent done, Holland had to answer criticism that the engineering budget was inflated, and had to overcome political interference even at the level of hiring a draftsman "who did not come from the Hudson County Democratic organization." By early 1924, the final cost was estimated to be $42 million, with larger plazas and improved ventilation systems contributing to a more-than-tripling of the cost over the preliminary estimate. The efficacy of the ventilation system was an ever-present concern. The new twin Liberty Tunnels through Pittsburgh's South Hills were opened prematurely in 1924, then had to be closed after only two hundred cars had passed through, because of dangerously high concentrations of carbon monoxide. Later that year, a trolley strike in Pittsburgh caused massive traffic jams in the tunnels, and the accumulated fumes caused scores of people to become ill in their cars. It would not happen under the Hudson, Holland assured New Yorkers and New Jerseyites, for his tunnel would have four times as many ventilation shafts, and traffic congestion would not be permitted to occur. In a long letter to the editor of *The New York Times* that read more like a scholarly article, Yandell Henderson, professor of applied physiology at Yale University, proposed obviating the fumes problem by fitting vehicles with vertical exhaust pipes that extended upward to a few inches above the tops of the cars and trucks.

Additional problems continued to arise. In early April, the tunnel sprang a leak and water rushed in, forcing thirty-five workers to flee for their lives.

Shortly thereafter, there was another sandhog strike, with increasing numbers of men complaining of the bends. In September, there was growing concern for the mound that was being thrown up on the river bottom by the shield driving the tunnel through the silt. The three largest ocean liners of the day, the *Berengaria, Leviathan*, and *Majestic*, had to follow a revenue cutter through "the only spot in the Hudson River where the great liners could pass" without danger of grounding. Though the mound would be dredged away once construction was completed, it meanwhile presented a distinct danger to port traffic.

A milestone in the construction project was to occur in late October 1924, when the two tunnel shields boring the north tube were to cease working and the remaining thickness of silt was to be blasted out, thus completing the connection between New York and New Jersey. Though careful and continual surveying made it virtually impossible for the two tunnel halves to be out of alignment, it was reported that "some of the engineers engaged in the work lie awake nights just the same worrying about it." Nevertheless, an elaborate ceremony was being planned; President Coolidge would press a button in his library in the White House connected via telegraph wires to the dynamite, thus setting off the blast at precisely nine o'clock on the evening of October 29. Newark radio station WOR was to broadcast the event live, its "impresario" being present "in the depths of the tunnel," along with the governors, mayors, commissioners, and U.S. senators of the two states. As soon as the blast was set off, a band would play "The Star-Spangled Banner."

The blast occurred nine hours early, and it was not President Coolidge but George Flinn, head of the construction company, who pressed the button. No mistake was made; acting chief engineer Milton H. Freeman deliberately directed preparations for blasting away the final eight feet of silt. After the tunnel was holed through, the two ends were found to be well aligned, to within three-quarters of an inch of one another. However, the engineers and workmen present in the tunnel at the time took the breakthrough as part of an ordinary day's work, in fact making every effort to avoid any appearance of ceremony. The sudden change of plans took place because, two days before the historic event, chief engineer Clifford Holland had died.

It had been said of Holland that he "started working on tunnels the morning after he received his degree" from Harvard. For eighteen years, he had worked under the rivers defining Manhattan Island, and the physically and emotionally stressful conditions appear to have taken their toll on his health. Three weeks before his death, he suffered a nervous breakdown, which prompted the joint tunnel commission to pass a resolution commending him for "his continuous devotion, night and day, during the past five years," to the Hudson River project, and to grant him a month's vacation with full pay. He had gone to Battle Creek, Michigan, for a rest, and that is where he died, of heart disease. According to the newspaper report describing the unceremonious completion of the north tube, Holland "had expected that his heart would fail him some day and so had kept ready for a quick ending. From the time he was first chosen to plan and construct the tunnel until he left on his vacation recently he had kept the plans in such shape that the work would go on after his death." It did, of course, and the tunnel, which would open in 1927, remains to this day an essential link in the traffic network of the New York metropolitan area.

News of Holland's death brought immediate calls to name what had till now been called the Hudson River Vehicular Tunnel after its chief engineer. Within about two weeks, the tunnel joint commission did pass a resolution bestowing the name Holland Tunnel, and *The New York Times* hailed the action. Their editorial admitted that engineers had some reason to complain that politicians took more care to associate their own names rather than engineers' with public works, but that this was less true than it had been. There was in the mid-1920s, according to the *Times*, "a general and cordial appreciation of the services rendered by engineers." If they did not get the proper publicity, it was "largely the fault of their modesty. As a rule they are the poorest of self-advertisers, and are, or profess to be, content with professional recognition of their achievements." Whether this is true generally, Holland was certainly thought to have been "the last to exploit his own merits," and so it was fitting that his greatest achievement be recognized by being named after him. However, the *Times* also feared that, "as he shared his name with the nation that ruled New York in its early days, the title of the tunnel will be understood by many, in the years to come, as referring to the nation rather than to the man." A device for preventing this misunderstanding remains elusive, however, and even today few realize the true origin of the tunnel's name, which is emblazoned in letters several feet tall atop the tollbooths that stretch across the New Jersey entrance plaza.

In 1924, construction on the Holland Tunnel had continued under Milton Freeman, who had almost twenty years of experience in tunnel work. However, after less than four months as chief engineer, he too died, of pneumonia, to which his resistance had been lowered by "long-continued overwork." *Engineering News-Record* reflected on the significance of it all: "Throughout time the building of great works for the use and convenience of man has claimed its victims. Not only gold and labor but also human life is wrought into what the builder creates." In a letter to the editor of the magazine, Robert Ridgway, chief engineer for the New York City Board of

Transportation, remembered Freeman as being Holland's "able lieutenant, working modestly and quietly, studying under the direction of his chief every one of the many details that made up the work." During crisis conditions, Freeman would spend countless hours on the job, "without going to the street for food," sleeping in his field office. Ridgway speculated that Freeman's modesty may have been carried too far, for few but his associates knew of his devotion and commitment to the projects on which he worked. "Mr. Holland used to say of him that he was an engineer who made the reputation of other engineers." Holland, like all the best chief engineers, would have known that anything named after him was a memorial to the many, like Freeman, upon whom he so depended.

The worries of the Holland Tunnel engineers may or may not have been aggravated by the "nice controversy" that rearose early in 1924, when both Governor Smith of New York and Governor Silzer of New Jersey proposed that the recently formed Port of New York Authority take over the tunnel. Among the primary purposes of the Port Authority was to simplify and rationalize the transfer of great volumes of goods between the rail freight yards in New Jersey and the factories, warehouses, and wharves in Manhattan. The tunnel, originally conceived of as "merely an underwater street," was certainly intended in part to carry trucks between New Jersey and New York, but in the years since its construction began, the development of the motor truck and private automobile had so changed the complexion of traffic on the roads around New York Harbor that the tunnel was likely to be overcrowded with automobile and truck traffic as soon as it was opened. To do its job better, the argument went, the Port Authority should control such an interstate traffic artery.

There was good reason to believe that other factors also motivated the proposal. Since the Authority had only a modest annual appropriation for administration, it had no money for capital projects. The tunnel, however, with the projected amount of vehicle tolls to be collected, promised to be an enormous source of revenue. If the Port Authority had control of the tunnel and its tolls, it would have the capital basis and continuing revenue for issuing bonds to finance other projects relating to traffic between New York and New Jersey. The Authority had been talking of building as many as five vehicular tunnels, and thus there was good reason to believe that it was indeed looking for ways to finance them. Even if trucking interests dreaded the prospect of the precedent of perpetual tolls for the use of a tunnel that had been intended to become free after about twenty years, there was growing public support for using public money to build bridges and tunnels whose toll revenue would in time repay whatever bonds were issued. In the 1924 election, for example, every county in New Jersey voted, by an overall margin of almost four to one, in favor of an $8-million bond referendum to continue support for building the suspension bridge between Camden and Philadelphia and the tunnel under the Hudson, both of which they expected would eventually be paid for by tolls.

What was curiously missing from the Port Authority's public plans was the construction of a bridge across the Hudson River. Since before World War I tunnels had been argued to be much less expensive than bridges, and they avoided the complications of providing high clearance for ships or the necessity of condemning large amounts of land to accommodate long approaches. However, with the final cost of the Holland Tunnel coming in at about four times the original estimate, and the growing vehicular as opposed to rail traffic in the New York metropolitan area, tunnels could no longer be argued to be the obvious economic choice. Relatively small capacity tunnels at different locations still had the advantage of diffusing traffic rather than concentrating it the way a single large bridge would, but which form of traffic communication should be picked in a particular instance began to involve arguments that amounted to deciding between six double tubes of one and a dozen lanes of the other. Such choices had long been debated.

4

In October 1908, an article in *The New York Times* had looked ahead to the bridges and tunnels that could be expected to lead into and out of Manhattan fifty years in the future. With the Queensboro and Manhattan bridges across the East River then under construction, the article predicted that "the next of the great viaducts will probably be the New York and New Jersey Bridge across the Hudson River." The midtown-Manhattan site of the "originally planned" bridge was reported to have been replaced by "a site further northward," with the vicinity of 181st Street mentioned as one possible location. In a bird's-eye view of Manhattan and its links to the west and east, only four Hudson River crossings were sketched in: the two pairs of McAdoo tunnels, the Pennsylvania Railroad tunnels that would lead to Pennsylvania Station, then yet to be designed, and a suspension bridge way up north, near 179th Street.

An interstate bridge commission had been formed as early as 1906 to determine whether and where a bridge for pedestrians, private vehicles, and trolleys could be built across the Hudson. Commissioners from both sides of the river had been looking at sites up and down the waterway, and the vicinity of 110th and 112th streets had also appeared to be a good choice

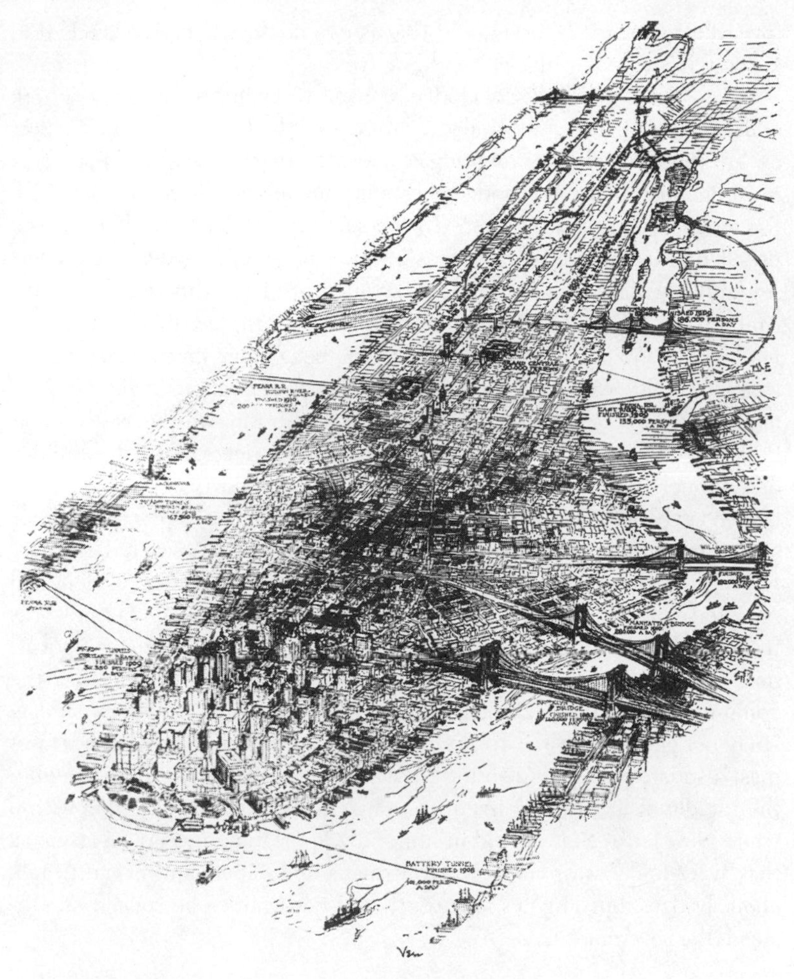

"Tunnels and Bridges of Manhattan Already Finished or in Process of Completion" in 1908, with an unnamed suspension bridge shown at about 179th Street

technically. However, that site "would seriously injure the property of Columbia University, St. Luke's Hospital, and the other eleemosynary institutions situated in that neighborhood," representing a total investment of some $30 million, according to Columbia's president, Nicholas Murray Butler, who summed up by calling any decision to build a bridge there "little short of vandalism." Whereas the New Jersey commissioners continued to remain favorable to a site near Columbia, the New York commissioners appear to have sympathized with President Butler's concerns; they leaned toward a site at 179th Street, and that is essentially what was sketched in in the bird's-eye view in *The New York Times*.

In the meantime, Boller & Hodge, consulting engineers to the New York and New Jersey Interstate Bridge Commission, had submitted their report on "the most feasible sites for bridges across the North River." The engineers recommended that three bridges be built: one at 57th Street, one at 110th Street, and one at 179th Street. (In the same report, Boller & Hodge also, presciently, identified the most feasible sites for bridges between New Jersey and New York's Staten Island: at Bayonne, at Elizabethport, and at Perth Amboy.) With regard to Lindenthal's 57th Street Bridge, though Boller & Hodge acknowledged its $75-million cost to be an impediment, they saw its location, in line with the Queensboro Bridge on the other side of Manhattan Island, to be a strong point. The engineers estimated the 179th Street bridge to be the least expensive to build, and they rejected the possibilities of a tunnel above 34th Street because they believed that the high cliffs known as the Palisades on the Jersey side made the idea impractical.

Among those interested in a bridge across the Hudson was Robert A. C. Smith, "the steamship and tobacco man," who had a summer residence in New Jersey. Smith thought the best view of the prospective bridge sites was from the deck of his steam yacht. Thus, on a sunny day in June, the governors of New York and New Jersey boarded the vessel as the guests of the commissioners, who had brought along their chief engineer, Henry Hodge: "In honor of Gov. [Charles Evans] Hughes, the *Privateer* carried at her foremast a square of deep blue bunting, with a white New York State shield in the middle of it. At her mizzenmast she carried another blue flag with a white New Jersey State shield in honor of Gov. [Franklin] Fort." As soon as they were under way, Hodge got out maps and papers and began to talk about bridges, but Hughes apparently did not want to be "talked at." Instead, the governor talked:

> The problem we want to settle right here is what is going to justify the building of the bridge and the choice of its site. You say that Fifty-seventh Street is a desirable location. But no one wants to go to Fifty-seventh Street per se. What class of people are going to use such a bridge, and how are they going to get from its termination at Fifty-seventh Street and Tenth Avenue to places downtown where they really want to go to? Will enough people use the bridge to make it pay from the taxpayers' viewpoint? If a ferry won't pay, a bridge won't pay.
>
> I understand that so few people of late years have been going over the Forty-second Street and Weehawken Ferry, for example, that its service has been practically abandoned. I rather like the 179th Street bridge plan,

on the other hand. It might not be so popular among New Jersey people. But it would connect New York City with the other New York State counties on the other side of the Hudson and help them to bring their farm produce into Manhattan.

According to the reporter, the *Privateer* was passing the 179th Street site at this time, and Hodge pointed out a castlelike structure on a wooded hill about a mile farther up the river. It had been built by an Italian immigrant named Paterno, who in twelve years had worked himself up from a day laborer to a wealthy building contractor, and who had hired Hodge to make plans for "first-class apartment houses all over the city." Hughes first wanted to know how long it would take to build a bridge, however, and only after Hodge told him about five years did the governor look up at Paterno's "white, hill-crowning 'castle,' with a thoughtful smile," presumably thinking of how the hills of New York's Orange and Rockland counties could one day be dotted with symbols of other success stories. The governor of New Jersey remained in favor of the 57th Street site.

Although these men of influence and power were talking on the deck of a steam yacht as if they were in a smoke-filled room, and although their interests were selfishly directed toward those of their respective states, the process was not without a redeeming social value. There was no single absolutely correct answer as to where a bridge should be located. Clearly, technical, financial, and economic issues—not to mention social, aesthetic, and metaphorical considerations—left plenty of permutations and combinations of assertion and opinion about what was the right thing to do.

Among the remaining open technical questions were the relative riverbed conditions at the competing sites. Information about the 57th Street site was in hand, but borings at the 179th Street site still had to be done in order to confirm that a bridge could be built there for the estimated cost of $10 million. In 1910, after boring had gone as deep as 180 feet, nothing but sand and mud was discovered where Hodge had hoped to found a bridge tower. Though the site was not described as abandoned, the office of Boller & Hodge announced that plans were being made to explore other possibilities. Within a week, however, George F. Kunz, chairman of the Geological Section of the New York Academy of Sciences and president of the American Scenic and Historic Preservation Society, disclosed the engineering report to the press. According to Kunz, bedrock was too far below the surface of the Hudson River to allow the building of a practical bridge south of about West Point, and he believed that ten tunnels could be built for the cost of one bridge. The geologist Kunz was not shy about estimating the cost of tunnels, and the conservationist Kunz was not disappointed that a bridge would not despoil the natural beauty of the Hudson. The day following Kunz's exposé, McDougall Hawkes, chairman of the New York section of the Interstate Bridge Commission, confirmed reports that, because of the depth of bedrock, it did not look as if a bridge could be built at a reasonable cost at any of the proposed sites.

Following these reports, Gustav Lindenthal wrote a letter to the editor emphasizing that it was indeed possible to build a long-span bridge without piers in the river, but it would cost on the order of $100 million when land and approaches were added to the basic price of the bridge. He believed that only a lack of capital, not of technical knowledge, had kept the project from proceeding, and he admitted that the Pennsylvania Railroad tunnels

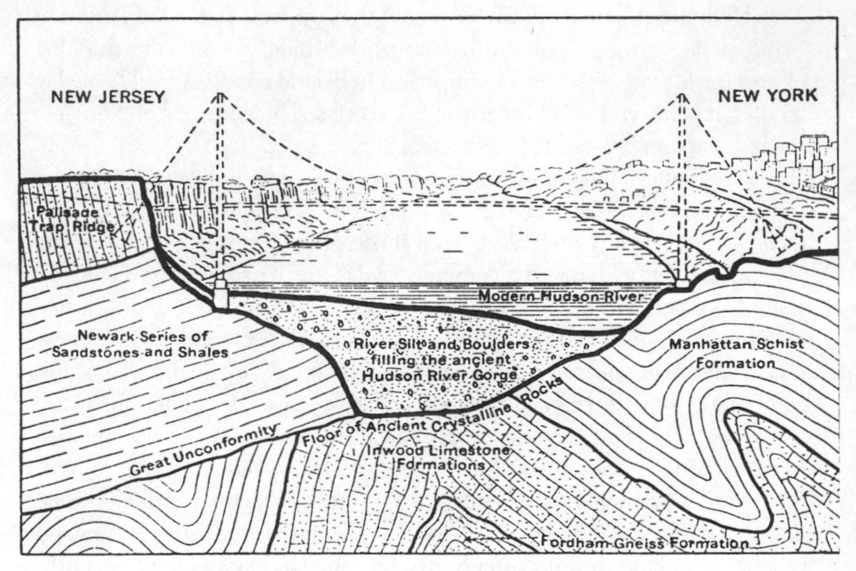

Geological cross section at the site of proposed bridge at 179th Street

then under construction "had ended for the present all prospect for a bridge across the North River," because he continued to see the railroads as an essential technical and financial component of a viable bridge scheme. In the meantime, Lindenthal's North River Bridge Company charter was running out, though after some opposition and debate in the Senate it would be extended for another decade.

Lindenthal continued to push his bridge plan, arguing in a long letter to the editor of *The New York Times* in late 1912 that tunnels were not enough, but he was losing credibility, and some readers at least were losing patience with his diatribes. John F. Stevens, who was identified as a "civil engineer," but whom everyone must have known to be the one who got the Panama Canal project back on track before Goethals took it over, immediately wrote a terse letter in response, calling schemes to bridge the Hudson "archaic" in a time when "the designing, construction, and operation of sub-aqueous tunnels have reached such a plane that in comparison a bridge is almost a joke." Not all engineers were so opposed to bridges, however, and though Hodge agreed with the feasibility of a tunnel at Canal Street, he also argued for the desirability of a bridge at 57th. Plans prepared by Boller & Hodge showed a braced-chain suspension bridge, with a 2,880-foot center span, that was estimated to cost $30 million. It would have trolley but not railroad tracks, and its cost would be offset by the "increased valuation of property and the increased population."

Though the bridge-tunnel debate was interrupted by World War I, the lower initial cost estimate of a tunnel, albeit one of limited traffic capacity, was in the end what led to the Holland Tunnel. However, Hodge had clearly become established, both technically and politically, as a prime candidate for the job of bridge builder when the time came, and he had shown himself to be more temperate and flexible than Lindenthal. Hodge would no doubt have been an obvious choice for the Port Authority to have looked to when that body became disenchanted with the tunnel solution in the early 1920s and reopened in earnest the question of bridging the Hudson. Henry Hodge would have been in his mid-fifties then, with just the right balance of experience and age to lead an ambitious bridge-building program like the one the Port Authority was to undertake. In 1917, Hodge resigned the position of public-service commissioner, to which he had recently been appointed by the governor of New York, to go to France and join General Pershing's staff as director of military railroads and bridges. Before the domestic bridge-building business recovered from the hiatus caused by the war, however, Hodge became ill with an embolism, and died after a six-week illness late in 1919. The untimely death of "one of the foremost American engineers in bridge building" left the field open to others.

Lindenthal made one of his more reasoned and objective arguments for his bridge in a paper read before the annual meeting of the American Society of Mechanical Engineers held in New York late in 1920. An abstract of his paper, published in *Engineering News-Record*, began with some hard, up-to-date data and continued with a fair and convincing argument:

> The Hudson River is a barrier which must be crossed daily by 700,000 passengers, by 3,000 freight cars and by 8,000 to 10,000 vehicles. Of the passengers about one-half come through six submarine tunnels, four of the

Hudson & Manhattan R.R. and two of the Pennsylvania R.R. All of the rest of the passengers, all vehicles and all railroad cars come over, as they did 80 years ago, on floating equipment along a river front of 12 miles. The congestion and the delays act as a deterrent to the spreading of population into New Jersey, where there is ample room within one hour's travel from New York for at least 4,000,000 people to live in suburban comfort. The two vehicular tunnels just started will, when completed in three or four years, add only four lines of traffic for vehicles, a mere drop in the bucket. The two tubes with approaches will be 10,000 ft. long and cost $28,000,000.

Compare with this backward condition the crossing facilities over the East River, for a population of 2,500,000 on Long Island. There are here besides numerous ferries four large municipal bridges (ignoring the Hell Gate railroad bridge) with, together, thirty-six tracks for rail cars and sixteen lines for vehicles, besides sixteen subway tunnels—all together, 52 lines of traffic, along a river front of only 6 miles. Two additional bridges and sixteen additional subway tunnels across the East River are under contemplation.

It is easy to see that if the 2,000,000 people in New Jersey are to be accommodated in the same proportion we must have across the North River, in addition to the above-mentioned railroad and vehicular tunnels, at least fourteen railroad tracks and twelve lines for vehicles. If put into tunnels these twenty-six lines would require twenty additional submarine tubes.

Let us assume that all these twenty tunnels would be about the same length and have the substantial construction of the Pennsylvania tunnels, or as proposed for the vehicular tunnels just started. That would make the cost of the twenty tunnels about $240,000,000.

But the higher shores of the North River above 23d Street would require most of these tunnels, with approaches, to be much longer than 10,000 ft. Some of them would have to pass under the Palisades on the New Jersey side, like the Pennsylvania tunnels, making them, with approaches, 15,000 to 16,000 ft. long, so that the total cost of these twenty tunnels would be nearer $400,000,000.

Lindenthal was being the consummate engineer, showing with hard numbers that the main objection to his bridge—namely, its cost—should in fact be a greater objection for tunnels of the same aggregate traffic capacity. Furthermore, he pointed out, his bridge would have a moving passenger platform, an idea of long standing. With renewed vigor, Lindenthal pushed for his bridge "in the vicinity of Fifty-ninth Street," which he estimated would cost only about $100 million, and it was to work on this new project that Othmar Ammann was called back from the clay mine. What Lindenthal did not allow for or calculate, however, was traffic trends and alternatives to a single bridge for combined rail and vehicular traffic.

5

As construction of the Holland Tunnel progressed, it continued to confirm the feasibility of vehicular tunnels, but it also revealed their true cost, even higher than Lindenthal's estimate. Although the financial climate was still not right for his ambitious plan, Lindenthal continued to seek support for his bridge-and-terminal project, whose total cost was to be on the order of half a billion dollars. And he would talk to whoever would listen about the frustrations of having economics stand in the way of a great engineering achievement, on one occasion telling a Lions Club audience in West Hoboken that "it was possible to bridge the Atlantic Ocean, but impossible to finance such an undertaking."

Perhaps Ammann saw a resolution to the financing impasse at the Hudson River in a less ambitious bridge, or perhaps he simply recognized the need for a bridge whose main traffic was vehicular. In any case, he began to work on such a plan independently of Lindenthal, who had not been able to retain him full-time. Ammann had his own stationery printed, identifying himself as "O. H. Ammann, Consulting Engineer," with offices at 7 Dey Street, which was also the address of the North River Bridge Company. In early January 1923, he was using this stationery with the Dey Street address corrected to 470 Fourth Avenue, the location of a separate office that had been loaned to him. On January 9, 1923, he wrote on this letterhead a personal note to newly elected New Jersey Governor George Silzer, whom he knew, of course, from the clay-mine enterprise, that there had been a very successful meeting with the Board of Freeholders of Bergen County regarding a bridge at Fort Lee, across the river from Manhattan's 179th Street. Ammann alluded to the governor's policy that no money be spent on new vehicular tunnels until the one under construction had proved itself, and he passed on the suggestion of the freeholders that a bill be introduced in the legislature supporting a bridge at Fort Lee.

The story of Ammann's behind-the-scenes involvement in the political machinations that were necessary to get a Hudson River crossing approved and financed has been told in considerable detail by the political scientist Jameson Doig, most recently in collaboration with the structural engineer David Billington. Ammann engaged in what can only be described as a deliberate campaign to promote his own plans for a bridge, a campaign not unlike the one Joseph B. Strauss was engaged in almost contemporaneously on the other side of the continent to promote his dream of a span across the Golden Gate. As late as 1923, Ammann appears to have been willing to work with Lindenthal, perhaps partly out of loyalty and partly out of pragmatism, for his septuagenarian mentor was still the pre-eminent bridge engineer in America, and any technical and financially reasonable plan supported by him would no doubt have been seen as technically credible by the politicians and financiers whose support was essential. In March, Ammann revealed his thinking to Lindenthal, whose quick response the younger engineer recorded in his diary:

March 22 / 1923

> Submitted memo. to G.L., urging reduction of H.R.Br. program, dated Mar. 21.
>
> G.L. rebuked me severely for my "timidity" & "shortsightedness" in not looking far enough ahead.
>
> He stated that he was looking ahead for a 1000 years.

Ironically, it was Lindenthal who was being shortsighted; New York, like the rest of the country, was witnessing the beginning of the era of the truck and automobile, which would ensue at the expense of the railroads, and a vehicular and light-rail bridge was an appropriate solution to the problem of communication between New York and New Jersey. Ammann would break with Lindenthal over the issue, and he worked independently on his plans for a more modest proposal. He had also begun to establish his own identity as an independent engineer.

Unlike the prolific Lindenthal and in spite of the praise heaped upon the Hell Gate Bridge report, Ammann had to this time published few pieces of technical advocacy, his two reviews of David Steinman's book on suspension bridges being the exception. Perhaps the attention Steinman's book had received or the notice those reviews had gotten Ammann prompted him to write a "think piece" on the "Possibilities of the Modern Suspension Bridge for Moderate Spans," which was published in *Engineering News-Record* in June 1923. He argued that, contrary to the conventional wisdom of bridge literature, according to which suspension bridges were suitable only for very long spans, recent years had seen the construction of a number of highway and light-railway suspension bridges with spans in the three-to-eight-hundred-foot range. He took a long historical view—albeit idiosyncratic, as his history was wont to be—of the suspension bridge, and discussed such questions as aesthetics, safety, stiffness, and economy. Whether designed to or not, the paper would give him credibility as a suspension-bridge designer somewhat beyond his years or his experience with the form.

The year 1923 was a busy one for Ammann, and he had had no time to communicate with his mother before the approaching holidays prompted him, in mid-December, to write and post a letter right away, so she might receive it in Switzerland by Christmas. In spite of his growing reputation, he explained that "the giant project" for which he had "been sacrificing time and money for the past three years" then lay "in ruins" because of Lindenthal, who remained unnamed:

> In vain, I as well as others, have been fighting against the unlimited ambition of a genius that is obsessed with illusions of grandeur. He has the power in his hands and refuses to bring moderation into his gigantic plans. In stead, his illusions lead him to enlarge his plans more and more, until he has reached the unheard of sum of half a billion dollars—an impossibility even in America.
>
> However, I have gained a rich experience and have decided to build anew on the ruins with fresh hope and courage—and, at that, on my own initiative and with my own plans, on a more moderate scale. It is a hard battle that I have already been fighting for six months now, but the possibility of success is constantly increasing, so that I do not allow myself to be frightened in spite of the great handicaps and my shrinking finances. I wait and hope that the New Year will finally bring my work to fruition.

Among the things Ammann had done during the year was to write to Governor Silzer, with a copy to Samuel Rae, who was not only president of the Pennsylvania Railroad but also associated with the North River Bridge Company. In his letter, after stating that he had "no desire to discredit Mr. Lindenthal," Ammann gave his views on the "scope of plan and size" of the 57th Street Bridge:

> It is over a year ago that I began to suggest to Mr. Lindenthal reduction of the stupendous scope of the plan and also of the size of the bridge. Not having succeeded to convince him and having noticed the discouraging effect of his policy upon the supporters of the project I wrote him a memorandum on May 21, 1923. . . .
>
> As regards the size or capacity of the bridge proper it is generally recognized that even for a light bridge a width of 1/20 of the span gives ample lateral rigidity. That would mean 160 feet for the Hudson River Bridge as against the 235 feet now provided for. But in such a long span with a deadweight of more than 50 times the probable greatest wind pressure even a smaller width would give ample rigidity.

> The width is therefore determined by traffic requirements alone. The plan now provides for 20 lanes of vehicles and 2-15 ft. promenades on the upper deck and for 12 tracks on the lower one. A roadway capacity of only 12 lanes of vehicles would provide for a possible traffic of 50 million vehicles per year, which is equal to twice the traffic now crossing the four East River Bridges and is much more than should at any time be concentrated at one point even with ample capacity of approaches. . . .

Such arguments regarding the interrelated issues of structural rigidity, traffic capacity, and cost vis-à-vis the width of roadway would have profound implications for the very near future of bridge building, but for the moment Ammann's objectives were less general. He argued that it would probably be ten to twelve years before Lindenthal's 57th Street Bridge could be completed, and that in the meantime traffic congestion was becoming "calamitous," especially in the upper part of Manhattan. He also pointed out that "tunnel advocates" appeared to be gaining ground because Lindenthal's proposal tipped the economic argument in their favor. Ammann felt the only way to keep further tunnel projects from being pushed forward "by popular demand" was to build a vehicular bridge at 179th Street. He had not told Lindenthal of this specific new proposal of his, as he made clear in the final paragraph of his letter to Silzer: "At the opportune moment I shall lay this proposition before Mr. Lindenthal and I trust that in case he agrees to take it up, it will also have your sanction and support. I shall not mention to him that I laid these matters before you."

Later in the year, the Port Authority, which had rejected Lindenthal's grand bridge plan, scheduled public hearings on building further tunnels beneath the Hudson. Governor Silzer was able to broaden the agenda to include the question of bridges as well. In the meantime, Ammann had come up with a proposal for a bridge at 179th Street that would cost only $30 million, just as Hodge had estimated a decade earlier for a remarkably similar bridge to be located farther down the river. The cost of Ammann's bridge could be reduced even further, to $25 million, if only vehicle traffic were carried. This proposal was certainly competitive with a tunnel, and Ammann even supported the Port Authority as the natural agency to build such a structure. He also suggested that they might need an expert bridge engineer, and that he would be happy to assume this position. Whereas Ammann may have painted a dark picture when writing to his mother, within a few days Silzer had forwarded Ammann's analysis of the problem to the Port Authority and issued a press release announcing what he had done. Lindenthal was to learn of these developments in the newspaper, and he penned his reaction to Silzer:

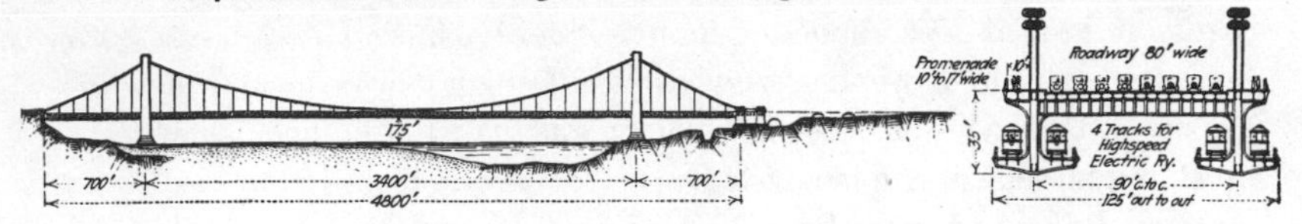

Othmar Ammann's 1923 proposal for a bridge across the Hudson River at 179th Street

> A bridge at Fort Lee of the kind described by [Ammann], cannot be built for $25,000,000. The estimate is far too low. It is the same old way to mislead the public either from design or from ignorance, just as the estimate of Gen. Goethals of $10,000,000 for the vehicular tunnel, which when fully completed will now cost $45,000,000. The public cannot judge of such vagaries in estimates of which engineers are constantly guilty.
>
> Mr. A. had been my trusted assistant and friend for ten years, trained up in my office and acquainted with all my papers and methods. But I know his limitations. He never was necessary or indispensible [*sic*] to me. Many other assistant engineers are very able and glad to fill his position. But one does not like to make changes and train up new men as long as it is not necessary.
>
> Now it appears that A. used his position of trust, the knowledge acquired in my service and the data and records in my office, to compete with me in plans for a bridge over the Hudson and to discredit my work on which I had employed him. He does not seem to see that his action is unethical and dishonorable.

In spite of Lindenthal's protests, the practicality of Ammann's plan attracted immediate interest. A sketch of his suspension bridge with a thirty four-hundred-foot main span appeared in the first issue of *Engineering News-Record* for 1924, but there was no protracted discussion of what to many may have looked like just another engineer's dream. However, in April 1925, that same magazine reported that, "almost unnoticed, the first step has been taken toward the ultimate construction of a bridge so far beyond any existing structure in its size as to rank virtually in a new order of magnitude." The occasion for the story was that the legislatures of New York and New Jersey had "appropriated a total of several hundred thousand dollars for the preliminary studies." After he submitted Ammann's plan to the Port Authority, Silzer had suggested that the agency might want to engage "such a man as Mr. Ammann, who is thoroughly skilled in this kind of work," but that was not to be for some time.

When the Port of New York Authority was formed in 1921, Benjamin F. Cresson, Jr., had been appointed its first chief engineer. Cresson was born in Philadelphia in 1873 and was educated at Lehigh University and the University of Pennsylvania. His early career was with the Lehigh Valley Railroad, but he soon joined Jacobs & Davies to work on plans for an East River railroad tunnel and on the first McAdoo tunnels under the Hudson. He had a well-established reputation in tunnel and harbor work when he was named chief engineer of the Port Authority. In that position, he would have had considerable influence on the nature of Hudson River crossings had he not died suddenly in early 1923 following an operation for appendicitis. The office of chief engineer was not filled until late September, by William W. Drinker, who had extensive harbor and terminal experience but was not a bridge engineer, which was what the Port Authority would soon need.

6

The 179th Street project was not the only bridge under consideration by the Port Authority. Since colonial days, travelers between New York and Philadelphia had passed through Bayonne, New Jersey, and Staten Island, using ferries to cross the Kill van Kull and the Arthur Kill, the bodies of water that separated Staten Island from the Jersey mainland, and whose designations derived from the Dutch word *kil*, meaning "channel" or "creek." First oar-propelled scows, then horse boats (whose side wheels were powered by horses on a treadmill), then steam ferryboats served the purpose, but with the growth of population and the attendant vehicle traffic in the 1920s, there were growing calls for bridges. The Port Authority came under some criticism for worrying about automobile traffic instead of the harbor-goods traffic that was said to be its principal charge; by 1924, legislation had been passed enabling the Authority to plan bridges. It did so, between Bayonne, which was at the bottom of a New Jersey peninsula approaching the north shore of Staten Island, and the towns of Elizabeth and Perth Amboy, which are across the kill that forms the western boundary of Staten Island and contains the New York–New Jersey state line. These were the same sites identified by Boller & Hodge years earlier, but it took time for politics and need to catch up with the vision of that engineering firm.

Two bridges across Arthur Kill were authorized first, and, in conjunction with the design of these structures, Silzer had written in a letter of support in May that "the Port Authority ought to avail itself of the services of O. H. Ammann," whom he understood to be available "just at the moment." In fact, by early November 1924, Ammann had actually submitted to the Port

Authority a bid for preparing plans for the Arthur Kill bridges and was "anxiously awaiting their decision." While he was waiting, on November 7, chief engineer Drinker sent a letter to "four or five prominent bridge engineers experienced in the planning of large bridges." After describing the conditions the bridges had to meet, Drinker wrote for the Port Authority: "We desire to secure the services of a bridge engineer and wish you to make a proposal for carrying on the work." Ammann's reputation seems not to have been sufficiently established for the young Port Authority, which was trying to establish its own reputation. Thus the task, which clearly involved more than bridge engineering, went to J. A. L. Waddell, whose fifty-year career was older than Ammann himself, and whose bearing and bemedaled appearance may have counted for as much in the politically charged circumstances as his expertise.

Since the Port Authority had to raise the construction cost of the bridges by bond issues backed by the earning power of the bridges themselves, "it was necessary that the bridges be planned by an engineer of recognized standing, to assure the confidence of prospective bond buyers, although only a small sum was available as compensation for this engineer." This was the situation, at least, as interpreted by *Engineering News-Record*, which called the announcement of the choice of Waddell, and the revelation of how it was made, "an unpleasant shock . . . to established views of the relation between the engineer and his client." Drinker's letter was described as follows:

> It calls for bids on price and time of delivery of expert engineering services—services involving a very high type of specialized engineering skill and trained judgment—in precisely the way that one would ask for price and time of delivery on a ton of coal or a thousand of brick. For the first time, thus, a responsible and well informed public authority has resorted to the bidding system of selecting a professional man to direct the expenditure of large sums of public money. . . .
>
> We doubt whether the Port Authority would think of defending a lawsuit on which its existence depends, or in which $10,000,000 is at stake, through counsel retained as the result of competitive bidding.

Though the journal admitted that competitive bidding for the services of engineers had increasingly been taking place in small communities, whose local staff not only was not qualified to carry out the work but also could not judge the qualifications of available engineers, the writer felt there was no excuse in the case of the Port Authority, which had "able and widely informed engineers on its staff." Yet the selection of Waddell remained, as did his designs for the two bridges, which were almost identical high-level cantilevers: the more southerly one, at Perth Amboy, was to have a main span of 750 feet and an overall length of over ten thousand feet; the main span of the bridge at Elizabeth was to be 672 feet. Even if undistinguished, especially in comparison with such cantilevers as the Firth of Forth, the Quebec, and the Queensboro, both bridges were major engineering projects, and Waddell could not do the work alone. Among those who also worked on the designs were William Burr and George Goethals. Ironically, however, Othmar Ammann is commonly misidentified as the designer of these bridges over the Arthur Kill, because of subsequent developments.

Ammann became associated with the Staten Island bridges when he was appointed bridge engineer for the Port Authority on July 1, 1925. Silzer had persisted in writing to officials of the Authority that Ammann had had a great deal to do with getting the 179th Street bridge project as far as it was, which in March 1925 meant that both states had passed legislation authorizing the Port Authority to build a bridge at the location. Shortly after that, and after an "encouraging interview" with Drinker, which took place at Silzer's suggestion, Ammann was hired to add some bridge-building experience to the staff. Although by that time an outside contract had already been let for the Staten Island bridges, in his new position Ammann oversaw their design and construction. The two bridges were completed six months ahead of schedule, and opening ceremonies took place simultaneously, on the same day in June 1928. The southern bridge was named Outerbridge Crossing—not because of its remote location but after Staten Island resident Eugenius H. Outerbridge, the first chairman of the Port Authority, who was guest of honor at the opening ceremonies. The other structure was to be named the Arthur Kill Bridge, but when General Goethals, the first chief engineer of the Port Authority, died shortly before its dedication, it was named the Goethals Bridge, thus making it, like the tunnel named for his critic Holland, one of the few civil structures named for an engineer.

The other bridge between Staten Island and New Jersey, across the Kill van Kull, was truly to be Ammann's design. There was to be a distinguished group of consulting engineers, including William Burr, Daniel E. Moran, Leon Moisseiff (as advisory engineer of design), and Joseph B. Strauss, who was at the time trying to generate support on the West Coast for his suspension bridge across the Golden Gate. For all the highly visible consultants, Ammann's experience with the Hell Gate Bridge had actually prepared him well to design, with assistant E. W. Stearns and engineer of design Allston Dana, a steel arch that would be two-thirds longer than the Hell Gate and, at 1,675 feet, actually the longest in the world, surpassing by five feet the span of the Sydney Harbour Bridge, then under construction in Australia.

(Ammann's arch would in fact not be surpassed until 1977, and then only by twenty-five feet, when the New River Gorge Bridge, at Fayetteville, West Virginia, was completed.)

The Sydney Harbour Bridge, known affectionately to some as "the coat-hanger," might have been a steel cantilever had not John Bradfield, chief engineer of the Public Works Department of New South Wales, visited America during its design. "Lindenthal's departure from the usual stark steel arch so impressed" him at Hell Gate that Bradfield quite consciously modeled the Sydney Harbour Bridge after it, but with some deliberate distinctions. Whereas Lindenthal's Hell Gate had an odd number of panels, so that a redundant diagonal member was necessary for symmetry and forms a steel X to mark the center of the arch, the Sydney Harbour Bridge was designed with an even number of panels, so that there is a more subtle visual transition between the two halves of the main structural element. As if to distinguish himself from his mentor's masterpiece, Ammann chose to give the Kill van Kull arch also an even number of panels. The Sydney Harbour arch, for which David Steinman served as a consultant, was built with another sharp distinction from the Hell Gate, the termination of whose top chord had been criticized for being continued visually into the masonry pylons. The steel of the top chord of the Australian arch ends abruptly some distance from the stone, emphasizing the true structural action of the arch's springing from the bottom chord.

The meeting of steel and stone in architect Cass Gilbert's early architectural drawings of Ammann's design for the Bayonne arch was actually somewhat ambiguous, but the question became moot when the steel framework at the ends of the arch was not encased in masonry as originally planned. Though the Kill van Kull bridge was dedicated as the Bayonne Bridge less than a month after the bridge across the Hudson at 179th Street, the details of the design and construction of the world's longest steel arch would forever be overshadowed by the contemporaneous design and construction of the great suspension bridge at 179th Street.

7

After the necessary legislation for a Hudson River bridge had been passed, only one great nontechnical obstacle to the four-decade-long struggle remained. Bonds had to be sold, of course, and when it was announced in December 1926 that $20 million was underwritten by National City Bank, *Engineering News-Record* reported on "the high regard of the banking world for the essential integrity of that unique body," the Port Authority, which

Cantilever design originally accepted for the bridge over Sydney Harbour

alone was responsible for the security of the debt. Even as the journal was praising the fiscal reputation of the Authority, however, it commented on the body's "metamorphosis" from a facilitator of freight handling in the port, which was its *raison d'être*, to an aggressive bridge builder. The trade journal concluded that "the Port Authority would do well not to delude itself into believing that building bridges was its job," but that would indeed appear to be the major activity during the next five years or so for Ammann and his engineering staff.

Early in 1927, five papers dealing with the suspension bridge at 179th Street were read before a local meeting of the American Society of Civil Engineers. Four were presented by Port Authority engineers: one by Ammann, who gave an overview of the project; one by Dana, who described some of the calculations involved in the structural design; and one each on surveying procedures and on traffic studies. A fifth paper, however, was presented by R. S. Buck, who "challenged the design as to a number of its major features, and called for a thorough reconsideration and restudy of the project." Buck may at first have appeared to have been in the long tradition of naysayers who oppose projects that go beyond state-of-the-art experience—as Ammann's certainly appeared to, even if that was being denied—but Buck's criticism was more reasoned than grudging. He suggested that a sense of urgency to get the bonds sold and the contracts let may have led to overconfidence, and spoke to the value of looking at competitive designs. His criticism focused on some specific points, including the question of wire cables versus chains, which was still open at the time, recalling the controversy that accompanied the design of the Manhattan Bridge. His final point related to the design of the towers; these were proposed to be steel-framed for the first stage of construction, which was to include only a single deck for vehicles, then later, when a second deck carrying light rail traffic would be added, to be encased in concrete and faced with stone. In Buck's opinion, such "pseudo masonry is out of place both esthetically and

Hell Gate Arch **Proposed for Sydney Arch** **As Built for Sydney Arch**

Comparison with Hell Gate Bridge of pylon-design details of the Sydney Harbour arch as proposed and as built

structurally." Only the "lateness of the hour" curtailed discussion of such matters, but they were not to lie buried in the minutes of an engineering meeting. As with large projects generally, early plans would be modified and solidified as design and construction progressed.

Formal ground-breaking occurred in September 1927, simultaneous ceremonies taking place on each side of the river as speeches were delivered from the steamer *De Witt Clinton* anchored midstream. Silzer, no longer governor of New Jersey but now chairman of the Port Authority, waited patiently, while a telephone cable was repaired, for the ceremonies to begin. This cable, which had been cut by a passing ship, was to carry the speeches from the steamer's saloon to WOR in Newark for broadcasting to those on the shore. After the forty-five-minute delay, the chairman began the proceedings and announced that "there are compensations for everything, and I made up my mind when this delay occurred that for every minute of it I would deduct a minute from my speech, and the result is that there is nothing left of the speech." Silzer, the consummate politician, had made a deal with himself that could please everyone.

Shortly after the ground-breaking ceremonies, the issue of how to suspend the deck was settled: the Trenton, New Jersey, firm of John A. Roebling's Sons submitted a wire-cable bid lower by about 10 percent than that for an eyebar design. Matters of price were not expected to be the only determining factors for securing engineers' professional services, but that was certainly not the case when it came to choosing one form of construction over another. As for the towers that would support the cables, their final appearance was to undergo a much more prolonged discussion, and how important a role economic considerations played in that decision is still debated. In the meantime, the contract for the steel superstructure for the bridge, with a single deck, was let to the McClintic-Marshall Company.

The design of the bridge was described in an August 1927 issue of *Engineering News-Record*, with bare steel towers shown in the construction drawings and masonry-clad towers in a perspective drawing of a completed bridge. The discussion began with the declaration that "esthetic considerations played a large part in the determination of all the general features of this bridge," and such considerations were "most commanding, perhaps, in relation to the towers." The architect Cass Gilbert, whose Gothic-style Woolworth Building was still, in the late 1920s, the world's tallest skyscraper, had been consulted by the Port Authority on the Hudson River Bridge "from the earliest stages," and his treatment of the towers was expected to prevail. The steelwork had to be designed and erected first, however, and thus Leon Moisseiff, as advisory engineer of design, would have a great deal to say about what was to support the architectural façade of the towers.

The structural design of the towers as a composite of steel and concrete to be clad in stone came in for considerable criticism of the kind R. S. Buck had raised; an early modification was to increase the capacity of the steel framework so that it could support the ultimate weight of the bridge and its traffic without any help from the concrete. Even then, the nature of the steel design was not simple, in that it resulted in what is known as a "statically indeterminate structure." As with the Queensboro cantilever, this, of course, meant that the distribution of forces among the various steel members was not just a function of applied loads and geometry, but depended upon the stiffness of the steel and how the structure was assembled and how it moved under the load. Because of possible uncertainties in construction and analysis, it was widely held that indeterminate structures were to be avoided, but Moisseiff supplemented intricate analysis with model tests and allowed for variations in the completed design.

Moisseiff's interest in the bridge towers went beyond his elegant structural analysis, however, as he revealed in a book review that appeared in *Engineering News-Record* late in 1928. "Esthetics of bridges has come to be so live a topic that the publication of a work on the subject by an engineer of standing and culture is timely," Moissieff began this review, of a new book by Friedrich Hartmann, professor at the Technische Hochschule in Vienna, and approvingly reported on its author's viewpoint:

> He fully realizes the difficulty of expressing the achievements of modern engineering in the esthetic habits of the past. He therefore believes that

> architects, as a rule, are distant from the conception of steel bridges. Structures which fitted in the landscapes of medieval towns do not meet modern requirements. Dogmas and slogans on the relation of the useful to the beautiful have lost their value. "Neither is the necessary always beautiful nor the beautiful always necessary. The reverse may be true in both cases." Engineers themselves should endeavor to develop the beauty of their structures.

Moisseiff explained that the modern "designer of bridges cannot always make use of stone and its substitute, concrete, to the continuous and solid texture of which man has been accustomed by ages. He is generally compelled by the demands of feasibility and economy to make use of the high resistance of steel." Moisseiff reported how Hartmann, "with a malicious pleasure . . . quotes diametrically opposed opinions of architects on the appearance of the Forth bridge," that great steel cantilever. Most important for Moisseiff, however, was that Hartmann had opinions about American bridges, both standing and under construction: "Naturally he is against masonry towers and prays for their omission in the Hudson River bridge."

The basic idea for the towers appears to have come from Ammann himself, even before he became associated with the Port Authority. One of his first open discussions of the plans he had sent to Silzer occurred at a meeting of the Connecticut Society of Civil Engineers in early 1924. After spending over half of his time making a case for a bridge based on developing traffic patterns, Ammann discussed various aspects of the proposed design. In the concluding section of the paper, he dealt with the towers, "another important part of a suspension bridge." While the Brooklyn Bridge towers were "a magnificent example of the old massive masonry type, unquestionably the one most pleasing to the eye," cost had become prohibitive for the great weight of stone needed just to support towers projected to be as high as forty-five-story buildings. Ammann continued by contrasting steel towers:

> The opposite extreme is represented by the slender flexible steel towers. With proper architectural treatment, as in the case of the Manhattan Bridge, this type may be rendered very pleasing to the eye, but often its construction suggests crude utility. This type is, in most cases, the cheapest in first cost.
>
> The "braced steel towers," as represented in the Williamsburg bridge, have a more massive appearance and certain advantages in erection but, unless well proportioned, look clumsy. They may be covered by a decorative shell of masonry as proposed in Mr. Lindenthal's design for the 57th Street bridge, but such shells involve a heavy additional expense.

Ammann's innovative compromise for the 179th Street bridge was to resort to "the combination and utilization, to the full extent of their strength, of the two modern materials available for structural parts in compression, namely steel and concrete, the former completely embedded in the latter." He was, in effect, proposing a sort of reinforced-concrete tower, something that would not be fully realized on such a scale until the Humber Bridge was built many decades later in England. In order to facilitate the construction of such a unique tower, Ammann proposed that the steel skeleton be designed to carry most of the load of the bridge itself, with the concrete and the steel sharing in carrying the added load of the traffic. Furthermore, the concrete would protect the steel from corrosion, thus reducing maintenance cost, and would provide a "more monumental appearance" than bare steel for a small additional cost. Finally, "the architecture of the towers has purposely been kept plain so as to be in harmony with the simple construction of the steel work. The monumental effect must be produced by harmony of mass and lines, and not by ornamentation."

The sketches accompanying the paper are captioned "Copyright by O. H. Ammann," but they may very well have represented the work of the architect R. A. Ruegg, whose "valuable assistance" the engineer acknowledged in closing his paper. The architectural argument may also have been Ruegg's, for Ammann seemed to be having some difficulty in establishing his own structural aesthetic at the time, or at least seemed willing to be swayed more by the weight of opinion than the towers themselves would be by the weight of the bridge. Whereas his later bridge towers would resemble visually the open portal design in this paper, albeit more slender and in steel rather than in massive concrete or stone, the 179th Street bridge would not.

In the tradition of large engineering projects, even after it was completed the final design of Ammann's Hudson River bridge remained to be described in considerable detail, in words and pictures. Such final reports were typically prepared for limited distribution, and thus were a relatively expensive undertaking. In the case of this great suspension bridge, the Port Authority chose not to print the report itself but, rather, appropriated funds for the estimated cost to the American Society of Civil Engineers; this organization issued a series of eight papers on the bridge as a volume of its *Transactions* for 1933, giving the Port Authority enough reprints to distribute as it saw fit. The society bragged about this publishing coup: "Not since December 1911, when Vol. 74 was mailed to the Membership, has TRANSACTIONS appeared on other than a very thin and transparent grade of Bible paper. In order to present the numerous photographs . . . to the best advantage, Vol. 97 is published on a good grade of English finish paper." How-

Four versions of Ammann's proposed Hudson River bridge (left to right, top to bottom): *the original 1923 proposal; a version with slender steel towers; a version with granite-faced towers and eyebar chains; and the bridge essentially as completed in 1931, with a single deck suspended from four wire cables and bare steel towers, but topped by observation platforms that were never added*

ever, since this was happening at the nadir of the Depression, the society found itself in the curious position of also having to justify such extravagance "in a year of unusual financial stringency." The explanation was that, "with the expenditure of relatively little more" than was given by the Port Authority, the society could "furnish all its members with a technical record of this milestone of engineering progress." The society also thus associated itself with the project in a distinctive way, and hence gave an extraordinary imprimatur to the project and its engineers.

The debate over whether the towers should or should not be encased in concrete and stone had continued throughout the construction of the bridge and remained an issue even in the final report. When his status changed from independent engineer to bridge engineer for the Port Authority, Ammann's association with architects seems also to have altered, and it was not Ruegg's but Cass Gilbert's architectural sketches, dominated by even more massive stone-faced towers (and approach viaducts on the Manhattan side), that then appeared to be the dominant feature of the bridge. In early Port Authority plans, the Gilbert stonework even overwhelmed the heavy eyebar chains that were chosen by the artist over wire cables. In his final report on the bridge, however, which dealt with its general conception and the development of the design, Ammann could not ignore the story behind the towers, which at the time remained the bare steel that Moisseiff had designed and had subsequently redesigned to be able to support even the double-decked bridge without the aid of concrete:

> There is no part of the design . . . which has called forth as much comment, favorable and unfavorable, on the part of engineers, architects, and laymen, as the towers. Indeed, as the writer has endeavored to show, the design of the suspended structure, the floor, and the cables, resolved itself largely in the application of natural and most simple structural forms which neither required nor permitted architectural treatment to satisfy aesthetics.
>
> The design of the towers, however, is not so well defined. There are widely different meritorious forms and the effect of the towers on the appearance of the entire structure is perhaps more pronounced than that of any other part. They may enhance or destroy the natural beauty of a graceful suspended structure. There are existing examples which illustrate both effects.
>
> It is futile to theorize about this question—it is largely a matter of aesthetic conception, which is intensely individual and changeable—nor can it be dealt with on general principles without regard to the local scenery or landscape. Moreover, the aesthetic treatment of a bridge, as that of any other engineering structure, is not always satisfactorily solved even by cor-

rect and honest application of engineering principles. The appearance of a structure so conceived may sometimes be materially enhanced by the addition or the architectural embellishment of certain structural parts, whether structurally required or not. The flanking abutments of an arch bridge, and the towers and the anchorages of a suspension bridge, offer opportunity for such enhancement. . . .

The writer, who has conceived and is primarily responsible for the type and general form of the design, considers the steel towers as they stand to represent as good a design as may be produced by a slender steel bent, and that they lend the entire structure a much more satisfactory appearance than he (and perhaps any one connected with the design), had anticipated. Nevertheless, he believes that the appearance of the towers would be materially enhanced by an encasement with an architectural treatment, such as that developed by the architect, Mr. Cass Gilbert. . . .

The writer is not impressed by the criticism, based solely on theoretical and utilitarian grounds, that the encasement would constitute a camouflage which would hide the true structure and its function.

Ammann may still have been unsatisfied with the bare towers, but not Moisseiff, who wrote an entire paper on the design of the towers for the special *Transactions* volume. Though he pointed out that "the fact that the towers were conceived to be ultimately encased in masory is important to the understanding of their design and their articulation," he spent few words on the controversy, other than to express his opinion that the towers "would probably remain without enclosure." The engineer was not to be alone in his view; the architect Le Corbusier would call the structure "the most beautiful bridge in the world," and "blessed." He would also write: "When your car moves up the ramp the two towers rise so high that it brings you happiness; their structure is so pure, so resolute, so regular that here, finally, steel architecture seems to laugh."

8

For all the debate over appearances, actual construction of the 179th Street Bridge had progressed very deliberately, and the bridge was finished six months earlier than had been thought possible. Yet final details of the towers remained uncertain even as the bridge was being dedicated, on October 24, 1931; the program for that event shows an observation platform that was never built. The program also contains a very brief popularized history of Hudson River crossings, and a description of the gigantic nature of the endeavor and enterprise. Not uncharacteristically, there is no mention of the

Photograph of the George Washington Bridge as completed in 1931, showing the extreme slenderness of its single deck

role that engineers like Ammann played, either in the technical design or in the political and financial machinations that were so necessary in the preliminary stages. The short program did, however, include a justification for the name of the structure, which, also not uncharacteristically, had been a matter of some controversy.

The site of the bridge, the program explained, was of historic interest: during the American Revolution, George Washington had led patriots against the British at this location. Thus it was fitting to bestow George Washington's name on the monumental structure. The name "George Washington Memorial Bridge" had actually been announced nine months earlier by the Port Authority, which had appointed a committee to select a name from the many suggestions that had been made. Though there was strong feeling that the simple title "Hudson River Bridge" was sufficient, the committee acceded to custom by agreeing that "something more historically significant was desirable." There was quick opposition to the "ridiculous name" that had been "born in a vacuum, warmed by patriotism but chilled by an utter absence of common sense." Since there was already a Washington Bridge over the Harlem River, barely a mile to the east, some argued that there could be confusion, but even the chairman of the Port Authority, John F. Galvin, admitted that people would continue to call the new bridge the "Hudson River Bridge," no matter what its official name. There were, naturally, alternatives suggested, including "Palisades Bridge," after the New Jersey cliffs that also served as the west abutment, and "Cleveland Bridge," after the U.S. president, who had been born in New Jersey and been a governor of New York. Within barely a week, the Port Authority commissioners, who wanted the bridge "named to suit the public," agreed to refer the matter back to the special committee that had recommended the name in the first place. It had become known that the George Washington Bicentennial Commission, which had been requesting public entities to take the occasion of the bicentennial of the birth of Washington to name "bridges, highways and other public monuments" after him, was a strong proponent of the name originally chosen. The public was now invited by the Port Authority to submit more names.

Hundreds of letters poured in with more suggestions. A man from New Jersey proposed "Verrazano Bridge," after the Florentine navigator who was claimed by some to have been the first European to cross New York Bay, in 1524, but who was "considered an uncertain quantity by many historians." "Columbus Bridge" and "Hendrik Hudson Bridge" had solid support, but the "volume of letters from school children and women's clubs" kept the

The Washington Bridge in 1889, when recently completed and known as the Harlem River Bridge

hope for honoring George Washington alive. In the end, the letters showed "a complete lack of unanimity of public opinion," and the Port Authority decided on "George Washington Bridge," only dropping the explicit mention that it was a memorial. Pleas for other names continued to be advanced in letters to editors, but to no avail. Though newspaper and magazine editorials expressed resignation to the official naming of the bridge, "what the millions who use it in the years to come will choose to call it is another matter." *Engineering News-Record* believed the bridge would "get its workaday name from the people, while its christen [*sic*] name is likely to remain unknown outside parish records and mortgage papers." How wrong these prognosticators were. The name "George Washington Bridge"—sometimes shortened in speech to "the George Washington," "the GWB," or, within the Port Authority offices, simply "the George"—is used to this day by New Yorkers, New Jerseyites, and travelers to and from all parts of New England.

At the dedication, New York Governor Franklin D. Roosevelt and New Jersey Governor Morgan F. Larson formally opened the bridge with a ribbon-cutting ceremony at the center of its span, where grandstands had been erected for five thousand guests. It was a day to celebrate and show unity. Roosevelt and ex-Governor Al Smith arrived together, and they received loud applause. "But probably the greatest and most spontaneous greeting was that accorded to" Ammann and Lindenthal, "who came to the grandstand in the same automobile." Unlike Lindenthal, who was described as "designer of the great Hell Gate Bridge and dean of American bridge builders," Ammann needed no identification in the newspaper story covering the ceremonies. His eclipse of his mentor was complete; Lindenthal was not even listed among the consulting engineers, although in the final report on the project Ammann would identify Lindenthal specifically as one who had "rendered special advice on design questions." Regardless of whether Ammann was setting the record straight or fulfilling a sentimental obligation to an aging colleague, the balance of influence and power had definitely shifted. Future bridge design in America would follow not Lindenthal's but Ammann's lead in pursuing longer, lighter, and more flexible bridges that were enormous but not Brobdingnagianly grotesque.

Among the many ceremonial touches that day was one that both recalled the past and foreshadowed the future:

> Defying the age-old rule that marching troops break step when crossing a bridge, columns of soldiers, sailors, marines and Coast Guard came swinging down the roadway from the Manhattan plaza. Those in the center of the bridge felt the gigantic span vibrate as if shaken by earth tremors, but the crowds were only amused at the strange sensation.

It had been possible to build the George Washington Bridge at a cost that made it politically acceptable because Ammann had used all the technical ingenuity at his command to reduce the dead weight of the structure; but this also made the bridge perceptibly flexible. He discussed the development of the suspension bridge in his 1923 article in *Engineering News-Record*, in which he described the evolution of the stiffening truss, the dominant feature of the roadway profile of such recent suspended spans as Howard Baird's Bear Mountain Bridge and Modjeski's Delaware River Bridge, each of which briefly held the record for the longest suspended span in the world. In his review, Ammann explained how, a century earlier, Telford had designed his Menai Strait Suspension Bridge with a flatter curve to its chains, which in turn required a flexible deck to accommodate the deflections of the chain due to traffic loads and temperature changes. The deck of the Menai was blown down in the wind in the nineteenth century, but in his characteristically selective and slightly skewed view of bridge-building history, Ammann did not mention this extreme effect of a too-flexible deck. Instead, he described how the Niagara Gorge Suspension Bridge could carry railroad trains because its engineer, whom Ammann misidentified as *Washington* Roebling, successfully incorporated a stiffening truss. Still, whatever the lacunae in his historical view, Ammann understood fully the importance of having a bridge deck stiff enough to withstand traffic loads vertically and wind loads horizontally, and what was true for spans of moderate length he saw could be extended even further for those of unprecedented length. The stiffness of a bridge deck and cables complicated the calculation of the manner in which the load was distributed among the various components of the structure, however, and the less precisely an engineer knew this distribution, the more conservative he had to be in his design. Conservatism in bridge design translates into using more steel to compensate for uncertainties, which in turn translates into a heavier bridge that costs more to build.

The required strength of a bridge deck depends not only upon the weight of the deck itself but also upon the weight of the traffic it has to bear. The requirements for Lindenthal's Hudson River proposals were all dominated by the extreme weight of heavy freight railroad trains, which was ultimately reflected in his prohibitive cost estimates. In the case of the 179th Street bridge, however, Ammann had begun from the premise that it would not carry any heavy railroad trains. As originally planned, the upper deck would be for cars and trucks only, and the lower deck would carry at most only light commuter rail traffic. When the project evolved into two stages, with the first stage having only motor vehicles carried on a single deck, Ammann had been able to make some drastic reductions in the weight of the deck itself, by reasoning about the nature of the traffic on the completed bridge.

Short-span bridges were typically designed by assuming that they might at some time be loaded with bumper-to-bumper truck traffic, which determined their strength requirements. As longer and longer spans came to be designed, however, it became clear that a solid line of trucks on a bridge deck was a very unlikely occurrence indeed; designing for that condition led to a very expensive structure that might have to support such a load very infrequently, if ever, and then for only a very short time. Since all structures were designed with some additional reserve strength anyway, in order to provide a prudent margin of safety, all bridges were in a sense overdesigned for the vast majority of the uses to which they were put. In his 1916 treatise on bridge engineering, Waddell gave currency to an idea that Modjeski had employed in the Manhattan Bridge—namely, that for the purposes of designing the bridge the unlikely extreme traffic condition be reduced by one-half. By the mid-1920s, such a practice had become somewhat standard in designing long-span bridges, such as the one then under construction over the Delaware River.

In working out plans for his own bridge in such a design climate, Ammann further reasoned that the mix of automobile and truck traffic at the northerly-Manhattan location would be such that he could make additional reductions in the assumed maximum load. One such reduction was possible because the bridge would have eight lanes, and it was unlikely that all lanes simultaneously would be equally loaded with truck traffic. In the final analysis, Ammann used only one-sixth of the maximum conceivable traffic load as a design load. Though this so-called live load may seem to lead to a drastic reduction in strength, the effect is greatly lessened by the fact that the dead weight of the bridge itself would dominate the total design load. In an insightful analysis, with Jameson Doig, of "Ammann's first bridge," David Billington estimates that, had the engineer used the standard reduction of one-half for the traffic load, more than $7 million worth of additional steel would have had to be added to the $23 million that did go into the bridge. Furthermore, according to Ammann, since "every dollar spent for steel in the flooring and stiffening trusses in a span of this length requires at least an equivalent expenditure for materials in the cables, towers, and anchorages to carry the floor steel, the total saving by the adoption of the flexible trusses is estimated to be almost $10,000,000." This 25 percent savings of the entire cost of the bridge may have made the difference in the economic attractiveness of Ammann's proposal to Silzer.

Ammann's calculations were supported by the "deflection theory" that Moisseiff had applied to the design of the Manhattan and Delaware River bridges. A consequence of that theory was that as the dead weight of a span increased, the stiffness of the deck could decrease. Since the weight obviously increased with length, longer bridges could be more flexible. This counterintuitive result was due in part to the necessarily heavier cables or eyebars, which alone would provide considerable resistance to deflection, whether caused by traffic or by wind. Whereas in many other suspension bridges, Ammann pointed out, "the stiffening system is heavier than the cables or chains themselves," in his bridge the weight of the stiffening system would amount to "only about one-eighth of that of the chains."

When the structure was completed in 1931, the slenderness of the deck of the George Washington Bridge was one of its most striking features, but some critics found fault with other aspects of its design. The American Institute of Steel Construction's award for that year's "Most Beautiful Bridge," for example, went not to Ammann's 179th Street suspension bridge but to his Bayonne arch. According to an architect on the jury, "The setting of the Washington Bridge with the lofty Palisades on one side and the low Manhattan shore on the other did not call for symmetrical treatment and might be filled better by an asymmetrical structure." But engineers by and large embraced fully Ammann's Hudson River achievement, especially as embodied in the structural aesthetic of its most slender deck.

At a special dinner meeting of the American Institute of Consulting Engineers held less than two weeks after the opening of the George Washington Bridge, the structure was described as a "great monument to Mr. Ammann," and to him had fallen the title "*Pontifex Maximus*" of New York. Amid the mutual admiration of engineers, the flexibility of light bridge decks came in for some light talk. After Ammann spoke about the long history of bridging the Hudson—and had his recollection of the date of the first tunnels corrected from the audience by their engineer, J. Vipond Davies himself—Ammann discussed the question of stiffening suspension bridges. Among his remarks was a reflection on the evolution of his ideas:

> In my preliminary studies for this bridge, it took me a long time to wean away from the rigid system and to find out what would be the required rigidity, but I finally came to the conclusion that if the bridge were built for highway traffic alone the weight would be sufficient to provide ample rigidity without any stiffening whatsoever; in other words, it was feasible to go back to the early English type of suspension bridge, that is, to a practically unstiffened floor suspended vertically from the cables.

What neither Ammann nor anyone else at the time seemed to recall about the "early English type of suspension bridge" was that many of those light decks were destroyed in the wind. Rather, the mood at the dinner was gleeful, the participants reveling in stories of people who feared the light bridges. After Ammann spoke, it was the turn of guest Colonel E. Vivien Gabriel, of the British Royal Engineers, "a man of vast and wide experience in India." Among his stories of designing and constructing suspension bridges in that country, he related how he sometimes "drove a flock of sheep or cattle over the bridges before we tried the cars, in order to convince" the local engineers. After listening to Gabriel's "delightful stories," Ammann ended the evening with some of his own "about the early bridges built by American engineers." Presumably recorded verbatim, Ammann's closing remarks suggest that he had less a sense of lightness in story than in bridges:

> There was one of these flexible suspension bridges in New England, and an engineer who visited that place stood in the middle of the bridge while a team of oxen went over, and he relates that he became perfectly seasick.
>
> In another instance, an American engineer had built a suspension bridge of one of those thin wires or ropes that the Colonel mentioned, and the Indians who lived in the vicinity were very doubtful about its strength. They were used to the substantial cantilever bridges built of solid wood, and they refused to go over it. Finally they held a council, and decided that before they would go over they would send their squaws over, and if the bridge stood up under them, they would go over, too.

These stories, told in a time before speeches were vetted for political correctness, must have gotten a laugh, not for their execution, but for their allusion to the George Washington Bridge, whose flexibility was of no concern to these worldly and sophisticated men of scientific engineering.

Of the forty-odd engineers present at the meeting, only Ammann himself would go on to build more daring bridges than the George Washington, but the success of that structure would also embolden engineers who were not in attendance. These included Leon Moisseiff and Joseph Strauss, who were engaged at the time in the design and construction of a bridge that would have a central span of forty-two hundred feet, a full 20 percent longer than the George Washington.

9

Talk of bridging the dramatic strait between San Francisco Bay and the Pacific Ocean known as the Golden Gate had begun in the nineteenth century, but the social and technical conditions of the early twentieth were needed to advance the project beyond talk. Bay Area movers and shakers were becoming increasingly annoyed by the ferry service between Marin County and San Francisco, and among them was the San Francisco *Call Bulletin*'s James Wilkins, whose engineering degree and daily ferry rides seem to have destined him to launch, in 1916, a new and relentless editorial campaign for a bridge. Another staunch proponent of a bridge was San Francisco's city engineer, Michael O'Shaughnessy, who had been hired a few years after the 1906 earthquake to rebuild an infrastructure that was still in shambles. To secure a long-term water supply for the city, O'Shaughnessy wanted to build—almost 250 miles away, in the Sierra Nevada Mountains—a dam across the Hetch Hetchy Valley, which was regarded as a natural setting equal in beauty to nearby Yosemite Valley. Opposition to the project by the environmentalist John Muir and the Sierra Club took the case all the way to the U.S. Supreme Court. When the dam was christened, in 1923, it was named after engineer Michael O'Shaughnessy, in recognition of his dedication to the scheme. Completing the supply line to the city took another decade. Half a century later, environmentalists were trying to have the dam torn down and the Hetch Hetchy Valley restored to a natural state, but San Francisco's residents have become as dependent upon the water it impounds as they have upon the Golden Gate Bridge.

Newspaper people and city engineers, no matter how determined, cannot merely will great water supplies or bridges into existence. To bridge the Golden Gate, a proposal that was sound both technically and fiscally was needed, and such a scheme would have to come from an engineer with the background, vision, and time to conceive and execute the first rough calculations on a dream. In the course of dealing with all kinds of infrastructure problems for the city, ranging from tunneling under hills to seeing that all was in order for the 1915 Panama-Pacific International Exposition, O'Shaughnessy encountered many an engineer with whom to talk of things that might be. Among those with whom he raised the question of a bridge across the Golden Gate was Joseph Strauss, one of whose patented bascule bridges with a massive concrete counterweight was the first such one in San Francisco. In addition to being responsible for a nondescript if not downright ugly Fourth Street Bridge, Strauss designed the amusement ride known as the Aeroscope for the 1915 World's Fair. The ride carried 118 fairgoers about two hundred feet in the air in the equivalent of a modest two-story house mounted on the end of a steel truss that was effectively a revolving bascule bridge. The view provided of the fair and the surrounding area must have been spectacular, for, as the ride progressed in a helical path over the fairgrounds, passengers could see "Alcatraz and the Angel Islands in the bay, and the Golden Gate and Pacific

Fourth Street Bridge constructed in San Francisco in 1916 by the Strauss Bascule Bridge Company

Ocean beyond." Engineer Strauss no doubt looked out at that view and dreamed.

Joseph Baermann Strauss was born in Cincinnati in 1870, the son of a noted portrait artist, Raphael Strauss. It was just three years since traffic had begun to move between that city and Covington, Kentucky, across the Ohio River, on the new suspension bridge designed and built by John Roebling. Though Joseph Strauss may not have grown up literally in its shadow the way engineer David Steinman would in the shadow of the Brooklyn Bridge, the dominant role such a bridge plays in the life of a city does not escape young men who want to realize certain kinds of great dreams and be remembered for them. As an engineering student at the University of Cincinnati, Strauss was well aware that his five-foot frame would not allow him to compete on the football field, and he was recalled as having become determined then "to build the biggest thing of its kind that a man could build." Ambition was apparent in his becoming class president and class poet by the time he finished school in 1892; in his graduation thesis, he proposed the construction of an international railroad bridge across the Bering Strait. Raphael Strauss gave the new graduate "$100 and told him to go out into the world and make it on his own."

Armed with this modest stake and his college degree, young Strauss moved to Trenton, the New Jersey town where the John A. Roebling's Sons Company had been founded in 1849 to manufacture wire rope, and where the Roebling family had become prominent. Strauss took a job as a draftsman with the New Jersey Steel and Iron Company, a bridge-building firm that dated from 1866 and was owned by the philanthropist Peter Cooper's New York–based Cooper, Hewitt & Company. After two years in Trenton, Strauss accepted an opportunity to return to the University of Cincinnati to teach for a year in the Department of Engineering, at the end of which time he married his college sweetheart, May Van. They moved to Chicago, where Strauss took a position as detailer with the Lassig Bridge and Iron Works. After gaining further experience as inspector, estimator, and designer with the Lassig firm, Strauss joined the Chicago Sanitary District, and advanced from designer to squad boss. In 1899, he became principal assistant engineer in charge in the office of Ralph Modjeski, which had designed many of the lift- and drawbridges over the Chicago River.

Among the problems confronting Modjeski, and other Chicago bridge firms, was the design of bridges with movable roadways. One relatively new type of drawbridge was the bascule bridge, which operated on a seesaw principle, but with the counterbalancing side greatly reduced in length by the use of massive counterweights. Though these bridges had the advantage of not taking up much space in crowded cities, they had the disadvantage of requiring a lot of expensive dead weight to function properly, and the added complexities associated with mechanical movements. In time, Strauss came up with a new scheme, which employed concrete for the counterweight. However, since this relatively inexpensive material was about two-thirds less dense than the cast iron that was being used in counterweights, Strauss's early designs had ungainly proportions. Another of his ideas, to use a new kind of trunnion bearing as the "element of movement," was reportedly also ridiculed by Strauss's superiors at Modjeski, and so he left.

In 1902, Strauss thus struck out on his own as a consulting engineer, and in 1904 he established the Strauss Bascule Bridge Company. Over the next decade or so, a series of significant patents were granted for movable bridges of Strauss's design, and his firm prospered somewhat, eventually becoming the Strauss Engineering Corporation, a consulting firm with offices in Chicago and San Francisco. Though the firm's early specialty was movable bridges, not all of which were by any means unattractive, it also designed and built such unusual structures as the Aeroscope, portable searchlights employed during World War I, and reinforced-concrete railroad cars. Strauss would go on to serve as consulting engineer on such significant bridges as the Arlington Memorial Bridge in Washington, D.C., whose center arch is in fact a double-leaf steel bascule bridge with concealed counterweights, as well as the Bayonne Bridge and the George Washington Bridge in New York.

O'Shaughnessy and Strauss began to talk seriously about the possibilities of bridging the Golden Gate in about 1919. The city engineer's office provided location maps, surveys, and the like, while Strauss's company worked on design. Soundings of the strait made by the U.S. Coast and Geodetic Survey in early 1920 were not encouraging. O'Shaughnessy shared the data not only with Strauss but also, unbeknownst to Strauss, with Frank McMath, who was with the Canadian Bridge and Iron Company in Detroit, and with Gustav Lindenthal. By mid-1921, Strauss had delivered a proposed bridge design to O'Shaughnessy, who did not make it public for some time, perhaps because of its unconventional appearance. It was an ungraceful hybrid, half cantilever and half suspension bridge, but it did carry the very attractive price tag of about $17 million. O'Shaughnessy must certainly have thought that Strauss's keen sense of economy counterbalanced at least in part the aesthetic shortcomings of the design—especially when the less cost-conscious Lindenthal eventually came in with a proposal requiring $60 million at a bare minimum, and other estimates had run in the $100-million range; McMath never did submit a plan. Thus Strauss's design came to be used "to help stimulate public interest in building a bridge and to determine what financial and political support really existed for such a project."

An attractive booklet titled *Bridging "The Golden Gate,"* prepared by Strauss and privately printed by O'Shaughnessey and Strauss in 1921, contains a concise and convincing argument that the project was technically and financially sound. The "new cantilever-suspension type of long span bridge" that was "conceived and patented" by Strauss combined the cantilever and suspension principles to reduce the length of cable required and at the same time increase the overall stiffness of the structure. The estimated cost of the bridge, laid out succinctly on a single page, was said to be "so reasonable" that it could be paid for either directly by San Francisco, Marin County, and others who would benefit from it, or operated as a toll bridge and thus would pay for itself, according to Strauss's estimate of revenue.

Strauss began to talk about his proposal before small groups in Marin County and northern California. Though he was apparently a terrible speaker, his idea for a bridge that would bring clear advantages in real-estate development and commerce gained increasing support. By 1923, enough momentum had been generated so that legislation was drafted au-

Joseph Strauss's 1921 proposal for a bridge across the Golden Gate

thorizing a Golden Gate Bridge and Highway District, which would create the first local tax district of any significance for the specific purpose of building a bridge. Within months, the legislation was in place. In the meantime, an engineer was added to Strauss's staff to work out the design of a bridge larger than any the office had ever tackled.

This engineer, Charles Alton Ellis, was born in Parkman, Maine, in 1876. His formal education consisted of an A.B. in mathematics and Greek from Wesleyan University, and he learned engineering on the job at the American Bridge Company. He learned so well that in 1908 he became an instructor in civil engineering at the University of Michigan, and he was a professor of structural and bridge engineering at the University of Illinois when Strauss hired him, in 1922, to work on the Golden Gate Bridge. While Ellis did the essential design work, Strauss continued to promote his bridge scheme, which had encountered the familiar challenge of presenting an ob-

A page from a promotional booklet showing Joseph Strauss's concise cost estimate for a Golden Gate Bridge

Estimated Cost, Proposed Golden Gate Bridge

The estimated quantities and cost of the structure are as follows:

Item	Quantity		Rate	Amount	Total
SUPERSTRUCTURE					
Cable erected	5,250 tons	@	$460.00	$ 2,415,000.00	
Structural Steel (Nickel steel)	74,000 "	@	150.00	11,100,000.00	
Rails	400 "	@	60.00	24,000.00	
Railing	500 "	@	200.00	100,000.00	
Lumber	1,675 M. ft. BM.	@	80.00	134,000.00	
TOTAL SUPERSTRUCTURE					$13,773,000.00
SUBSTRUCTURE					
Concrete	114,000 cu. yds.	@	12.00	1,368,000.00	
Reinforcement	1,460 tons	@	60.00	88,000.00	
2 caissons		@	260,000.00	520,000.00	
2 abutments including steel (anchorage, etc.)				440,000.00	
Miscellaneous				61,000.00	
TOTAL SUBSTRUCTURE					2,477,000.00
TOTAL COST					$16,250,000.00
ENGINEERING, SURVEYS, GENL. MISC.					1,000,000.00
GRAND TOTAL					$17,250,000.00

stacle to navigation. Final War Department approval was to come slowly, and in the meantime Strauss engaged Professor George F. Swain of Harvard University and Leon Moisseiff to serve on a board of consultants. Both endorsed Strauss's plans as workable, if not graceful, and at his request Moisseiff prepared plans for a comparable suspension bridge for the site. His design, with a four-thousand-foot center span, represented more than a doubling of the suspension-bridge length, itself a record, that was soon to be realized in the Moisseiff-designed Delaware River Bridge, and was to be an even greater leap in the state of the art than Ammann's Hudson River Bridge, then proceeding within the secure financial base of the Port of New York Authority. Though cost estimates for both Strauss's and Moisseiff's versions of a Golden Gate Bridge remained attractive at under $20 million, some supporters were beginning to have second thoughts about providing financial guarantees to the project and going ahead without further studies. By the end of 1928, however, the Golden Gate Bridge and Highway District was incorporated.

With incorporation, the district could get down to business, which meant fixing on a design that would be approved by the War Department, going to the voters for their approval, issuing bonds to raise capital, and letting contracts for the actual construction. Although Strauss was indeed chief engineer of the district, his or any other design still required the approval of the politically appointed members of the district's technical board. Among the engineers who were approached anew for designs in 1929 were Ammann, Lindenthal, and Modjeski. Needless to say, Strauss was not in their league when it came to long-span suspension bridges. Indeed, he had not designed or built a single one. Furthermore, questions that had arisen about foundation conditions made it increasingly doubtful that the heavier cantilever design, even if modified, could be supported. In order to bolster his credibility for the job, Strauss had engaged Ammann and Moisseiff as consulting engineers to him as chief engineer. They were already engaged in the building of the George Washington Bridge, of course, which not only put them in the forefront of bridge building but also made their full availability for a West Coast project somewhat questionable. Ammann had accepted Strauss's offer reluctantly, in part because he was unsure about the organizational structure that had him working for the chief engineer rather than for the district board, and he did not want to appear to be competing with consulting engineers in private practice. No doubt he avoided quitting the project outright because he must have realized that Strauss was not only determined but also politically effective enough to get the world's largest bridge built, and he did not want to be left out of participating in it. In the final arrangement that was worked out, Ammann and Moisseiff continued to be associated with Strauss, as members of a board of engineers appointed by the Bridge and Highway District, and he was to be associated with them as consulting engineer on the Bayonne and George Washington bridges. The Golden Gate board was chaired by Strauss, who was in fact also chief engineer of the project. O'Shaughnessy, who was so instrumental in getting the bridge idea off the ground in the first place, was disappointed, to say the least, at being left out entirely; he would eventually join the forces completely opposed to the bridge.

The newly printed letterhead of the Office of the Chief Engineer listed Strauss in that position, with Charles Ellis next in line as designing engineer. Three consulting engineers were also listed: Ammann, Moisseiff, and, as the local representative, Charles Derleth, Jr., dean of the College of Engineering at the University of California at Berkeley. Derleth had been chief engineer of the Carquinez Strait Bridge at Vallejo, California, designed by David Steinman and William Burr, and had also been a critic of Strauss's early design. Dominating the letterhead, however, was not the names of the engineers but the seal of the district, showing a combination cantilever-suspension bridge that strongly resembled the chief engineer's original design. Yet Strauss's ambition was greater than his pride: he came to acknowledge that a suspension bridge of the kind suggested by Moisseiff not only would be lighter and less expensive to build, but also could be completed in less time. Because of the Depression, cost estimates for construction had fallen even since Strauss's original estimate, and he had made his bid for the job more attractive to the Bridge and Highway District by agreeing to terms that represented "the lowest ever written for a bridge job of such magnitude." His firm's fee was to be 4 percent, which compared rather unfavorably with the 7 percent recommended by the American Society of Civil Engineers. Though this meant that Strauss would have to run a bare-bones engineering-design office for the project, he was apparently willing to compromise in many ways to get a chance to pursue his dream bridge. (Strauss had set his fee with the understanding that the bridge plans would be prepared sequentially as construction progressed. When the district directors ordered the plans prepared all at once, in order to go for lump-sum bids, Strauss sued for an additional 1 percent fee and won.)

Charles Ellis was put in charge of the detailed design at the Strauss Engineering Corporation, with Moisseiff checking all the calculations. Early plans called for the Marin pier to be two hundred feet out in the water, but a decision was made to put it adjacent to the shore, thus reducing costs while at the same time increasing the length of the main span of the bridge to forty-two hundred feet. The south pier and anchorage were to be the focus of controversy, however. Berkeley Professor Andrew C. Lawson had been retained as consulting geologist, and he had serious doubts that rock conditions near the San Francisco shore could support a tower of the size required. This sparked prolonged debates about how sound the rock really was and how the bridge would survive movement of the tower and anchorage during an earthquake. (The span was to be built parallel to the San Andreas Fault, only six miles away.) It was Ellis who assured critics on the latter point. The major design challenge Ellis faced, however, was that of the towers themselves. The decision was that they should be of steel and that their design was of prime importance not only structurally but also aesthetically. Heavy diagonal bracing was to be confined to below the roadway, with the upper portions of the towers braced only with horizontal members. Strauss suggested that the skyscraper towers have "the stepped-off type of architecture" that was so prominent a feature of the contemporary Chrysler and Empire State buildings, and this was followed. The first consulting architect retained by Strauss was John Eberson, an Austrian-born electrical contractor who had taught himself architecture and become a prominent theatrical-set designer. He produced a design that had archlike openings above the roadway and at the tower top, thus concealing with architectural embellishments two of the four strong horizontal structural members that Ellis and Moisseiff knew were needed to provide the proper stiffness.

As the design of the bridge proceeded, it was learned that the San Francisco approach roads had been realigned by federal highway authorities, requiring a new arrangement for the toll plaza and thus further architectural work. Since the New Yorker Eberson had proved to be too expensive for Strauss's budget, and since Strauss knew that a locally prominent architect might be more helpful in gaining political and financial support for the bridge, he hired Irving F. Morrow as the new consulting architect. Born in Oakland, California, in 1884, Morrow studied architecture at Berkeley and at the Ecole des Beaux-Arts in Paris. Though he had no prior experience in bridge work, he was a daily commuter by ferry between his office in San Francisco and his home in Berkeley, and thus had observed the ever-changing conditions of light, shadow, and color in the Golden Gate. It would be Morrow's Art Deco design treatment of the towers, incorporating such façade details as the vertically faceted aluminum fascia panels to cover the heavy steel structural horizontals, that would give the bridge much of its dramatic effect in the changing light. He would also play a key role in helping to decide on such details as bridge railings and paint colors that would make the Golden Gate Bridge one of the most distinctive in the world.

Before the bridge could be decorated, however, it had to be financed and built, and chief engineer Strauss presented his structural and architectural report to the Bridge and Highway District Board on August 27, 1930, just two weeks after a construction permit was signed by the secretary of war. A referendum was announced, and on November 4 the voters approved a $35-million bond issue by more than three to one. The successful outcome was due in no small part to the fact that the construction of the George Washington Bridge had gone so smoothly, technically and financially, its opening now only a year away. Even before that occurred, however, *Engineering News-Record* would remark that "the Golden Gate Bridge is a fact today because the Fort Lee bridge was built yesterday. It was the latter project that attuned the public mind to the possibility of financing such huge enterprises." The sale of the Golden Gate bonds would go no more smoothly than the continuing design, but construction would begin within three years.

The greatest tensions in the bridge's design proved to be not in the cables or anchorages but in the interpersonal relationship between Strauss and Ellis. With the bond issue passed, there was great optimism and pride in the project, and there were audiences wanting to hear all about it. Among these was the first West Coast conference of the National Academy of Sciences, to be held on the Berkeley campus. The university's president, Robert G. Sproul, proposed to Dean Derleth that the scientists be told about the significant engineering problems that were being overcome in designing the local bridge, and Derleth arranged for Ellis to make the presentation. Though he acknowledged that Strauss was the boss of the project, when it came to the design of the bridge itself, Ellis stated, "Mr. Strauss gave me some pencils and a pad of paper and told me to go to work." Derleth himself, at a luncheon hosted by skeptics about the design where Strauss expected him to silence the critics, appealed to the authority not of Strauss but of Ellis, "who stands high as a structural bridge engineer." The dean went on to describe how Ellis was in charge of the first design and how Moisseiff's theory was used to lighten the trusses employed. Finally, Derleth assured the audience that neither Ellis nor Moisseiff had made any errors in their lengthy and complex mathematical calculations, "even if Mr. Strauss and Mr. Ammann and I do not know anything about the subject." Thus revealed to be just another manager, Strauss was not happy.

Strauss had indeed put Ellis in charge of the design, and the staff of Strauss Engineering was to assist him, especially with calculations for the tower. But since Ellis had found at least one of Strauss's staff unprepared to work on "a problem of this nature," and others were busy on a potentially very profitable bascule-bridge job, he took on much of the work himself, having it checked by Moisseiff and his own staff. As the bridge seemed to be associated more and more with Ellis, Strauss tried to rein him in by pushing for milestones and deadlines. Ellis defended his reluctance to promise when things would be completed or to delegate calculations by describing the structure as one requiring extrordinary research and theoretical work. Yet times were tough; Strauss's corporation apparently felt financial pressures, and he was little interested in paying for work he thought to be a frivolous academic exercise. In late 1931, Strauss wrote to Ellis that he should turn the work over to his assistant, Charles Clarahan, Jr., and take a vacation. This left one person in the office working on the bridge; before his vacation was over, Ellis received a letter from Strauss effectively laying him off and instructing him to turn over the job to Clarahan. According to Strauss, "the structure was nothing unusual and did not require all the time, study and expense" that Ellis had been devoting to it.

It was the Depression, of course; Ellis remained out of work until he joined the faculty of Purdue University in 1934. In the meantime, he continued to work on the tower design, writing to the consulting engineers about his concerns. Eventually, Ellis "vexed" even Moisseiff, who wrote to Derleth that he and his staff found nothing wrong with the methods used by the Strauss Engineering Corporation. The tower design had been verified by means of a stainless-steel model at Princeton University, and at a program there to describe the tower design and testing, Moisseiff described how the stiff frame design, "without diagonals, was adopted for the sake of appearance and how this complicated the analytical design." Though he also commented on the "ingenious method of analysis" worked out by Ellis, he credited Frederick Lienhard of his own office with the computations that checked with the model tests.

While Ellis would certainly appear to have been badly treated by Strauss, he also seems not to have been able to maintain a sound perspective on the matter at the time. Strauss eventually replaced him with Clifford E. Paine, a onetime student of Ellis's at the University of Michigan and an associate of Strauss who once quit his employ "in protest over his boss's interference with his work." He was hired back with assurances that he would be left alone. Paine's strong personality and central position on the Golden Gate project were subsequently demonstrated again in the fact that Strauss Engineering Corporation became Strauss & Paine, Inc., in 1935, about halfway through the Golden Gate construction. In a brochure available from the Bridge and Highway District, Paine is identified as "principal engineer during the design and construction" of the bridge. When it was completed and a plaque was to be prepared, Paine wanted to be identified as "Assistant Chief Engineer," but Strauss preferred that the credit read "Assistant to the Chief Engineer." Today, the plaque, on the southeastern face of the eastern leg of the San Francisco tower, identifies

Paine as "Principal Assistant Engineer." Others receiving credit are Russell G. Cone, as "Resident Engineer"; Charles Clarahan, Jr., and Dwight N. Wetherell as "Assistant Engineers," and the consulting engineers Ammann, Derleth, and Moisseiff. There is no recognition whatsoever of Charles Ellis here, or in the listing of the engineering staff in Strauss's final report to the board of directors.

The Golden Gate Bridge opened on May 27, 1937, with an event known as Pedestrian Day. By 6 a.m., about eighteen thousand people had assembled on both sides of the bridge to be the first to walk over it. Over the course of the day, an estimated two hundred thousand strolled across the miraculous span. In 1987, as a means of commemorating the Golden Gate's fiftieth anniversary, another Pedestrian Day was announced, but so many people showed up before the opening ceremonies that they pushed onto the bridge and pre-empted the usual political speeches. It was estimated that about a quarter-million people were crowded onto the Golden

The Golden Gate Bridge, in its dramatic setting

Gate at one time, thus testing it as it had never been before. The bridge, which has come to be known among engineers for its flexibility in the wind, for having been stiffened since construction, and for having been found structurally unsuitable to carry an extension of the Bay Area Rapid Transit system into Marin County, had the graceful arc of its center span flattened out under all the people in 1987, and there was some concern for its safety.

It is unlikely that the centennial of the bridge will be celebrated with another uncontrolled Pedestrian Day, unless substantial structural retrofitting work is done in the meantime. This may indeed occur, for after the 1989 Loma Prieta Earthquake, extensive plans were prepared to make the bridge capable of surviving a quake registering as high as 8.3 on the Richter scale, to meet standards set by the state of California. The Bridge and Highway District's $128-million project was first delayed by an environmental study required by the Federal Highway Administration, and then by disagreements over liability between the district and the engineering firm of Steinman, Boynton, Gronquist & Birdsall. This successor firm to David Steinman's practice felt it was being asked to provide more security against financial loss than the district itself had felt prudent, and so refused to sign a contract for the retrofitting job. The irony of this development is only evident in the context of events subsequent to the completion of the Golden Gate Bridge, when Steinman and Ammann disagreed over the causes and cures of movements of their own suspension bridges in the wind. But that was still in the uncertain future when the Golden Gate Bridge was opened on Pedestrian Day, 1937.

On that day, forty-five years after he had proposed a bridge connecting the continents of Asia and North America over the Bering Strait, the class poet Strauss commemorated the opening of the golden bridge that he had built across the Golden Gate with a new poem, which began:

At last the mighty task is done;
Resplendent in the western sun, . . .

Though the poem did not involve Strauss directly, except as author, neither did it celebrate the assistant and other engineers who had labored over and checked calculations for the foundations, towers, and cables. The bridge itself is the hero of Strauss's poem; it deals mythically with the setting and the structure, which is anthropomorphized in titanic proportions. The bridge was destined to be, "for Fate had meant it so," and only in the fourth stanza did the poet begin to hint at the human toll such an engineering project could take:

Launched 'midst a thousand hopes and fears,
Damned by a thousand hostile sneers,
Yet ne'er its course was stayed;
But ask of those who met the foe,
Who stood alone when faith was low,
Ask them the price they paid.

. . .

An Honored cause and nobly fought,
And that which they so bravely wrought,
Now glorifies their deed;
No selfish urge shall stain its life,
Nor envy, greed, intrigue, nor strife,
Nor false, ignoble creed.

Strauss's vision of the battle for the bridge appears to have been of one among the human emotions that dominate the political and financial design, rather than among the mathematical and material forces that must be reconciled to make the structure work. On the other hand, Ellis seems to have fought the fight entirely on the technical level, and he certainly must be considered one who paid a price. He may never have read Strauss's poem, and it might not have mattered to him if he had. It is known that Ellis, who lived until 1949, never visited the bridge; that may be a fair indication that his satisfaction lay in the designing rather than the building— or that he was nursing a mighty grudge.

Later in his own career, David Steinman would take up writing poetry, adding to his list of accomplishments and awards, and he too would versify on his own poems in steel. However, for every Strauss and Steinman, who seemed to wear their accomplishments as Waddell wore his medals, there have always been and no doubt always will be uncounted and unheralded engineers like Ellis designing and building bridges equally worthy of poetry. Many such engineers were glad to have jobs in a worsening economic climate; few were as contentious and uncompromising as Ellis with their chief engineers.

10

The sense among bridge engineers generally in the 1930s was one of supreme confidence in their theoretical capabilities, which Ammann himself articulated in an article in *Civil Engineering* in 1933: "When Telford planned the Menai Bridge [in the 1820s] he developed the major forces largely on models. Bridge failures of that day resulting from inadequate design might be excused on the ground of insufficient knowledge; today the designer has no such alibi." Ironically, Telford's bridges might themselves have served as models of the kinds of failure that could still befall bridges, and the designer, whether modern or not, indeed had no alibi for being unaware of such things. But designers in the 1930s certainly seem to have been oblivious to the power of the wind, as developments soon would show.

In 1930, Ammann rose from bridge engineer to chief engineer of the Port Authority, and as such he oversaw the planning and construction of the Lincoln Tunnel, which entered Manhattan at 39th Street and thus provided, when it opened in 1937, the long-dreamed-of midtown crossing of the Hudson. That same year, Ammann assumed simultaneously the position of director of engineering of the Port Authority. With the growing use of automobiles, other New York City transportation needs had been developing between the boroughs, but those intracity projects were under the control of the master planner Robert Moses, who was chairman of the Triborough Bridge Authority. This autonomous and yet highly political entity was named after its first large project, which was to connect the three boroughs of Manhattan, Queens, and the Bronx with a system of bridges and viaducts known collectively as the Triborough Bridge, whose centerpiece was a suspension bridge across the Hell Gate just south of and parallel to Lindenthal's famous railroad arch. With his success at the Port Authority well established, Ammann was asked by Moses to bring his (and Allston Dana's) experience to the troubled Triborough Bridge project, whose Tammany engineers prized granite towers over traffic lanes. Since considerable design work had already been done, the bridge and its squat towers could not be completely reshaped to look like an Ammann structure. Nevertheless, the bridge, completed in 1936, remains a major achievement in facilitating traffic movement within the city.

From 1934 to 1939, while continuing in his position with the Port Authority, Ammann served also as chief engineer of the Triborough Bridge Authority. One of the projects that Moses promoted toward the end of this period was a bridge between Brooklyn and the part of lower Manhattan known as the Battery, with a central anchorage near Governors Island. The proposed design consisted of two suspension bridges in tandem, not unlike the San Francisco–Oakland Bay Bridge, then recently completed on the West Coast; Moses's bridge was so convincingly drawn on an aerial photograph of New York Bay that one would swear that the bridge was a reality. A tunnel at the location had been proposed at least a decade earlier, and the New York City Tunnel Authority had been created to construct it. Moses was not interested in giving up any opportunity to control another great project, however, and while the Tunnel Authority was looking for financing during the Depression, the bridge alternative was proposed. Considerable opposition arose to an above-water crossing, not only by the War Department but also from citizens who did not want to see the spectacular and world-famous view of the lower-Manhattan skyline hedged by the enormous elevated approach roads that would have had to accompany a bridge.

The tunnel project, under the direction of chief engineer Ole Singstad, had been estimated to cost about $65 million, but Moses, who had earlier made a deal with Mayor Fiorello La Guardia that a federal loan covering part of the cost would be supplemented by funds from the revenue-rich Triborough Bridge Authority to help construct the tunnel and connecting highways, put the cost at $85 million. This inflated figure gave Moses an excuse to renege on his agreement with the mayor, arguing that the bridge could be built for about half the cost of a tunnel. However, Moses's figure for the bridge did not square with the cost of other spans then under construction, and Ammann, the designer of the structure, was asked for his opinion on the matter. The engineer was apparently caught off guard, and, torn between his engineering integrity and his loyalty to Moses, who had given him opportunities to continue to design and build great spans, Am-

A controversial bridge designed in the mid-1930s, and strongly advocated by Robert Moses fifteen years later, convincingly drawn onto a photograph of its dramatic proposed location between Brooklyn and the Battery, in lower Manhattan

mann hemmed and hawed enough to lead representatives of a citizens' group to ferret out the true costs of each complete project; this comparison favored the tunnel.

The Brooklyn-Battery bridge thus became a footnote to Ammann's portfolio of designs and dreams, usually not even mentioned by biographers. Though the general slowdown in large projects toward the end of the 1930s must certainly have played a role, the uncomfortable incident relating to lower Manhattan must also have influenced, if not triggered, his departure from civil service, and he went into private practice the year the bridge-tunnel controversy came to a head, 1939. However, before he left the employ of Moses, Ammann did, as chief engineer of the Triborough Bridge Authority, oversee the design from scratch and the realization of a major suspension bridge between the Whitestone section of Queens and the Bronx. Planning for that bridge had begun in 1935, and the Bronx-Whitestone Bridge was to be completed in time to serve the traffic influx expected for the 1939 New York World's Fair. Since the site for the bridge did not constrain the lengths of the main or side spans as the Palisades location did those of the George Washington Bridge, Ammann had free rein to design a structure whose proportions were chosen principally for economic and aesthetic reasons. The latter was extremely important in the case of the Bronx-Whitestone, for its entire profile, including anchorages and approach spans, was to be in clear view, "so that the structure as a whole [would] be visible to an extent that [was] true in no other case." No conflict between architectural and structural considerations was necessary here; in a report published in *Civil Engineering* to coincide with the completion of the bridge in April 1939, Ammann wrote of how modern engineers like himself could view such matters when unencumbered by the constraints of the past:

> It is now well established that long-span suspension bridges for modern highway traffic may have a relatively flexible stiffening system, and that the degree of flexibility has a material effect upon the economy of the design. In this respect the Bronx-Whitestone Bridge marks another radical departure from past theories and practice. Its stiffening girders have greater flexibility in relation to span length than any other suspension bridge built in recent years, except the George Washington Bridge in its present state with only a single unstiffened highway deck. The latter bridge, however, with a present roadway capacity equivalent to that of the Bronx-Whitestone Bridge, has a suspended dead weight per linear foot 2½ times greater, a center span 56 per cent longer, and side spans somewhat shorter, all of which factors contribute to the greater rigidity of the unstiffened cables.

> It was the aim of the writer, on esthetic as well as structural and economic grounds, to restrict the height of the floor structure to a minimum, to avoid trusses, and to keep the top at such an elevation above the floor as not to obstruct the view of the landscape from passing vehicles. A depth of 11 ft for the stiffening girders was found to be sufficient and to fit best into the floor structure. This is only $\frac{1}{210}$ the length of the center span and only $\frac{1}{70}$ of the side span.

Ammann here laid down the prevalent philosophy of suspension-bridge building in the later 1930s. The solid plate girders defining the deck profile and the "design of the towers as rigid frames without any diagonal cross bracing" contributed to "the graceful appearance and structural simplicity of this type of bridge," according to Ammann. Even the anchorages and approach viaducts were "reduced to a minimum as to materials required for strength and stability," and were "devoid of extraneous architectural embellishments." Furthermore, all these factors "in no small degree contributed to the unprecedented speed in the construction of the bridge." Ammann's apparent about-face with regard to "architectural embellishments" was more a turning from the traditional view that masonry provided the texture of choice for monumental works to the modernist view, pre-eminently expressed by Le Corbusier, which saw beauty in steel. The outline of the towers of the Bronx-Whitestone were in fact not unlike those of the early sketches for the George Washington, when Ammann had worked with the architect Ruegg. After the unrealized Cass Gilbert treatments of that bridge, Ammann must have been happy to be collaborating on the Bronx-Whitestone with yet another architect, Aymar Embury II.

Embury had already established himself as an architect to the elite, having designed a classic dormitory at Princeton University and a good number of estates on Long Island, when Robert Moses persuaded him to work on parks in New York. In a remarkable series of articles in *Civil Engineering* in early 1938, Embury seemed to be single-handedly trying to heal the rifts that had developed between architects and engineers. He acknowledged help from the likes of Ammann, Steinman, Waddell, and, "in particular," Allston Dana, who was engineer of design on the Bronx-Whitestone as he had been on the George Washington. Indeed, Embury wrote that he "had the good fortune to work in close association with" Dana. Embury went on to say that he and Dana "had a fairly free hand, although, of course, the designs were always subject to Mr. Ammann's criticism and never out of his control. We were, in a sense, his instruments, and were guided by his desires as to the lines along which we should proceed."

The Bronx-Whitestone Bridge as completed in 1939, with an anchorage in the foreground

Embury came to Ammann's aid late, however, and it was only on the design of the Bronx-Whitestone anchorages that he worked with Dana *de novo*. They "wanted the anchorages to look like an anchorage and like nothing else," Embury reported, in keeping with Ammann's desire that "the whole bridge should be kept smooth, sharp, and clean." The design for the anchorages was "made without reference to any precedent," and the final result was one of simplicity and appropriateness for the foundation conditions. In his paper on the aesthetics of bridge anchorages, Embury also showed alternative designs for the anchorages of the Triborough Bridge and proposed designs for the anchorages of the George Washington Bridge. Then, five years after the opening of that great bridge, the completion of the architectural treatment of its New York anchorage was still on hold "until traffic conditions should necessitate construction of the lower deck." In the meantime, Cass Gilbert had died, and his design was being criticized as "an anachronism in a modern steel bridge."

But Embury did not have only positive things to say about engineers in his articles, for he spoke of "design by drawing instruments," and wondered "how often do engineers, because their triangles are 45, 30, and 60 deg, use one of these slopes for diagonal members?" Architects also came in for criticism for following "the easiest way," and Embury called for more of a meeting of the minds: "Engineers should be good architects, and architects good engineers!" He very graciously closed the second in his series of articles with a note on his collaboration, in which "a reversal of function" surprised those involved in the project: "As a rule it has been the architect who has suggested and the engineer who has acted as the artistic critic." Yet modern steel suspension bridges still presented immensely difficult problems in both structural and architectural design, not least of which was the problem of massive steel superstructures in close proximity to massive masonry anchorages. And solutions to such problems were often ineffable. In the end, Embury admitted that, although engineer and architect had in their anchorage design "a shape that pleases us both, we do not know why."

Among the many pithy phrases in architect Aymar Embury's conciliatory works, he wrote that "it is always easier to remember than to invent." It was here that he and the engineers did part ways, though inadvertently. As much as Ammann and his contemporaries referred to early-nineteenth-century suspension bridges, like Telford's Menai, as aesthetic models, they did not find it easy (or think it relevant) to remember how the light decks were tossed about in the wind. This professional amnesia was endemic by the mid-1930s, and the effects of it began to manifest themselves in a big way. As construction neared completion on the Bronx-Whitestone and preparations were being made for attaching the floor, the deck began to swing when the wind blew at certain angles, and it moved back and forth lengthwise between the towers. Though this "pendulum action" was unexpected, it does not seem to have excessively alarmed Ammann and his associates, who simply installed short guy cables at the bridge's midspan and braking devices at the towers. There was "marked improvement" in the behavior of the bridge, but the first cables had to be replaced by heavier ones. These slight modifications to control the motion of the new structure were apparently taken in stride, and the Bronx-Whitestone opened on April 30, 1939, in time for the World's Fair.

Contemporary suspension bridges had been designed to the same structural aesthetic as the George Washington and Bronx-Whitestone, and some of those other bridges were also beginning to exhibit excessive motion. Even the Golden Gate Bridge, which at forty-two hundred feet had of course surpassed the George Washington as the longest suspended span in the world, was more flexible than anticipated. In heavy winds, the Golden Gate moved sideways by as much as fourteen feet, but engineers calculated that this movement stressed the bridge less than did the expected variation of temperature. Though the deck of the Golden Gate was stiffened by a

The Bronx-Whitestone Bridge after stiffening trusses were added in the mid-1940s

conventional deep truss, it was extremely slender relative to its great length, and this resulted in great flexibility. Two much shorter suspension bridges, designed by David Steinman and stiffened like the Bronx-Whitestone with plate girders, also exhibited considerable motion in their decks. The eight-hundred-foot Thousand Islands International Bridge, opened in 1938 over the St. Lawrence River between New York and Ontario, and the 1,080-foot Deer Isle Bridge, opened in 1939 over Penobscot Bay in Maine, behaved similarly to the Bronx-Whitestone and were likewise fitted with restraining devices, albeit of a different nature. Although considerably stiffened since their initial opening, these bridges remained flexible. In 1978, for example, hundreds of people, including Joan Mondale, the wife of the then vice-president, were stranded on Deer Isle for several hours when the deck of the two-lane bridge began "swelling." A local resident characterized the bridge as normally having "a certain amount of play in it," but on that day the sway in the light breeze was far greater than during the seventy-mile-per-hour winds of the previous winter.

Such vagaries of behavior of their bridges in the wind led to considerable rethinking about calculations and testing of models by engineers in the late 1930s and into 1940, but most must have thought as Ammann did: "We have had to deal with very small movements, and would have felt no concern about them had they not tended to produce discomfort in some persons under unfavorable conditions." The movements were in most cases "very small" relative to the size of the structure (on the order of one foot in a thousand, for example) and were akin to the swaying of a skyscraper today. Although such motions are not thought to be life-threatening to the structure or its inhabitants, they can be psychologically distracting and can have adverse economic implications if people do not want to occupy a swaying skyscraper or cross a swinging bridge. In 1940, however, few seemed to be excessively worried about such things.

Among the official consulting engineers on the George Washington Bridge project were two who, with Ammann, were responsible for the very flexible bridges that followed from the aesthetic imperative. These were, of course, Joseph Strauss and Leon Moisseiff. Just as Strauss was a consultant for the George Washington Bridge, so Ammann fulfilled a similar role for the Golden Gate. Such interlocking relationships, common among the engineering elite, explain to a considerable degree how the state of the art can advance almost in lockstep, so that many different structures, in this case suspension bridges, can share the same aesthetic characteristics—and the same behavioral flaws.

It was Moisseiff's development of the deflection theory that enabled all the slender and flexible bridges to be designed in the first place. Leon Solomon Moisseiff was born in Latvia in 1872, when it was part of the Russian empire. He attended the Baltic Polytechnic Institute in Riga for two years, but his student political activities reportedly led his family to emigrate to the United States in 1891. They settled in New York City, where Leon worked as a draftsman before enrolling in Columbia University, from which he graduated in 1895 with a degree in civil engineering. He worked as a draftsman for the New York Rapid Transit Railroad Commission, as a design engineer with the Dutton Pneumatic Lock and Engineering Company (doing work on drydocks, gates, and lock improvements proposed for the Erie Canal), and as a draftsman with the Bronx Department of Street Improvements, before joining the New York Department of Bridges in 1898 as chief draftsman and assistant designer. It was in this position that he worked on the Williamsburg, Queensboro, and Manhattan bridges and met Gustav Lindenthal. In 1910, Moisseiff became engineer of design for the Bridge Department, and in 1915 he struck out on his own as a consulting engineer. In 1920, he was appointed chief designer of the Delaware River Bridge, which remained his office's major project until the record span was completed in 1926. Thereafter, through 1940, he consulted on virtually every major suspension-bridge project in the United States, including many an engineer's dream that did not materialize. Even if biographical sketches and memoirs have to be taken with a grain of salt, since they often rely so heav-

Leon Moisseiff

ily on the word of friends and relatives, sometimes they do contain a grain of truth: "Although he did not always receive formal credit, Moisseiff was the principal designer of the George Washington, Bronx-Whitestone, Tacoma Narrows, and Mackinac bridges."

II

The twenty-eight-hundred-foot main span of the Tacoma Narrows Bridge made it the third-longest suspension bridge when it was completed in 1940. In keeping with the engineering aesthetic and economic thinking of the times, the bridge deck was stiffened with plate girders. However, although the Tacoma Narrows' main span was five hundred feet longer than that of the Bronx-Whitestone, only eight-foot-deep girders were used to stiffen the roadway, because it was much narrower than that of the New York span. The combination of a longer span with shallower depth and narrower width made the Tacoma Narrows Bridge more flexible than any other. Nevertheless, the two-lane crossing of the Narrows, about thirty miles south of downtown Seattle, provided a reasonable highway alternative to taking ferries between Seattle and the Olympic Peninsula, across Puget Sound. As soon as the Tacoma Narrows opened in July, drivers noticed how flexible it was, the wave motion of its roadway bringing cars that were ahead of a driver on the bridge alternately into and out of view as the pavement rose and fell. Rather than scaring toll-paying customers away, however, the bridge became affectionately known as Galloping Gertie and attracted even more traffic as an unintended amusement ride. Although some of the riders were reported to have become seasick, traffic over the bridge in its first two weeks of operation was twice what engineers had expected.

A bridge across the Narrows had been proposed as early as 1933 by the Tacoma Narrows Bridge Company, which had obtained a franchise and was then seeking capital. However, the growing sentiment for publicly owned bridges and utilities led to a competing application by Pierce County, on the peninsula. In 1937, no doubt fueled by the success of bridges like the George Washington and the almost complete Golden Gate, the State Legislature created the Washington Toll Bridge Authority, which took over the Pierce County initiative and its application for a construction grant from the federal Public Works Administration. As was typical, various conceptual designs had been considered, including a cantilever bridge and a multiple-span suspension bridge such as had been recently completed across San Francisco Bay. By mid-1938, the State

The Tacoma Narrows Bridge, when it was opened in July 1940

Highway Department had made a preliminary design for a suspension bridge with a single main span of twenty-six hundred feet, and two side spans of thirteen hundred feet each, all resting on a stiffening truss twenty-two feet deep. The structure was to carry a twenty-six-foot-wide roadway and two four-foot sidewalks. The total width of the bridge, including the stiffening truss, was to be only thirty-nine feet—a remarkably narrow deck relative to the length of the bridge.

As consulting engineer, Moisseiff was asked to study the Highway Department's design, and he submitted his initial report in July 1938. His first criticism addressed the unequal tower heights. Though these had been chosen to accommodate the unequal elevations of the two ends of the bridge, they meant that the entire roadway of the preliminary design had an upward incline toward the higher, Tacoma shore, and the consulting engineer criticized it in no uncertain terms:

> Unless there are very valid reasons which compel the making of the towers of unequal heights the towers should be of identical design and fabrication. Economic fabrication and good appearance demand it. The symmetry of the structure should be adhered to.

Moisseiff's solution was to "raise the west end of the bridge by 19.5 ft." With regard to the twenty-two-foot stiffening truss of the Highway Department design, the consultant found that it could not "effectively stiffen the bridge except at great cost." He proposed eight-feet-deep plate girders, which would not only "result in a neat and pleasing appearance" but also "be about one cent per lb. less than for a truss." And he reported that his studies showed it "best to attain rigidity by shortening the side spans and by a reduction of sag ratio" in the cables. This was "not only a better solution but also the cheapest," he concluded. In a second part of his report, Moisseiff recommended further that the spacing of the suspenders supporting the roadway from the main cables be increased from thirty to fifty feet, not only to achieve a more pleasing appearance but to effect a further savings of about $35,000 out of the total estimated cost of $6 million. Ironically, he also argued for keeping the "height of the towers to a minimum due to the relatively great effect of transverse wind pressure."

In September 1938, an application to secure a federal loan for the bridge project was submitted by the Washington Toll Bridge Authority to the Reconstruction Finance Corporation, which funded numerous projects of the Public Works Administration. As was standard procedure, the application was referred to the Legal, Finance, and Engineering divisions of the Administration, but it was a review for the Reconstruction Finance Corporation that raised the strongest concerns about the soundness of the project. Theodore L. Condron, advisory engineer to the bond purchaser, was a septuagenarian consulting engineer best known for designing the steel structure for the seventy-two-bell carillon in the Rockefeller Memorial Chapel of the University of Chicago. In his report on the application Condron identified the board of consulting engineers as consisting of Charles E. Andrew, bridge engineer of the San Francisco–Oakland Bay Bridge, and chairman of the board; Luther E. Gregory, a retired navy rear admiral and resident of Olympia, Washington; and R. B. McMinn, a bridge engineer with the U.S. Bureau of Roads in Portland, Oregon. The consulting engineer Moisseiff, and his associate Frederick Lienhard, were identified as "New York engineers" associated with the design of the San Francisco–Oakland Bay and Golden Gate bridges.

The report of the board of consulting engineers on the Moisseiff-modified plans had found them to be "in satisfactory shape for receipt of bids," even though the board did not examine the project in detail. Time did not permit a checking of stresses in the cables or stiffening system, for example, but the board had "full confidence in Mr. Moisseiff," considering him "to be among the highest authorities in suspension bridge design." With this endorsement, Condron may have thought that the approval of the project before him would have been a routine matter; the more he looked at the plans, however, the more doubts he seemed to have. In particular, Condron had serious reservations because of the extremely narrow width of the proposed bridge relative to its main-span length. When he compared this ratio with that of recently completed suspension bridges, the Tacoma Narrows was definitely more slender than any of them, and thus Condron could not see it as just a routine application of bridge-building experience. Even the Golden Gate Bridge, then the longest suspension span in the world, was not nearly so slender as the Tacoma Narrows design, as Condron's tabulation showed:

Bridge	Span (ft.)	Width (ft.)	Ratio
Delaware River	1,750	89	1:19.7
Ambassador	1,850	59.5	1:31.1
[Bronx-]Whitestone	2,300	74	1:31
San Francisco Bay	2,300	66	1:35
George Washington	3,500	106	1:33
Golden Gate	4,200	90	1:46.7
Tacoma Narrows	2,800	39	1:72

Advisory engineer Condron may well have known of the surprising flexibility of the Golden Gate Bridge, and he had heard that "certain tests had been made on models of suspension bridge spans" at the University of California. When Condron could find no published reports on those tests, he went to Berkeley to confer with Professor R. E. Davis about concerns over the deflection of the very slender Tacoma design, whose construction loan was awaiting approval. Condron reported that Davis "felt reasonably confident that the lateral deflections of the Tacoma Narrows Bridge as designed and determined by Mr. Moisseiff would be in no way objectionable to users of the bridge." As if to document as best he could the authority of Moisseiff and the deflection theory, Condron quoted from a 1933 report on the accuracy of calculation that the theory permitted: "Moisseiff and Lienhard have presented a method which is closely accurate for determining lateral deflections of truss and cable stresses in the truss due to lateral forces." Whereas Condron had gone to Berkeley with questions about vertical as well as lateral deflections, he appears to have been reassured only about the latter, however, and he seems to have gone to lengths in his report to make that point clear. The problems with the bridge would not, of course, be with the lateral deflections.

Condron continued to have doubts about the design, and even a letter from Moisseiff to him could not put them to rest. When Moisseiff wrote that, considering the slenderness of the bridge, its stiffness was "rather satisfactory," Condron pointed out that "there seems to be some question even in his mind as to whether the obtained stiffness is other than *rather satisfactory*." In the end, however, the consulting engineer to the Reconstruction Finance Corporation acceded to authority and expertise:

> In view of Mr. Moisseiff's recognized ability and reputation, and the many expressions of approval and comment of his methods of analyses of stresses and deflections in the designs of long span suspension bridges, particularly as expressed by the engineers who participated in the discussion of the paper presented before the American Society of Civil Engineers by Messrs. Moisseiff and Lienhard entitled "Expansion [*sic*] Bridges under the Action of Lateral Force," I feel we may rely upon his own determination of stresses and deflections.

The Freudian slipping of "expansion" for "suspension" into the title of the paper by Moisseiff and Lienhard may have indicated Condron's fundamental unwillingness to concede that the Tacoma Narrows Bridge was stiff enough. Nevertheless, the weight of evidence presented by experts in the discussion of the key theoretical paper was too much for the lone advisory engineer to refute. In that discussion, the University of California models were repeatedly referred to, and Dean Charles Derleth found that their confirmation of the theory was "gratifying." He pointed out that the paper of Moisseiff and Lienhard "had its inception in the early debates on the Golden Gate design," with the authors "seeking a convincing argument to justify shallow stiffening trusses and slender wind-bracing for a 4000-ft. span." Though "engineers of considerable accomplishment" had argued that deck widths approaching two hundred feet might be necessary, Moisseiff and Lienhard's method of analysis had justified a ninety-foot roadway and was considered sufficient to "silence all arguments for unnecessary floor widths." Derleth also reached beyond mathematical analysis, to a "poetic rather than a mechanical figure of speech," to emphasize how important the cables were relative to the deck of very long suspension bridges, saying he liked to "describe the theory of Messrs. Moisseiff and Lienhard as assigning to the floor system the nature of a kite in the wind, with the cable and suspenders acting as the restraining strings and tail." Few but the likes of Condron seem to have worried more about the kite than the strings of the "different species in a genus of suspension bridges" that had been evolving toward the Tacoma Narrows Bridge in the wake of Moisseiff and Lienhard's theory.

For all the expertise that was assembled against him and to which he felt obligated to defer, however, Condron could not bring himself to unqualified approval in the conclusion to his report:

> With regard to the super-structure, I do not pretend to be qualified to analyze and check the design of the long span suspension bridge, but I have studied this design in connection with the designs of other bridges, which have been successfully erected, and are in successful operation. I also have great confidence in the ability and integrity of the Consulting Engineers under whose direction the computations and design drawings for the super-structure have been made. Moreover, these engineers have earned a very enviable reputation as experts in this field, as evidenced by the commendation from other suspension bridge experts which they have received in technical publications.
>
> I therefore, [*sic*] feel that with the exception of the unusual narrowness of this bridge with reference to its span length, the super-structure design is technically sound. It is probably technically sound notwithstanding its narrowness, but there are several reasons why it would be of material advantage if the bridge could be widened at a reasonable increase in the cost, and therefore, I recommend that serious consideration be given to the possible increase in the width of this structure, before the contract is let or work begun.

To Condron, the extreme narrowness of the deck of the Tacoma Narrows design forced him to conclude that "it would be advisable to widen the super-structure to 52 ft.," which would give the bridge a width-to-span ratio of 1:53.8—still very narrow, but a less radical departure from experience. Had Condron's recommendation been followed, it is very possible that the Tacoma Narrows Bridge would have been stiffened enough that, even had it exhibited some degree of flexibility in the wind, that might have been within tolerable limits and thus subsequently correctable, as it was to be in other contemporary bridges. Even if the course of suspension-bridge development had gone that way, however, this is not to say that some subsequent slender-bridge design would not have been proposed and approved without the reservations of so conscientious and perceptive an advisory engineer as Theodore Condron.

Condron could not have made his case more rationally or emphatically, unless perhaps he had appealed to the experience earlier that year of Russell Cone, resident engineer of the Golden Gate Bridge, who had observed not only horizontal but also vertical deflections of that span. According to Cone, during a windstorm on February 9, 1938, "the Bridge was undulating vertically in a wavelike motion of considerable magnitude." He went back to his office to get his camera and record the motion that "appeared to be a running wave similar to that made by cracking a whip," but when he returned that motion had stopped, and soon the wind died down. Neither Condron nor the board of consulting engineers, however, seems to have been aware of or excessively concerned about the behavior of the Golden Gate Bridge at the time the Tacoma Narrows was being designed. In any event, Condron's warning about the width of the Tacoma Narrows Bridge was not heeded, and the report of the consulting engineers prevailed:

> It might seem to those who are not experienced in suspension bridge design that the proposed 2800-foot span with a distance between stiffening trusses [girders] of 39′ and a corresponding width of [*sic*] span ratio of 72, being without precedent, is somewhat excessive. In our opinion this feature of the design should give no concern.

The board emphasized its conclusion by asserting that it believed the span could even be "materially increased if it were necessary, keeping the same width without any detrimental effect." With such an endorsement, the Toll Bridge Authority received a loan for about $3 million and a grant of a like amount from Pierce County. Construction bids were received by October 1938, and the bridge was completed less than two years later.

The Tacoma Narrows Bridge executing its fatal oscillations in November 1940

Even before the bridge was completed, however, engineers were surprised by its large movements; these were being studied on a model at the University of Washington, by Professor F. B. Farquharson, when a new twist developed in November. Until that time, the bridge deck had moved up and down in waves, and various checking cables and devices had been applied to it, as they had to Ammann's Bronx-Whitestone and David Steinman's Deer Isle bridges. However, on November 7, 1940, the clamps holding one of the checking cables at center span slipped, and the bridge began to move in a new way, twisting about its centerline in a wind of about forty miles per hour. The motion became so severe that the bridge was closed to traffic, and Farquharson went to see what was happening.

Camera equipment from a nearby shop was taken to the bridge, and so its twisting through a total of nearly ninety degrees was caught on the most famous film footage in structural-engineering history. A reporter's car was the only vehicle on the bridge, abandoned when it could not be controlled, and only Farquharson, the reporter, and his dog felt the full heaving of the steel-and-concrete deck. An attempt to get the dog out of the car was also abandoned in the increasingly violent motion, and the reporter and Farquharson were captured on film crawling, staggering, and climbing back toward the bridge tower and terra firma. Farquharson, apparently knowing more about structural vibrations than the reporter, walked along the centerline of the bridge, which as a nodal line was almost motionless, while the reporter fought along the heaving curbline. Not long after they reached safety, the bridge deck twisted itself apart and fell into the water. The motion had been so violent that the massive steel towers were permanently bent out of shape and had to be dismantled before a replacement bridge could be built—with very deep trusswork providing not a terribly slender profile but a very stiff deck.

The collapse of the Tacoma Narrows Bridge revealed a classic case of hubris, for the success of bridges like the George Washington and its close antecedents and descendants had given the coterie of major suspension-bridge engineers almost unlimited confidence and license in their designs, even as these were beginning to sway and wave in the wind. Because the new breed of engineers believed they were calculating, with the deflection theory, stresses and strains more accurately than nineteenth-century engineers like Telford and Roebling, their classic works were conveniently taken as aesthetic rather than structural models. The new field of aerodynamics, which was being applied to the development of the airplane in the 1930s, was seen to be largely irrelevant to designing and analyzing generally static structures like bridges.

There was, however, at least one civil engineer in the mid-1930s who "felt an obligation to make available to the civil engineering profession" the results of tests and theoretical studies being carried out by aeronautical engineers at that time. W. Watters Pagon knew, for example, that the principle of the wind tunnel was valid, because a powered structure flying through the quiescent air is equivalent to wind blowing over a stationary body, and in 1934 and 1935 he had published a series of eight articles on aerodynamics in *Engineering News-Record*, in which he discussed wind forces and their action on structures. The first article, entitled "What Aerodynamics Can Teach the Civil Engineer," opened with a recitation of how much was unknown about how structures behaved in the wind, including why a building had twisted in the recent Miami hurricane, but the whole series seems largely to have been ignored by the bridge builders. Only after the collapse of the Tacoma Narrows Bridge were Pagon's articles described as "must reading." Such a turnabout was prompted in no small measure by a letter that appeared in *Engineering News-Record* shortly after the bridge collapse. The letter, from Theodore von Kármán, director of the Daniel Guggenheim Aeronautical Laboratory at the California Institute of Technology, presented a very concise and convincing mathematical analogy between the twisting of an airplane wing and that of a bridge deck in the wind.

Although half a century later von Kármán would be identified on a commemorative U.S. postage stamp as an aerospace scientist, no doubt in part for his efforts to advance rocketry from "an eccentric study into a reputable discipline," his training and background were in engineering. Von Kármán was born in Hungary in 1881, and received a mechanical-engineering degree from the Budapest Royal Technical University in 1902. After a year of military service, he returned to Budapest to teach for a while, but left before long to take a position as a mechanical engineer with a machinery manufacturer. Two years later, he went to Berlin to study mechanics at the University of Göttingen, from which he received his Ph.D. in 1908. He became prominent in Europe in the newly established field of aeronautics, and in the late 1920s divided his time between the University of Aachen, in Germany, and Caltech, in Pasadena. In 1930, he accepted the position of head of the Guggenheim Laboratory and moved permanently to the United States, where he came to lead the country's first jet-propulsion and rocket-motor program. Von Kármán was in the process of establishing a model supersonic wind tunnel at Caltech when the Tacoma Narrows Bridge collapsed.

Von Kármán was one of three engineers appointed by the Federal Works Agency to investigate the failure of the Tacoma Narrows Bridge. He was joined by Glenn B. Woodruff, the consulting engineer from San Francisco who had been the engineer of design for the San Francisco–Oakland Bay Bridge, and, not surprisingly, Ammann, who had, of course, dominated suspension-bridge design and had investigated the failure of the Quebec cantilever bridge at the beginning of his career. The committee's report, issued less than five months after the collapse, concluded that "the Tacoma Narrows Bridge was well designed and built to resist safely all static forces, including wind, usually considered in the design of similar structures." In other words, the constant sideways push of the wind had been taken into account in the standard way for engineers of the time, according to what is known as the state of the art, and there was no blame to be placed on them. They were simply taken by surprise by the structure's "excessive oscillations," made possible by the "extraordinary degree of flexibility." Ignorance, and not incompetence, was to blame: "It was not realized that the aerodynamic forces which had proven disastrous in the past to much lighter and shorter flexible suspension bridges would affect a structure of such magnitude as the Tacoma Narrows Bridge."

That the report read as it did should perhaps not have been surprising, given the composition of the board of engineers, but their relationship and conclusions must have evolved over the months they worked together. Von Kármán was somewhat of a maverick, a confirmed bachelor who seemed as likely to be found posing with a buxom blonde or a world leader as with a wind tunnel, if his bombastic autobiography published a quarter-century later is a fair representation of the man. In the book, written "with" a free-lance writer in the manner of a celebrity, von Kármán related how he followed the news reports of the Tacoma Narrows collapse, only to be startled by a news item the following day reporting that the governor of Washington had announced that "the bridge was built correctly and that a new one would be built according to the same basic design." That evening, the engineer "took home from Cal Tech a small rubber model of the bridge" that one of his mechanics had made for him and demonstrated in his living room with an electric fan and the model an "instability which grew greater when the oscillation coincided with the rhythm of the air movement from the fan." As he had suspected, "the villain was the Kármán vortices," or the whirlpools of air, named after the investigator himself, that were shed in the wake behind the moving model and thus buffeted it. Von Kármán wrote to the governor, to Farquharson, and to *Engineering News-Record* about his discoveries and concerns, an initiative that could not have hindered his being placed on the investigatory board.

In von Kármán's recollection of board meetings during the investigation, he mentioned that he was surprised at the "long standing of the prejudices of the bridge engineers," as embodied in their consideration of static as opposed to dynamic forces and their difficulty in seeing how "a science applied to a small unstable thing like an airplane wing could also be applied to a huge, solid, nonflying structure like a bridge." This all led to "some definite undercurrents of rivalry"; Ammann was portrayed as especially reluctant to accept such suggestions as wind-tunnel testing of bridge designs.

In the final analysis, von Kármán may have thought it best to let the bridge engineers worry about bridges, for which they were paid. He admitted that they had won him over on one "difference in thinking" between them and him. Though he was prepared to serve for his standard government consulting fee of fifty dollars a day, the other engineers "bargained for a sizeable percentage of the value of the bridge, which after all was insured for six million dollars." This indicated to him that aeronautical engineers acted in consulting positions as if they were only "elevated laborers," and he "learned a good deal in a nonengineering way from this experience" at the bargaining table from the less flamboyant if not outwardly shy bridge engineers.

Ammann's thoughts during the investigation could not have been very far from his George Washington and Bronx-Whitestone bridges, to whose design Moisseiff had contributed so much. Woodruff, who had also been associated with Moisseiff in conjunction with the design of the San Francisco–Oakland Bay Bridge, must not have been predisposed to believe that the engineering of the Tacoma Narrows was faulty. In fact, five years earlier, as part of an issue of *Civil Engineering* focusing on that project, Woodruff had written a short article, "From the Viewpoint of the Bridge Designer," in which he spelled out the advantages bridge designers then had over their predecessors. However, in spite of the "more complete theory, an immense amount of experimental data, and more reliable materials, as well as the accumulated experience of past years," he warned that there was the danger that bridge design would come to be considered routine.

At the same time, Woodruff wrote, as if anticipating von Kármán's amazement and frustrations at the meetings that would occur in December 1940 in Seattle, bridge building was becoming so highly specialized that there was the "danger of losing contact with the other branches of engineering and with allied sciences." This all would seem to have predisposed Woodruff to be more of a finger-pointer when it came to blaming engineers, but he did not do so. Rather, he closed his article with a quote from another engineer: "The most perfect system of rules to insure success must be interpreted upon the broad grounds of professional intelligence and common sense." That these were the words of none other than Theodore Cooper, whose own intelligence and common sense were seriously called into question in the wake of the collapse of his Quebec Bridge in 1907, suggests that Woodruff was not one to hold engineers culpable for not foreseeing problems of an extraordinary kind.

Though Ammann and Woodruff believed that intelligence and common sense required the designer to "analyze all the assumptions made, estimate the possible errors in them, and also make a careful study of the properties of materials to be employed," they also appear to have believed that doing this to the best of the designer's ability satisfied his obligation and freed him of any guilt. Doing all one knew to do was, after all, the best that could be expected of an engineer of bridges or of rockets. It is very likely that, before the final report was drafted by Ammann, the rocket scientist von Kármán came to see this point of view of the bridge engineers.

If the engineer Moisseiff, along with the profession of engineering, was clearly exonerated by his colleagues, the precise physical causes of the motion of the bridge and the final disaster were left somewhat ambiguous and vague by the failure report. Vertical oscillations of the bridge were "probably induced by the turbulent character of wind action," but there was "no convincing evidence" that such vertical motions were unstable or even dangerous to the bridge. It was "reasonably certain" that a cable band that had been installed to check some of the deck motion had slipped, and this "probably initiated the torsional oscillations," which brought the span down. Among the more general conclusions of the report were, not surprisingly, that "further experiments and analytical studies are desirable to investigate the action of aerodynamic forces on suspension bridges." The report also concluded, however, that, "pending the results of further investigations, there is no doubt that sufficient knowledge and experience exists to permit the safe design of a suspension bridge of any practicable span," without reference to how wide such a span might be. Such a conclusion might have been subject to ridicule in less turbulent times.

The onset of World War II would no doubt have interrupted bridge building much the way World War I did even if the Tacoma Narrows collapse had not occurred. In any case, there was less urgency to following up on the report than there might have been. Many unanswered questions about the aerodynamic behavior of bridges remained, however, and it fell largely to Professor Farquharson, in the Structural Research Laboratory of the Department of Civil Engineering at the University of Washington, to continue throughout the 1940s to work on and pull together the results of laboratory and mathematical studies on the stability of suspension bridges in the wind conducted principally at his institution and at Caltech. Farquharson's work was sponsored mainly by the Washington Toll Bridge Authority, which needed to replace the bridge that had collapsed, in cooperation with the Public Roads Administration of the Federal Works Agency. The Authority's consulting board included Woodruff and von Kármán, who was identified as "aerodynamicist" and who would on occasion identify himself as "representing the wind." Ammann, as a consulting engineer to the Port of New York Authority, represented that body on the Advisory Board on the Investigation of Suspension Bridges.

Whereas Ammann's selective use of history had been symptomatic of the myopia that characterized suspension-bridge building in the 1930s, Farquharson's report opened with a broad, inclusive historical survey of the dynamic behavior of suspension bridges. Farquharson began this survey by noting that the collapse of the Tacoma Narrows Bridge "came as such a shock to the engineering profession that it is surprising to most to learn that failure under the action of wind was not without precedent in the history of suspension bridges." He then proceeded to describe trends and recount the main events of that history, which he summarized in a table that listed "bridges severely damaged or destroyed by wind" between 1818 and 1889, plus the 1940 Tacoma Narrows disaster. After that last collapse, "much old information long forgotten was once again made available to the profession."

Among the first to bring such information to the fore was J. Kip Finch, professor of civil engineering at Columbia, whose article "Wind Failures of Suspension Bridges, or, Evolution and Decay of the Stiffening Truss," appeared in *Engineering News-Record* about four months after the collapse. His article concluded with a section headed "The lesson is plain," and the lesson was: "History, in short, has been repeating itself, although this fact has, apparently, not been known even to engineers who have made a specialty of this type of construction." After this indictment of ahistorical engineers, Finch concluded optimistically, but perhaps without complete conviction, "This time the problem of preventing undulations in suspension bridges will undoubtedly be solved." Two weeks later, in a letter to *Engineering News-Record*, no doubt prompted by letters to it, Finch took pains to assure readers that he had not meant to infer that "the modern bridge engineer . . . was remiss in not anticipating" what had happened. He argued that it was "unbelievable" that the eight-thousand-ton center span of the Tacoma Narrows Bridge could be lifted by the wind as easily as the 460-ton deck of the Wheeling Suspension Bridge, which had been destroyed in 1854, or that the thirteen-thousand-ton deck of the Bronx-Whitestone or the fifty-six-thousand-ton deck of the George Washington could be compared to the lighter fabrics of old.

Finch's own interpretation of history was proved to be a bit questionable when he added that no engineer, so far as he knew, had "recognized in the twistings of some of these earlier failures, a characteristic aerodynamic phenomenon." Though John Roebling might not have used such terms, he had written as early as 1841 of the problems of bridges in the wind. Indeed, before that, the Scottish engineer J. Scott Russell had written, in the wake of the 1836 collapse of the Brighton Chain Pier (which was in fact a multispan suspension bridge out to sea), about how the wind can set structures like bridge decks into oscillation as surely as a bow does a violin string. An engineer's knowledge and use of history was a touchy point after the Tacoma Narrows collapse, however, and Finch concluded that "it is asking too much of the human mind to suggest that the engineer should have anticipated the Tacoma failure." He then articulated what many other engineers were saying and would continue to say about the issue that had been thrust upon them:

> It is also a mistake to assume that failures should never occur. The engineer cannot wait until he knows "all" about a device before he builds it. In general theory follows practice—the theory of heat followed the first use of the steam engine, truss analysis came after the truss, etc.—and there has, as yet, been no substitute developed for experiment—even in science. A history of bridge building which contained no records of failures would be a history almost devoid of progress. Man must ever struggle to bring into being the children of his imagination, for through such creation progress is possible. It is thus inevitable that, in daring to do bigger and better things, there will always be some failures. Failures, in fact, are a sure sign of progress. While a failure is always materially wasteful, it

> is always a stimulant to increased knowledge. We may rest assured that the engineer will not make the same mistake twice.

Argue as Finch and others might, engineers had indeed made the same mistake twice, but Finch's reasoning, questionable though it was, would serve to console engineers who could not bear to be wrong. They could get on with picking up the pieces, turning them around in their hands and their minds, and going on to the next project with more experience and judgment. As for Moisseiff, who had to deal with the collapse of the Tacoma Narrows more directly, engineers by and large must have thought, "There but for the grace of God go I." Moisseiff continued to work on engineering projects, including plans for the reconstruction of the Brooklyn Bridge and "assisting in the solution of the problem forced upon the profession by the Tacoma Bridge failure," but his heart may not have been in them—or may have been in them too much.

Moisseiff died of a heart attack less than three years after the disaster, and neither his obituary in *Engineering News-Record* nor the unusual number of letters to the editor paying tribute to his memory gave much more than passing mention to the Tacoma Narrows incident. Only a letter from Ammann, who had not merely relied on Moisseiff but also derived a measure of his own reputation for success from the work of the late engineer, even dared address the matter. "The one great disappointment in Mr. Moisseiff's career," Ammann wrote, "was the failure of the Tacoma Narrows Bridge, the design of which he had originated and guided." Yet, Ammann continued, "it would be improper for his fellow professionals to put the blame for that failure entirely upon Mr. Moisseiff's shoulders," for "he followed a trend in long span suspension bridge design which appeared justified" at the time. Ammann may well have been speaking of himself. He returned to this theme in the memoir of Moisseiff, which, written with his associate Frederick Lienhard, appeared in the *Transactions of the American Society of Civil Engineers* in 1946. Moisseiff was called "one of the best informed of engineers," and his activities at the end of his career were described as those of "a consultant to consulting and executive engineers." Ammann himself was a consummate example of these latter two categories—no engineer could have accomplished single-handedly what he did in his career.

12

Ammann left the Triborough Bridge Authority in 1939 to go into private practice. In that capacity, he worked on a variety of bridge and other projects, including studies for a suspension bridge across the York River in Virginia and one across the Delaware River at Wilmington, in addition to participating in the Tacoma Narrows investigation. In 1946, when he was already past what in those days was generally considered retirement age, Ammann entered into partnership with Charles S. Whitney, who as an engineering student at Cornell had worked under Ammann during vacation periods, and who had gone on to establish himself in Milwaukee as a specialist in reinforced-concrete structures. Together, they formed the firm of Ammann & Whitney. Since few bridges were being built at the time, the engineering firm worked on projects involving large airfield hangars, long-span buildings, and highways.

After the war, bridge-planning activity picked up in New York and elsewhere. In 1955, a joint report of the Port of New York Authority and the Triborough Bridge and Tunnel Authority was prompted by "the unprecedented increase in the ownership and use of automobiles, trucks and buses since the end of World War II," which had "forced accelerated nationwide planning and construction of our arterial highway system." The main results of the joint study were to recommend for construction: (1) a lower deck on the George Washington Bridge; (2) a suspension bridge between Brooklyn and Staten Island across the water known as the Narrows; and (3) a suspension bridge between Throgs Neck in the Bronx and Little Bay in Queens, across the water variously known as the East River and Long Island Sound. The consulting engineer for these projects, as well as for studies for another Hudson River Bridge, at 125th Street, which was not recommended at the time, was the firm of Ammann & Whitney. Thus Ammann, the partner whose expertise was suspension bridges, was once again to be involved in the kind of work he dreamed about.

The second deck of the George Washington was, of course, part of Ammann's original design, and he not only would live to see it realized but would direct work on it himself. The New York Narrows bridge was a project that had been developed in the offices of the Port Authority two decades earlier. In his autobiography, the engineer Clarence Whiting Dunham recalled being asked by Ammann in the summer of 1936 to drop his work on the Lincoln Tunnel, then under construction, "to help him with a special project," which "was to be kept secret." This involved "an intensive preliminary study of a suspension bridge across the Narrows." Working directly and intensively for six weeks with Ammann, Dunham made drawings of the elevation of the bridge and sections showing its proposed construction, along with approaches relative to existing streets. According to Dunham, though city officials liked the plan, federal authorities rejected it because the destruction of the bridge during wartime might block access to the

Brooklyn Navy Yard. The plans were then shelved, only to be dusted off when, in the post–World War II years, air power had diminished the importance of the navy yard. The final bridge design and location were to be remarkably close to those proposed in 1936, but in the meantime another New York project would also occupy Ammann.

The Throgs Neck Bridge was constructed within sight of the Bronx-Whitestone, which in the wake of the Tacoma Narrows collapse had been slated for the addition of a stiffening truss. Materials shortages during the war delayed that modification work until 1946, at which time Ammann described the retrofitting in an article in *Civil Engineering*. "While the truss members will undoubtedly detract somewhat from the extreme simplicity of the original design, with its plain shallow girders, they will not be sufficiently conspicuous to mar the graceful appearance," he wrote, perhaps with not a little disingenuousness. In part for ease of construction, this truss was to be added to the top of the deck, which it was now admitted had "inadequate vertical stiffness," but the resulting superstructure obstructs a dramatic view of the Manhattan skyline from the bridge's roadway. In the process of stiffening, the traffic capacity of the bridge was also increased, by eliminating the pedestrian walks on either side of the roadway, thus reducing the possibility that people would feel how flexible the bridge did in fact remain.

The lines of the eighteen-hundred-foot main span of the Throgs Neck Bridge, to the east, were not to be so sleek as those of the original Bronx-Whitestone. Instead of plate girders, more conventional open trusswork was used to stiffen the wide deck, and the towers were to be rather squat-looking. But this bridge project did not bring Ammann renewed public attention in the early 1960s; that was to come with the opening of the lower deck of the George Washington Bridge.

Ammann's original concept for the George Washington was, of course, that it would eventually have light rapid-transit railroad tracks on a lower deck, but travel habits and traffic patterns had changed considerably since the bridge, with its single vehicular deck, was opened in 1931. By the mid-1950s, motor-vehicle registration in the region had more than doubled, to three and a half million, and annual crossings of the Hudson River through the Holland and Lincoln tunnels and over the George Washington Bridge had about quintupled. Almost thirty-five million vehicles were using the bridge alone on an annual basis. Thus, when the second deck of the George Washington Bridge opened in August 1962 for the exclusive use of motor vehicles, it was hailed as a "masterpiece of traffic relief."

The ceremonies marking the opening of the lower deck were attended by politicians from both sides of the river, and the formalities were highlighted

Othmar Ammann at the dedication of his bust at the George Washington Bridge, shaking hands with Governors Richard Hughes of New Jersey and Nelson Rockefeller of New York, with the completed lower deck of the bridge visible in the enlarged photograph in the background

by the unveiling of a bust of Ammann by Governors Nelson Rockefeller of New York and Richard Hughes of New Jersey. But Ammann, who was remembered by his daughter to be at that time a "small man," only five feet six inches tall, and a "sandy-haired, slightly frail octogenarian—then 83—who stood as majestically as the giant structures that he had fathered," was not conspicuously present at the ceremonies. According to the newspaper account, Ammann was not standing with the politicians, and "it took a few minutes to locate the designer who was sitting back in the crowd, to get on with the unveiling."

The bust of Ammann is now on display in the bus terminal at the foot of the bridge on the Manhattan side, but it is scarcely noticed by the travelers and commuters who pass it each day, and it is never seen from the cars of those who drive back and forth across the bridge. The inscription on the bust reads simply, "O. H. Ammann / Designer / George Washington Bridge." It is ironic that there is no mention on the pedestal of Ammann's being an engineer, but perhaps that was his choice. According to his son, Werner Ammann, an engineer himself and then a member of Ammann & Whitney, the shy old man agreed to the public honor only because it would "reflect favorably on the entire engineering profession." For Ammann to be identified as *the* engineer of the George Washington Bridge may have seemed arrogant; for him to be designated its designer, conceiver, and dreamer was just stating the obvious.

When the *New York Times* editorialized on "Mr. Ammann's Work of Art," it acknowledged his insistence that "no one man designed the bridge," yet went on to admit the public reality: "We shall think of Mr. Ammann, however, every time we look at the George Washington Bridge. He was the dreamer, he was the artist, he was the solid and reliable planner who made this beautiful structure possible and durable." Yet, as a letter to the editor several days later pointed out, nowhere in the editorial was Ammann identified as what he really was, "one of America's outstanding engineers."

Othmar Ammann was said to insist that "anyone who would take exclusive credit for bridge design" was "an egotist." It would certainly be even more egotistical to declare oneself *the* engineer of such a structure, especially to the deliberate exclusion of assistant and design engineers who had done much of the work. Joseph Strauss, without question the driving force behind the Golden Gate Bridge, had done just that to his assistant Charles Ellis, of course, and the statue of chief engineer Strauss that was installed at the bridge plaza must have irked many a subordinate. Though first authorized by the Golden Gate Bridge and Highway District late in 1939, a public expenditure for a bronze statue was successfully challenged by the Taxpayers Defense League, and, like many such a monument, this one was dedicated, in 1941, only after funds were provided by Strauss's widow.

The Strauss statue was originally located conspicuously at the toll plaza, atop an ostentatious pedestal, obstructing the view of the bridge itself, according to some. It was subsequently relocated to a less ornate base and less prominent setting, in a small plaza between the bridge's gift shop and parking lot. In New York, the perhaps overly modest pedestal and bust of Ammann, and its imprecise inscription, soon became so obscure in its location in an urban bus terminal, and forgotten by public and profession alike, that the issue of ego seems long ago to have become moot.

13

The final great bridge design that Ammann was to be credited with was the one he had had Clarence Dunham work on secretly in the summer of 1936. A bridge crossing the Narrows between Brooklyn and Staten Island was proposed as early as 1910 by the New York engineer Charles Worthington. His design consisted of a twenty-five-hundred-foot arch made up of hollow voussoirs of nickel steel that would be erected by a novel method devised by Worthington so that no falsework would obstruct the entrance to the harbor during construction. The proposed arch was to be 260 feet above high water, subject to the approval of the War Department, and the $15-million bridge was expected not only to provide a monumental gateway to New York from Staten Island but also to open up the island to "mercantile development."

The arch plan was shelved, and so were many later suspension designs, but by the time Ammann's bust was unveiled at the George Washington Bridge, his Narrows bridge had long been under construction. Work had begun in earnest in the mid-1950s, and Milton Brumer, its chief engineer, recalled that 125 engineers in the Ammann & Whitney design office were assigned exclusively to the project, and there were another seventy-five field engineers, not to mention the thousands of construction workers: "Every one of them has a good reason to say 'I played a part in building that bridge.' There's honor in a project like that, and it should be shared." Brumer did indeed play a part in the Verrazano-Narrows Bridge, as he had in many other of Ammann's designs.

Milton Brumer, born in Philadelphia, was a 1923 graduate of Rensselaer Polytechnic Institute and a classmate of Werner Ammann. Beginning as a junior engineer with New York's Interborough Rapid Transit Company, Brumer subsequently held various engineering positions, including assistant engineer for the Port of New York Authority on such projects as the Outerbridge Crossing, Goethals Bridge, George Washington Bridge, Bayonne Bridge, and the Lincoln Tunnel. He joined Ammann's firm in 1944 and became a partner in Ammann & Whitney in 1949. Having served as chief engineer for the Throgs Neck Bridge, he no doubt paid more day-to-day attention to that project than did Ammann. Of all the engineers at Ammann & Whitney, Brumer was most closely associated with Ammann on bridge and highway projects, "and it was on Milton Brumer that O. H. Ammann placed great reliance for the final execution" of the Narrows bridge project. Charles Whitney died before it was completed, and Brumer then became executive head of Ammann & Whitney, which had grown to have eight partners and a staff of about five hundred.

Though Ammann may not have been constantly bent over a drawing board, bridges were never out of his sight or very far from his mind. At Ammann & Whitney daily, he could be found in "his simply styled office surrounded by renderings or photographs of some of his bridges." A drawing of the Brooklyn Bridge was the only exception, "his way of paying tribute" to

John Roebling as pioneer suspension-bridge builder. Ammann, who lived in New Jersey, also kept an apartment in Manhattan, on the thirty-second floor of the Carlyle, a hotel located at Madison Avenue and East 77th Street. From this position at the more or less geographical center of the island, he had a view of all the New York bridges that had defined his career and reputation. Confined to his apartment with a cold on the day before his eighty-fifth birthday, on March 26, 1964, he was able, with the aid of a telescope, to look at the 680-foot-tall Brooklyn tower of the bridge across the Narrows, twelve miles away and still under construction. From the bedroom window was visible his favorite bridge, and the one he considered his "greatest achievement," the George Washington. From another window, he could see the Hell Gate, Triborough, Bronx-Whitestone, and Throgs Neck bridges. Finally, his living-room window afforded a partial view of the Bayonne and the Verrazano-Narrows bridges.

Ammann's bridge views, described in the newspaper article occasioned by his birthday, are reminiscent of those from the Brooklyn flat where a bedridden Washington Roebling watched the completion of the Brooklyn Bridge of which his late father had dreamed. John Roebling, of course, had had the total misfortune of having his foot crushed by a ferryboat while he was laying out the alignment of his bridge and of contracting tetanus. Other bridge engineers had had the bad luck to see the crowning achievements of their careers collapse, as did Cooper and Moisseiff, near Quebec and Seattle, respectively. The reporter recounted that Ammann had worked on the investigations of each of those colossal bridge failures and that Ammann's bridges had "known no tragedy through his own engineering miscalculations." Ammann conceded that he was "lucky."

Luck or no, Ammann was hailed as "the most respected engineer of his time." A further opportunity to lionize him would be provided at the opening ceremonies of the bridge across the Narrows, in November 1964. First, however, the bridge had to undergo the initiation rite of having its chosen name challenged. The name Verrazano-Narrows Bridge had been decided upon early enough for a commemorative stamp to be issued by the U.S. Post Office. The design of the five-cent stamp was unveiled "at a thinly disguised Democratic campaign rally on the steps of Brooklyn's Borough Hall," four weeks before it would be issued, which was to be on the day of the bridge opening. The crowd heard President Lyndon Johnson "praised by virtually every speaker," including one who spoke of his personal recognition of the "tremendous role which has been played in our national life by Americans of Italian heritage," and there was also the first public performance of "the Verrazano Bridge song," which began, "In 1524, he opened up the door; that Verrazano man, whose name is on the span." Italy also issued a stamp commemorating the opening of the bridge, but with the spelling "Verrazano." There was opposition to the name up to the very end, with ridicule for the honor thus paid to "a brave vagrant who is believed to have poked the nose of his ship through the Narrows."

The name Verrazano-Narrows Bridge would nonetheless remain, although the hyphen symbolizing the tension it had generated would often later be dropped or forgotten. But there would be plenty of comparisons with the Golden Gate, whose forty-two-hundred-foot main span was now surpassed by the 4,260-foot New York bridge. Other "statistical details" of the San Francisco landmark were also bettered by the Verrazano-Narrows, which in the wake of the Tacoma Narrows collapse included the newer span's support of a 75-percent-greater load. The aesthetic of light and slender had been replaced with one of strong and solid, and the ever-popular numbers whose publication accompanied the completion of great structures stressed that the Verrazano-Narrows Bridge was so large that the tops of its towers were more than an inch farther apart than the bottoms, merely because of the curvature of the earth. Another oft-repeated statistic was that the main span, which on average was 230 feet above the water, would be twelve feet lower in summer than in winter, when the lower temperatures caused the steel to contract.

One chronicler of bridges has written that "the success of an engineering project may often be measured by the absence of any dramatic history," but what may appear to be undramatic from one perspective can be very traumatic from another. To build a bridge from Brooklyn to Staten Island across the Narrows that ferryboat services, including one begun by Cornelius Vanderbilt in 1810, had plied for centuries required an enormous amount of land for approaches. Robert Moses called the bridge "the most important link in the great highway system stretching from Boston to Washington, or, if you please, Maine to Florida." However, to close this link, especially on the Brooklyn side, meant disrupting long-established neighborhoods, and this was at least as difficult to accomplish as any engineering aspect of the problem.

Since the main span of the Verrazano-Narrows was only sixty feet greater than that of the Golden Gate, the engineering choices were certainly nowhere near so dramatic as those made thirty years earlier in designing the George Washington, which, of course, had roughly doubled the then longest span. In his "preface" to a book about the building of the Verrazano-Narrows, Ammann referred to the "engineering phase in the construction" of such a great bridge as "essentially the application of scientific and technological progress in many fields." Though he was not saying that engineering was merely applied science, he was stating that, in this case at least, rational experience was a sure guide. One bit of experience that Ammann insisted on applying to the Verrazano-Narrows was the construction of both upper and lower decks at the same time, even though there was no expectation of an immediate traffic demand for the lower deck, a decision no doubt made to eliminate any possibility that the bridge would not be stiff enough in the wind. Another supposedly impersonal engineering and economic decision was to design the Verrazano-Narrows Bridge without a pedestrian walkway, but one can speculate as to whether this may have been dictated by a social or psychological concern that pedestrians, more easily than drivers and passengers in vehicles, would sense the flexibility of the record span—or merely to eliminate the bother of people on the bridge. Whatever the case, this limitation is now overcome at least once a year, when the beginning leg of the annual New York City marathon is run across the span.

The opening ceremonies for the bridge were held on November 21, 1964. Music was provided by the Department of Sanitation band, and Robert Moses rode in the first of fifty-two black limousines that brought official

The Verrazano-Narrows Bridge, shortly after it was opened in 1964, with Brooklyn and Manhattan in the background

guests. In an editorial on the occasion, *The New York Times* spoke of the completion of the bridge as crowning "the careers of two men to whom New York already owes a colossal obligation," Ammann and Moses, and recalled the Triborough Bridge and Tunnel Authority chairman's "resolve to conquer the staggering obstacles" to the bridge's construction, which had "resulted in a masterpiece he rightly ranks second only to the works of Shakespeare in the durability of its beauty." The bridge that the engineer was said to have designed to "last forever" was not just a critical link in a traffic artery, however, and "the realization that all this grace is merely an instrument for the insensate rush of endless ribbons of cars, trucks and buses is too depressingly mundane to contemplate in this moment of magnificent birth." The *Times* had to resort to the poet Hart Crane's lines about the Brooklyn Bridge to close its paean:

> *Through the bound cable strands, the arching path*
> *Upward, veering with light, the flight of strings,—*
> *Taut miles of shuttling moonlight syncopate*
> *The whispered rush, telepathy of wires. . . .*

The newspaper's reporter, Gay Talese, had more of an ear and eye for the immediate than the editorial writer, however. Talese had written a book about the design and construction of the Verrazano-Narrows, and now he reported how the motorcade proceeded to the Brooklyn approach to the bridge, where five pairs of gold scissors awaited, respectively, Moses, Governor Rockefeller, Mayor Robert Wagner, and the borough presidents of Brooklyn and Staten Island. The ribbon-cutting was delayed somewhat while politicians fought through the crowd of "generals, admirals, politicians, women in mink coats, business leaders, pretty girls." Talese also noted Ammann's arrival, not in the first but in the eighteenth limousine: "A quiet and modest man, he was barely recognized by the politicians and other dignitaries at the ribbon-cutting ceremony. He stood in the crowd without saying a word, although occasionally, as inconspicuously as he could, he sneaked a look at the bridge looming in the distance, sharply outlined in the cloudless sky."

The motorcade resumed to carry the dignitaries to the other side of the bridge, where Moses was to be the master of ceremonies. When it came time to introduce the engineer, Moses said, "I now ask that one of the significant great men of our time—modest, unassuming and too often overlooked on such grandiose occasions—stand and be recognized." The engineer removed his hat and stood, and Moses continued: "It may be that in the midst of so many celebrities, you don't even know who he is. My friends, I ask that you now look upon the greatest living bridge engineer, perhaps the greatest of all time." Unfortunately, Moses never mentioned Ammann's name, and the engineer resumed his seat, "again lost in the second row of the grandstand."

Later that evening, when Ammann was home with his family, the phone rang and his wife answered. She turned to Ammann and announced, "It's Ed Sullivan. He wants you to appear on his TV program tonight." Ammann is reported to have said, "Tell him, 'No, thank you.' " After his wife hung up, the engineer asked, "Who is Ed Sullivan?" Whether Ammann actually knew who he was or not, the story serves to carry one step further the image of this engineer as quietly devoted to his job, oblivious to everything else in the world. But one could also interpret the story as one shy engineer's private retaliation for his anonymity at the ceremony dedicating the bridge that owed so much to him.

Among the onlookers at the dedication ceremonies for the Verrazano-Narrows Bridge was a college freshman named Donald Trump, who was attending the event with his father. Sixteen years later, after the younger Trump had established himself as a real-estate developer in his own right and been credited with "reshaping the skyline of Manhattan," he recalled to a reporter having had "a sudden realization, an epiphany," at the ceremony that "always remained with him, shaping the way he made his fortune in real estate in New York City." Though his recollection of the day, which Talese at the time described as cloudless, may have undergone a certain embellishment, Trump did remember correctly some of its more salient points:

> The rain was coming down for hours while all these jerks were being introduced and praised. But all I'm thinking about is that all these politicians who opposed the bridge are being applauded. Yet, in a corner, just standing there in the rain, is this man, this 85-year-old engineer who came from Sweden [*sic*] and designed this bridge, who poured his heart into it, and nobody even mentioned his name.

Trump's epiphany, "then and there," was "that if you let people treat you how they want, you'll be made a fool." The realization that he "would never forget" was that he "didn't want to be made anybody's sucker." Regardless of what he forgot or did not, Trump's recollection of the way Moses omitted Ammann's name at the Verrazano-Narrows ceremony is a poignant reminder of the fate of the engineer. The record shows Moses's speeches not to be terribly well crafted, and so it seems likely that his verbal slight of Ammann was unintended. When Moses said candidly that, among celebrities, the engineer was not likely to be known, he was only speaking the truth.

It has often been said that engineers get their satisfaction not from personal recognition but from the recognition of their works. Whether this is the collectively shared rationalization of often moody personalities who tend to be more comfortable engaged in problem solving than engaged by crowds, many an engineer appears to have subscribed to it. And Ammann seems to have been no exception. He accepted honors but seems to have sought out none other than, or at least none so deliberately as, the honor of being the engineer among engineers to lead the building of great bridges. In this he was not shy, and for this he would be remembered.

Othmar Ammann died in 1965, after a long and active life ranging from lonely work to shared glory. When Robert Moses spoke of him at the dedication of Othmar Ammann College at the State University of New York at Stony Brook in 1968, the engineer was recalled as a "dreamer in steel" who was more than an individual, a paragon among engineers. "The Ammanns represent not merely mathematics, materials and stresses and strains, but character which can't be mined, fabricated and molded, but has to be there from the beginning. You have it or you don't, and Othmar Ammann had it." Though not every one of the engineer's contemporaries may have agreed, Moses's praise was, in broad terms, very well deserved indeed. Ammann had dreams of great bridges, and he had the exceptional inclination and talent to realize those dreams in his distinctive style.

STEINMAN

Many children have grown up in the shadow of a bridge, especially in a city like New York. In the late nineteenth century, the Lower East Side of that city was crowded with small children and one large bridge—that leading to Brooklyn—and life among the ever-present but ever-changing shadows cast by its approaches, abutments, decks, and towers was hard and squalid. Escape via the automobile to the suburbs was for many not yet even a realistic dream, and one did what one could do with what one had.

The stone towers of the Brooklyn Bridge had risen ever so slowly during the first half of the 1870s, and its cables had been spun equally slowly during the second half of the decade. As the bridge deck was hung in the early 1880s—piece by piece, like laundry from a line—the shadows cast by the structure lengthened and thickened over the tenements of New York City. There was a bright day of celebration and a still brighter evening of fireworks when the bridge opened in May 1883, and the central promenade that John Roebling had so thoughtfully designed above the traffic provided a welcome escape route from the heat and closeness of the tenements, if only for the hour or so it took to walk to Brooklyn and back, perhaps stopping midway to look out at New York Harbor.

Perhaps some found the bridge or its shadows oppressive, but the great structure provided an alternative to the ferries that so many people had daily to take back and forth across the East River. Others discovered in the bridge a new prosperity, with the uniting of the formally separate cities of

The New York approach to the Brooklyn Bridge, near where David Steinman spent his childhood

New York and Brooklyn providing new opportunities for commercial growth and real-estate development. On the other hand, some residents of the Lower East Side may not have thought much about the bridge at all, especially if they were busy raising large families in small apartments, as so many of the immigrant factory-worker families in the neighborhood were. However, at least one child growing up in those apartments became obsessed with bridges of all kinds.

David Barnard Steinman was born on June 11, 1886, and his childhood might have seemed unremarkable even to himself had he not lived in the shadow of the Brooklyn Bridge, which was to him not cold but warm. He grew up with the dominating bridge, whose towers looked over the city day and night, and whose arms of traffic reached far into the city, bringing and taking people and goods in unprecedented volume and with unprecedented speed. Not only was the Brooklyn Bridge there and functioning as a great communicator between what young David knew and what he could discover, but as he grew so did another bridge—the Williamsburg—under construction a mere mile or so away. For him, the Brooklyn and Williamsburg bridges became almost surrogate parents.

Eve and Louis Steinman were the real parents of David and his six siblings, but his hagiolatrous biographer, William Ratigan, no doubt acceded to Steinman's wishes in keeping them as much out of his later life as possible. Ratigan's 350-page biography has young David recalling only that his immigrant parents were lonesome, that "his father lashed him with a cat-o'-nine tails for wearing out his shoe leather" exploring Manhattan, and that "his mother wept." Among Steinman's few recorded recollections of his mother was of her "softly weeping" for the "cottage and the fields, the streams and meadows, of her native land," which remained nameless. Neither Steinman's mother nor father appears in the index to the biography, and there are no pictures of them or of his nameless siblings, from whom he learned the alphabet and the numbers. His first alluring taste of school was at five years old, when he was taken by his older sister to her teacher and principal so that he could show them his prowess in mathematics: "He could rattle off the powers of two: 2, 4, 8, 16, 32, up to a million." He was tested with mental multiplication problems, like 17 times 19 and 27 times 43, and he was rewarded with candy and visits to teachers' homes, which "were a glimpse of another world." A boxed charlotte russe brought home from one of those visits was such a treasure that it was nursed for nearly three weeks, being kept fresh on the fire escape because the Steinman apartment had no ice box. There is a mythic quality to Steinman's childhood, to his finding solace mainly in the cradle of the Brooklyn Bridge's cables and stays, and in the promise and reward of education.

Talk about building a new bridge across the East River to Williamsburg had, of course, begun even before the Brooklyn Bridge was opened, but a debate ensued as to whether the next crossing should be farther north instead, at Blackwell's Island. Before young David Steinman had reached his tenth birthday, Theodore Cooper had prepared plans and specifications for a "steel wire suspension bridge, stiffened by a longitudinal girder," between 59th and 60th streets, and Leffert Buck had had his plans approved for a new suspension bridge with four cables, each three inches larger in diameter than those of the Brooklyn Bridge, so that the elevated railway could be extended from the Williamsburg section of Brooklyn into New York. By the time David was twelve years old, the controversy had died down over the bare steel towers and the stepped-truss design of Buck's bridge, and work on the foundations and anchorages had begun. The precocious and studious youngster took a keen interest in the construction project and began to seek opportunities for further education.

Like many a child of immigrants, without money or access to established private colleges, Steinman began attending the City College of New York. In fact, because of his precocity, he began taking college classes even while he was still in high school. Eventually, the ambitious and conscientious young student, with the help of one of his teachers, was able to obtain a pass to enter the Williamsburg Bridge construction site. He climbed upon the steelwork and proceeded across the catwalks set up for the cable-spinning operation, thus seeming to follow an inexorable pull toward a life of, on, and about bridges. Steinman had to work to put himself through college, but he graduated *summa cum laude* in 1906 with a bachelor-of-science degree. Since he wanted a degree in engineering, he applied to Columbia University, whose School of Mines had been established in 1864, just two years after the Morrill Land Grant Act had promoted an expansion of engineering schools around the country.

Steinman's application was read by Professor William H. Burr, who endorsed it with a personal note: "The most deserving case I have known in all my years at Columbia." The aggressive and assiduous young man was eventually able to piece together enough scholarships, fellowships, and nighttime teaching jobs at City College and Stuyvesant Evening High School to complete three degrees at Columbia. In 1909, he was awarded the A.M. and C.E. degrees, having written for the latter an engineering thesis entitled "The Design of the Henry Hudson Memorial Bridge as a Steel Arch." Twenty-five years later, the Henry Hudson Bridge, connecting the uppermost tip of Manhattan and the Bronx, would be built by Steinman's firm, substantially as he had designed it in his thesis.

Even while still a student, Steinman engaged in miscellaneous engineering work, including projects involving subways, elevated railways, and aqueducts for New York City, and in 1910 he accepted an offer to become the youngest professor of civil engineering in the country—at the University of Idaho. Although he was far from the Brooklyn Bridge, Steinman's thoughts were not far from bridges. The following year, while continuing to teach in Moscow, Idaho, Steinman received his Ph.D. degree from Columbia. His dissertation, a comparative study of cantilever and suspension bridges, was of less urgency in the wake of the collapse of the Quebec cantilever, but nevertheless did treat a topic of keen interest to engineers. The work, *Suspension Bridges and Cantilevers: Their Economic Proportions and Limiting Spans*, was soon published under the same title as a textbook in the Van Nostrand Science Series, and a second edition appeared two years later. Steinman had an instinct for writing and publishing, especially on new, significant, and controversial topics, and he exploited it to the hilt. While teaching in Idaho, he also translated two books from the German: the highly mathematical *Theory of Arches and Suspension Bridges*, in which Josef Melan expounded the deflection theory that Moisseiff had introduced in the design of the Manhattan Bridge, and Melan's *Plain and Reinforced Concrete Arches*. Such a prolific output was fast establishing Steinman as a successful academic, but he by no means neglected practical engineering—or practical self-promotion. It is difficult to imagine how otherwise

Professor William H. Burr, who strongly endorsed David Steinman's application to Columbia University

Engineering News in 1913 carried an article on a timber cantilever bridge "built by a troop of Boy Scouts over the Potlatch River, Idaho, after a sketch made by Prof. D. B. Steinman." The magazine could hardly have been expected to send a reporter to Idaho to cover the story, collect a sketch of the design, and take a photograph of the Boy Scouts flanking their engineer leader.

I

David Steinman was not satisfied with building timber bridges with Boy Scouts, and he wrote to Gustav Lindenthal about the possibility of working with him on the Hell Gate Bridge, whose construction was then beginning in New York. Preceded by the credentials of his translation of Melan's books on arches and suspension bridges, which had been published in 1913, the young engineer began working in New York as special assistant to Lindenthal, second only to Ammann, on July 1, 1914. Steinman personally calculated the internal loads, including those due to temperature changes, and visible deflections associated with the erection of the Hell Gate arch, and he supervised teams of engineers who measured the actual strains and displacements at key points on the structure. The design and construction of the great arch was based on theoretical calculations; the measurements on the actual bridge confirmed not only the validity of the specific calculations for that structure but also the basic validity of the theory itself, thus advancing the confidence of engineers to apply it to still larger structures, such as the Bayonne arch, in the future.

Steinman reported the results of his calculations, and their comparison with the measured values, in a paper presented at the same meeting of the American Society of Civil Engineers at which Ammann presented his paper on the design and construction of the Hell Gate Bridge. Next to Ammann's global paper, which put the great project in historical perspective, Steinman's appeared to be that of an engineer with his nose pressed to the drawing board, looking so closely at the details and how to calculate and check them with measurements as to lose sight of the bigger picture. There was, however, in Steinman's paper a brief outline of the "growing movement" to supplement and check theoretical calculations with experimental measurements. Lindenthal and Steinman, at least, knew that the bigger picture was in jeopardy, as it had been in Quebec, without close and personal attention to details that rested solely on theory. At the end of his synopsis of his paper, Steinman gave "special acknowledgement" to Lindenthal, "who undertook to make these measurements in furtherance of engineering science." Lindenthal, in his discussion of this "able paper," explained that he

David Steinman (seventh from right) *and Boy Scouts on the wooden cantilever bridge they built across a stream in Idaho*

had "wanted to ascertain what, if any, bending stresses remained in the trusses after erection," recalling that such stresses had been significant enough to cause one of the steel tubes in the Eads Bridge to need replacing after it broke when the arch was closed. Lindenthal concluded his remarks with the confident assertion that, thanks to Steinman, "there are no unknown stresses in the Hell Gate Arch structure" to cause any cracking or breaking.

At the end of his paper, where acknowledgments were more traditionally made, Steinman mentioned those who had helped him with the new extensometer, or "strain gauge," that was employed, and those who had helped with some of the calculations. Finally, he also thanked Ammann for unspecified "suggestions." Ammann apparently could not graciously leave his involvement at that, however, and in a written discussion he expressed caveats about the generality of Steinman's work: "The analysis of the painstakingly recorded stress measurements, made by Mr. Steinman, may lead the uninitiated reader to overlook the important fact that he has to do with an extremely special case, which may not repeat itself in the history of bridge construction." Ammann also pointed out that "one important object" had not been accomplished by Steinman—namely, that the "actual stresses," as opposed to the secondary and erection stresses referred to by Lindenthal, remained indeterminate. But, so as not to appear to be contradicting Lindenthal, Ammann added that "the expense for such further investigation is too heavy for an individual engineer"—an allusion to Lindenthal, who himself had assumed the costs of the study. Ammann suggested that a government agency or an engineering society in cooperation with the railroads should sponsor such a project.

In his closure to the discussions of his paper, Steinman pointed out that, contrary to Ammann's suggestion that he was claiming more generality than his work allowed, there was "but one paragraph" in the entire paper that was "not a rigorous deduction from the results of the investigation," and it was a simple statement of considered judgment as to how other structures might behave. With undertones somewhat at odds with the usual gentlemanly exchanges among members of a professional society discussing one another's papers, Steinman wrote that he "would like to ask Mr. Ammann to point out anything in the summary of conclusions which can possibly be regarded as too far-reaching a deduction from the results of the investigation." That there was a tension and competition between the two engineers was thus evident in the discussion of this early paper, and perhaps one aspect of it was highlighted in Steinman's closure, in which he appears to have deliberately introduced titles before the names of the discussants, referring to Mr. Ammann and Dr. Lindenthal. Although Lindenthal's doctorate from Dresden was honorary and Steinman's from Columbia was earned, they shared a title of which he no doubt wished to remind Ammann.

Acknowledging Ammann at all may have been somewhat begrudging on Steinman's part, because were it not for Ammann, Steinman might have been in charge of the entire project and thus the logical person to write the more comprehensive paper. When the war called Ammann back to Switzerland, Steinman had assumed responsibilities for the connecting-railroad project for which the Hell Gate Bridge was the engineering centerpiece. However, despite his increased responsibilities, Steinman continued to be paid his initial salary of $200 per month. It was only as a "wedding present," when Steinman married Irene Hoffmann on June 9, 1915, that his salary was raised to the $225 which Ammann had been receiving. After Ammann returned from his stint in the Swiss army, Steinman was no longer to be second in command to Lindenthal: he seems clearly to have preferred the European-trained Ammann to the American Steinman, who seemed to want to forget his European roots.

David Steinman had also played an important role in the design and analysis of the Sciotoville Bridge, Lindenthal's other technologically significant project of the period. As a result of the new methods Steinman had developed in the course of this work, *Engineering Record* "commissioned him to write a series of articles presenting his new design methods." According to Steinman's biographer Ratigan—a World War II correspondent and a writer of "stories and adventure serials"—when Ammann returned from Switzerland he reportedly persuaded Lindenthal to curtail his rival's articles, although the impending consolidation of the journal with *Engineering News* may have been a less insidious factor. In any case, there was clearly a lot more than technical know-how to being a successful engineer—and to letting the world know about it.

Lindenthal reportedly called the younger engineer into his office one day and told him, "Steinman, bridge engineering is easy. It is the *financial* engineering that is hard." A major part of Lindenthal's complaint, which no doubt centered on his continuing frustrations in finding backers for his Hudson River Bridge proposal, was that bankers added millions of dollars in financing costs to bridges after "engineers had sweated and strained to secure the most economical design." These words were evidently taken to heart by Steinman, an inveterate student who seemed to measure his life by his documented degrees, honors, and achievements, and almost to lust after any recognition or achievement that he did not yet have. Not that Steinman did not work for what he got. Upon recognizing that he had no formal training in the important aspect of engineering that Lindenthal had discussed, Steinman enrolled in a correspondence course in business administration, which he found invaluable in his later career. Even if this aspect of the engineering endeavor seemed subsequently not to have been Steinman's favorite, he learned to talk the language of and to work with not only bankers and investors but also public officials and other nontechnical people essential to getting a large engineering project off the drawing board, and also off the ground. Lindenthal, on the other hand, for all his understanding of the importance of financing, saw no room for compromise in his plans. When the war put a stop to engineering projects, especially of the kind that he had dreamed about, Ammann may at least have given the impression of being more sympathetic to the elder engineer's technical resolve. In any case, it was Ammann who was kept on Lindenthal's payroll, however indirectly, and Steinman who was let go.

Years hence, when the two rivals would approve, if not compose, their own curricula vitae for *Who's Who in Engineering*, Ammann would list his service under Lindenthal as extending from 1912 to 1923, not mentioning that some of those years were spent in exile at the New Jersey clay mine in which Lindenthal had an interest. Steinman's record of service under the master extended only from 1914 to 1917. Their respective entries in the 1959 edition of this biographical dictionary tell a good deal more than factual details, however; they also tell a lot about the personalities of the engineers.

Ammann's entry occupies only one column, though this could reflect either his relative shyness and modesty or his sense of security in his significant accomplishments. After the standard identification of his origins, education, and marital status, there is a chronological listing of his principal engineering projects, giving in parentheses the dollar value of the most significant ones. The entry concludes with a list of memberships in professional and other organizations.

Steinman's *Who's Who* entry offers a sharp contrast. Following a long list of engineering projects, but without any mention of their dollar value, there is a much longer list of honors, awards, and memberships, seemingly citing every organization from which he had ever received a certificate of membership or a statement of dues. More curious than what is included is what is omitted from the very beginning of Steinman's entry. In a biographical dictionary whose entries customarily begin with a description of a person's origins, Steinman's contribution omits entirely any mention of his parents, as if he had maintained a firm resolve to suggest that his beginnings were in the stone and steel of a mythic bridge rather than in the flesh of immigrants. After his place and date of birth, the entry goes immediately to his education—including the number of medals, scholarships, and fellowships he won as a student pursuing his various degrees—as if to record that he did it all himself.

He was not merely keeping personal matters out of a professional biography, for his marriage is recorded, as are the names of his three children. It is hard to escape the conclusion that Steinman wished to obscure if not forget his origins, which were, according to a 1958 *New York Times Magazine* profile, "in the slums, in the shadow of the Brooklyn Bridge," giving a different twist to the bridge's inspiration for his career. On the other hand, he was immensely proud of his marriage to Irene Hoffmann, daughter of a former member of the Faculty of Medicine at Vienna, who not only approved of his daughter's marrying a young man with a Ph.D. but also encouraged her to do so, that he might have a son-in-law with whom he could discuss Kantian philosophy. Such dichotomies would naturally lead to tensions in Steinman's later life, at which the biographical dictionary could only hint. In the mid-1950s, for example, he would be identified as one of many significant personalities who had begun life as a Jew and had made things happen. Yet, during the same period, Steinman himself was turning away from those roots, telling reporters that he was "active in Presbyterian affairs."

In spite of whatever unresolved personal tensions he experienced, Steinman accomplished a great deal in his life and career. And in spite of the niggardly professional recognition Ammann gave him in discussions and reviews of his work, Steinman's reputation became established through his books. In 1917, he accepted an appointment as professor of civil and mechanical engineering at his alma mater, City College of New York, which then was organizing a school of engineering. One day, in the spring of 1920, while Steinman was still head of the engineering school, he received a telephone call from "a man who modestly introduced himself as H. D. Robinson." The two met, and Holton Robinson described to Steinman an international design competition for a bridge at Florianópolis, to connect that capital city of the off-coast island state of Santa Catarina with the Brazilian mainland, and proposed that they join in an effort to produce an entry.

2

Steinman had dreamed of actually building bridges of his own, but until that time, with the exception of directing the Boy Scout troop in building a modest cantilever, he had worked exclusively on others'. Now Robinson made it possible for him to participate as an equal partner in a major bridge project. It was a rare opportunity, for there was little work for bridge engineers in the early 1920s, and Steinman jumped at the chance. He went into private practice as a consulting engineer, renting a desk in the office of a friend for ten dollars a month and working on jobs for fees as small as five dollars. He soon got larger jobs, such as writing a survey for $250 and inspecting forty railroad bridges for a fee of ten dollars each. With business picking up, he was able in 1921 to move into his own office and hire assistants and draftsmen. Steinman invited Robinson to share this office, and the older engineer thus moved from "a drafting table in his home, where he did all computing and drafting himself," including solving "difficult three-span catenary problems by suspending a fine chain against the wall and measuring the ordinates." With Robinson's practical experience and Steinman's theoretical talents, technical traits that complemented each other as nicely as did the engineers' different personalities, the partnership of Robinson & Steinman would be able to compete successfully for major bridge projects for many years to come.

Holton Duncan Robinson was a generation older than Steinman, having been born in 1863 at Massena, New York, near the Canadian border. He was the son of Ichabod Harvey and Isabelle McLeod Robinson, and his Scots-English ancestry included Sir Alexander Mackenzie, the Canadian explorer after whom the river is named. Robinson grew up on the family farm beside Robinson Bay, which is located on the St. Lawrence River, and from childhood he was "outstandingly shy, modest, and retiring." He attended a local college, St. Lawrence, in nearby Canton, New York, and studied liberal arts and sciences, receiving a bachelor-of-science degree in 1886.

Young Robinson entered the engineering field through his uncle, the bridge builder George W. McNulty, who was associated with Leffert Buck. Buck and McNulty in turn had begun in engineering under Washington Roebling on the construction of the Brooklyn Bridge, and had started their own firm after that project was completed. Robinson began working on survey crews for Buck and McNulty and studying engineering at home. He slowly gained a variety of experience, being sent to the sites of various bridge projects, including one in the small town of Suspension Bridge, New York, where he took charge of repairs on the stiffening truss of John Roebling's aging Niagara Gorge Bridge. After a few years as draftsman and assistant engineer in the chief engineer's office of the New York Central & Hudson River Railroad Company, Robinson accepted an offer to return, as chief draftsman, to work under Buck, who was then chief engineer planning the Williamsburg Bridge. Robinson eventually became assistant engineer in charge of cable construction on the bridge, remaining so when Lindenthal became bridge commissioner. (Perhaps Holton Robinson and young David Steinman may actually have passed each other on the catwalks.) In 1904, after the Williamsburg Bridge had opened, Robinson was transferred to the Manhattan Bridge project and placed in charge of design

Holton Robinson, when he was engineer in charge of construction of the Williamsburg Bridge

and construction—again under Buck, who was consulting engineer to Othniel Foster Nichols, an 1868 graduate of Rensselaer Polytechnic Institute who was chief engineer of New York's Department of Bridges from 1904 to 1906, in which position he oversaw the redesign of the Manhattan Bridge after Lindenthal's departure from the position of commissioner.

Robinson left the employ of the city in 1907 to join the Glyndon Contracting Company, fabricator of the cables for the Manhattan Bridge. Besides designing the machinery to effect the spinning of the twenty-one-inch cables, then the largest ever, during his tenure with Glyndon he produced an unsuccessful design for a suspension bridge to cross the St. Lawrence River at Quebec, where the great cantilever had failed. He left Glyndon in 1910 to build, as an independent contractor, a suspension bridge near his hometown; this structure was completed in about six months at a cost of $40,000, 50 percent lower than the lowest bid that had been received by the town of Massena. Over the next several years, Robinson worked on a variety of bridge, tunnel, and navy war projects; his experience was broad and deep by the time Steinman met the "modest, distinguished-looking man" in 1920.

In 1922, Robinson was appointed consulting engineer for cable construction on the Delaware River Bridge between Philadelphia and Camden,

New Jersey. He assured the joint commission that, instead of four smaller ones, two cables of thirty-inch diameter could be spun, thus simplifying construction, and he supervised their design before resigning as consulting engineer to work for the contractor, the Keystone State Construction Company, which was to make the cables. The office of Robinson & Steinman, in turn, was given responsibility for designing the temporary work and machinery needed to accomplish the task. In 1926, at a joint meeting of the Franklin Institute and the Philadelphia section of the American Society of Civil Engineers, Robinson, then in his early sixties, presented his first technical paper, "Construction of the Cables of the Delaware River Bridge." For all his accomplishments, his boyhood shyness had not left him; "he suffered excruciatingly from stage fright and the experience so unnerved him that he vowed he would never repeat it."

He may have eschewed public speaking, but Robinson did not shy away from the physical challenges bridge engineers had constantly to face. According to Steinman,

> Even in his last years, Mr. Robinson was active, agile, and fearless in his outdoor work on bridges. He would climb on high steelwork or walk the cables of a suspension bridge with greater ease than most younger engineers. In 1941, during the investigations that followed the Tacoma Narrows . . . disaster, he was retained by the insurance companies, and made a personal examination of the cables of the wrecked structure. Although seventy-eight years old at the time, he calmly walked out over the 17½-in. cables, each 5,900 ft long and 450 ft high at each tower, to examine the condition of the wires and to cut out samples of the wire at midspan. His feat was rendered more difficult and hazardous by the fact that the hand ropes in the main span had been wrecked.

Thus, in spite of his social reticence, Robinson suffered no fear in the face of technical or physical challenges. Steinman, on the other hand, was, ostensibly at least, as comfortable before large audiences as he was on tall bridges. The partnership of Robinson & Steinman, extremely complementary and compatible, would last for a quarter-century without a written contract between the men.

The project that brought them together, the Florianópolis Bridge, was a success, thanks in large part to an unusual and distinguishing structural design by Steinman that had the eyebar chains doubling as the curved upper chord of the stiffening trusses, which resulted in a very economical structure. When it was completed in 1926, the Florianópolis Bridge, with a main span of over eleven hundred feet, was the largest in South America, and the

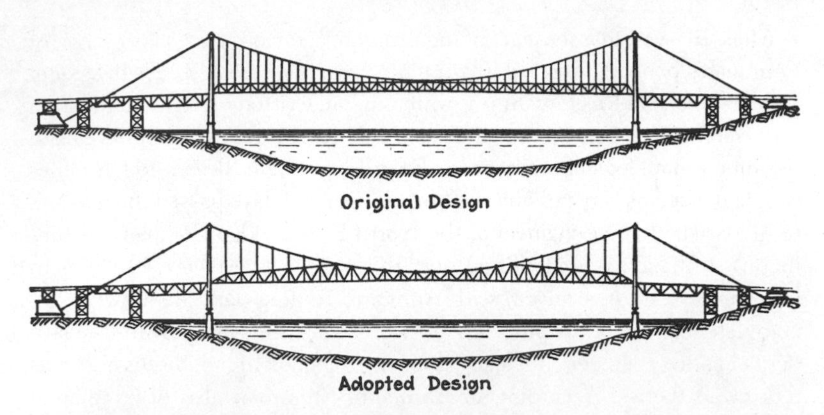

The Florianópolis Bridge, as originally designed by Robinson & Steinman, and as altered to suit the client

largest eyebar suspension bridge in the world. Steinman's article on the design of the bridge appeared in *Engineering News-Record* late in 1924, and he explained how the bridge was originally "designed along conventional lines," which meant a wire-cable structure that looked very much like the Williamsburg Bridge, with which Robinson was so familiar. When a decision based on economy was made to use eyebars rather than cables, however, this led to a reconsideration of the truss, into which the eyebars then became incorporated. According to Steinman, first sketches showed a "most pleasing outline" for the truss, which curved as it did at the towers, but straight chords were employed in the final design, "in deference to the preference expressed by our client." Such compromises might not be made by an engineer like Lindenthal, but Robinson and Steinman were more interested in establishing their firm's reputation for economical and reliable work than in making an engineering or artistic statement.

The new truss-eyebar arrangement produced a very stiff bridge with less material, and such an economical solution was something which other suspension bridge engineers would now have to take into account. It presented a realistic alternative to the stiffened cables or stiffened eyebars, such as the kind Lindenthal had proposed for his North River Bridge, which were not integrated with a deck truss. An immediate response to Steinman's article came in the form of a letter to the editor from Leon Moisseiff, who took exception to Steinman's claim that his structure was the first to incorporate a bridge's chain or cable into a stiffening truss that continued for the entire length of the main span. Moisseiff included a drawing of his 1907 design for a bridge over the Kill von Kull, which, "for better appearance," continued the line of the truss through the towers. But a drawing is not a bridge.

Moisseiff was, in a sense, echoing Ammann's review, two years earlier, of Steinman's book, *A Practical Treatise on Suspension Bridges: Their Design, Construction and Erection*. Since Ammann had written little for publication on suspension bridges up to that time, Lindenthal would actually have been the much more logical reviewer for *Engineering News-Record* to have chosen, and it seems very possible that he may indeed have passed the book on to his assistant chief engineer at the North River Bridge Company, as Ammann's affiliation was identified over the review. The review itself might best be described as mixed, with Ammann finding parts done with "fair completeness" and thus providing a "useful manual, especially for the student or young engineer," but also criticizing the book for not discussing matters of aesthetics. According to Ammann, Steinman also discriminated "against the eyebar chain" on technical grounds, but he was in fact ever flexible in his thinking, as the Florianópolis Bridge was to demonstrate.

Steinman, both with Robinson and independently, began to get more and more significant commissions, and the younger partner wrote about them with facility. The Carquinez Strait Bridge, located about twenty-five miles northeast of San Francisco, was one such project. Consisting of two main spans of eleven hundred feet, it became the second-largest cantilever in the United States and the fourth-largest in the world when it was completed in 1927. The chief engineer of the bridge project was Charles Derleth, Jr., with William H. Burr as consulting engineer and Steinman as design engineer.

But Steinman's real ambition was to build world-class suspension bridges that would also be recognized as things of beauty. Though the Florianópolis Bridge was a major structure, its oddness of type and compromised lines, not to mention its location, put it in a category almost by itself. A new opportunity arose, albeit still off the beaten track, with the Mount Hope Bridge, which Steinman designed, and whose construction the firm of Robinson & Steinman supervised "to take the Island out of Rhode Island." The total length of this bridge was over a mile, and its twelve-hundred-foot main span put it almost in a class with the major suspension bridges of the day. Its cross-braced towers suggested a Gothic arch above the roadway, and the bracing was topped by a crown of smaller crosses, this latter feature echoing somewhat the tower tops of several contemporary suspension bridges, including Modjeski's Delaware River Bridge, whose towers Pennell so disliked. Steinman's Mount Hope towers have a balanced look, however, and are in good proportion to the uniformly deep truss of the roadway. The bridge received the 1929 Award of the American Institute of Steel Construction as the most artistic new long-span bridge in America.

At the same time, the firm of Robinson & Steinman was designing the St. Johns Bridge over the Willamette River at Portland, Oregon. With a twelve-hundred-foot main span supported from rope-strand cables, which for such a distance were found to be somewhat more economical than parallel-wire cables spun in place, this bridge was then the longest suspension type west of Detroit. According to Steinman, perhaps responding to Ammann's criticism, "the desire to secure a beautiful public structure was a governing consideration" in the design, and the towers were the result of "extensive architectural studies," although he identified no particular architect or style. The unique towers have battered (i.e., slightly inclined) sides, spires, and, in a more extreme fashion than the Mount Hope Bridge, Gothic arches above and below the roadway. The stiffening truss, however, is undistinguished, and there does not seem to be a successful integration of towers and deck. Although the bridge was described in the Robinson & Steinman firm's brochure, *Bridges Lasting and Beautiful*, as "a poem stretched across the river" and "a symphony in stone and steel," the aesthetic success of the towers and the overall bridge can be debated. The towers were designed to echo and harmonize with the dramatic scenery of evergreens, mountains, and clouds, visible through the four-hundred-foot-tall structures, but they seem too unintegrated into the natural setting. In something of a departure from tradition, the bridge was painted a pale green. In 1931, "on a gusty, rainy day," Robinson and Steinman, whose firm had complete charge of design and construction, gave the newly completed crossing its final inspection from the open cockpit of a stunt plane which Tex Rankin, "northwest flying ace," flew around the towers and over and under the roadway. Both engineers were thrilled by the experience, and with the bridge.

In his memoir of Robinson, who died in 1945, Steinman described him as being "professionally connected with the construction of almost every notable suspension bridge built during his lifetime," a fact that "was his chief pride." Without detracting from Steinman's eulogy of Robinson, this could be said of a number of the great bridge engineers; indeed, it almost naturally followed, because great engineers wanted to be associated with great bridges, whose designs in turn relied on a variety of engineers who had a variety of experience with the unique and specialized design and construction problems that were faced. Sometimes, of course, as in the case of the Tacoma Narrows Bridge, the greatness of the engineers has come to seem more important than the design itself.

In any event, who would consult on what bridge had a lot to do with who was the chief engineer, of course, and who had the dominant reputation or the most correct politics at the time. When plans for the George Washington Bridge were being finalized in the mid-1920s, for example, Lindenthal, because of his relationship with Ammann, was a problematic choice as a consulting engineer. On the one hand, he was the engineer who had been most visibly associated with such a Hudson River project; on the other, his inflexibility and prior relationship to Ammann put him in a special category. Robinson, because of his extensive experience, was a natural choice, but his then recent association with Steinman may have presented problems for Ammann. As for Steinman himself, for all his writing about bridges, he was only just beginning to gain his first experience with their design and construction. In some accounts, however, Robinson and Steinman, in particular, "helped to design" the bridge, as consultants on the erection of the steel superstructure, even though they are not listed among the consulting engineers in the dedication program.

3

The 1930s were a heyday of large-bridge construction, with the George Washington opening in 1931 and the two great bridges connecting San Francisco with Marin County and with Oakland under construction simultaneously. Of these, the bridge to Oakland was actually completed first, in 1936, but it was to be permanently overshadowed by the Golden Gate, completed six months later, in 1937, with its world-record span of forty-two hundred feet between towers. The beginning of construction of the lesser-known structure took place in mid-1933, with President Franklin Roosevelt setting off a blast by remote control from Washington, and the first earth being turned up with a golden spade. At this ceremony, Herbert Hoover called the San Francisco–Oakland Bay Bridge "the greatest bridge yet erected by the human race," yet until the 1989 earthquake it remained largely unknown outside the Bay Area, where it serves such an important transportation role. Among the reasons for its relative obscurity must also be counted the fact that this bridge had no single prominent and dominant dreamer like a Roebling, Lindenthal, Ammann, or Strauss serving as executive director and providing a visible personality to the project. Even its official name—San Francisco–Oakland Bay Bridge—is impersonal and awkward; it has often been abbreviated to the Transbay, or simply the Bay Bridge, the name by which it is best known locally.

In spite of these differences, the Bay Bridge, like all great engineering projects, did encompass a long history of dreams and dreamers. Talk of having a bridge between San Francisco and Oakland began shortly after the Gold Rush and continued throughout the latter part of the nineteenth century. The 1906 earthquake distracted attention from a bridge, since the city had to be rebuilt, and in the meantime a ferry system carrying four million vehicles and fifty million passengers a year developed. Agitation for a bridge again arose, only to be suppressed by the world war. In the decade after the war, numerous applications for bridge-building franchises were filed, only to meet continuing opposition by the War Department, especially for a bridge north of Hunters Point, across the bay from Alameda. By the end of

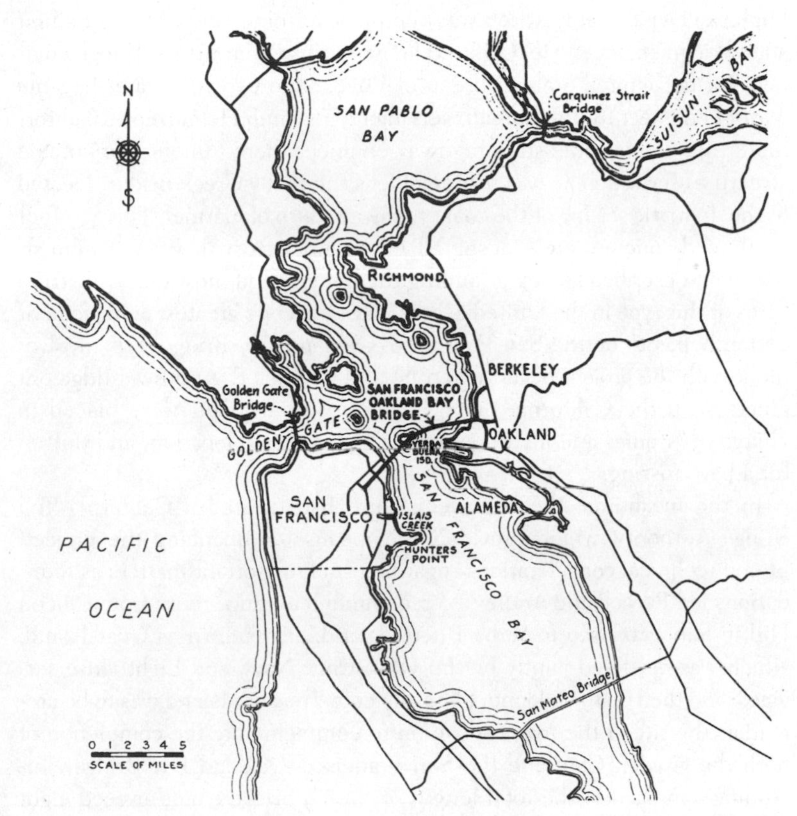

The San Francisco Bay area, showing the locations of the Carquinez Strait, Golden Gate, and San Francisco–Oakland Bay bridges

that decade, the progress of the Port of New York Authority in financing and constructing the 179th Street bridge across the Hudson had led to calls for a West Coast bridge supported by revenue bonds. A San Francisco Bay Bridge Commission was appointed by President Hoover, which seems ultimately to have made the objections of the War Department less absolute; the state highway engineer Charles H. Purcell was appointed as secretary.

Purcell was born in 1883 in North Bend, Nebraska, and attended Stanford and the University of Nebraska, where in 1906 he received his B.S. in civil engineering. He began working for the Union Pacific Railroad in Wyoming, then held positions in Nevada, New York, and Peru, with smelting, refining, and mining companies, before returning to construction and railroad work in the Pacific Northwest. In 1913, he joined the Oregon State Highway Department, which was then just forming, and became its first state bridge engineer. He accepted an appointment in 1917 as bridge engineer for the United States Bureau of Roads, and two years later became district engineer for the bureau, serving in Portland. He moved to California in 1927, to become state highway engineer there. Among the notable structures for which he was responsible is the Bixby Creek Bridge, located in the dramatic setting of the coast highway south of Carmel. This 330-foot reinforced-concrete arch, designed in conjunction with F. W. Panhorst, has been described as being "among the lightest and most graceful structures of this type in the United States." But Purcell's greatest achievement certainly has to be the San Francisco–Oakland Bay Bridge. His involvement with the project began when he and Charles E. Andrew, bridge engineer with the California State Highway Department, were placed in charge of "studies and investigations of engineering, location, and traffic" for a bay crossing.

In the meantime, the state legislature had created a California Toll Bridge Authority, which provided the means for financing the project. Sound technical considerations regarding such important matters as foundations led Purcell and Andrew to recommend a bridge route from Rincon Hill in San Francisco to Yerba Buena Island, also known as Goat Island, which was occupied jointly by the U.S. Army, Navy, and Lighthouse services, and then on to Oakland. (The adjacent Treasure Island was to be created as the site of the 1939 exposition to commemorate the completion of both the Golden Gate and the San Francisco–Oakland Bay Bridge.) Including approaches, the total length of such a bridge would exceed eight miles, half of which was over the bay; to cross each of the two stretches of water, engineers would have to devise independent structural solutions as great as any single major bridge then extant or under construction. After preliminary designs and underwater borings were made in 1930 and 1931, a San Francisco–Oakland Bay Bridge Division of the Department of Public Works was set up, with Purcell as chief engineer, Andrew as bridge engineer, and Glenn B. Woodruff as engineer of design. The board of consulting engineers comprised Ralph Modjeski, the chairman, who with J. Vipond Davies had made a preliminary survey for such a project a decade earlier;

Engineers making final inspection of San Francisco–Oakland Bay Bridge (left to right): *Charles Derleth, Jr., Glenn B. Woodruff, Leon S. Moisseiff, Henry J. Brunnier, Charles H. Purcell, Carlton S. Proctor, Ralph Modjeski, and Charles E. Andrew*

the partners Daniel E. Moran and Carlton S. Proctor; Leon Moisseiff; Charles Derleth, Jr.; and Henry J. Brunnier.

In an article in *Engineering News-Record*, subtitled "A Review of Preliminaries," Purcell, Andrew, and Woodruff described some of the site and design alternatives they had considered. Though they confessed that it would be impossible for them, in this paper, to "consider the large number of tentative designs that were made," they did discuss several, which included cantilevers longer than the Quebec Bridge and suspension bridges almost the equal of the Golden Gate. They admitted that the forty-one-hundred-foot suspension design "presented strong temptations" for its acceptance: "It required fewer departures from past practice than any alternate layout, reduced the number of piers to be constructed and was a more monumental structure." However, it did present some drawbacks: it would have required a large amount of material to construct the San Francisco anchorage and to stiffen the truss against the wind. Furthermore, the longer span would have provided inferior clearance for shipping, required the destruction of some piers, and cost about $3 million more than the adopted design.

Detailed considerations of the many alternative design possibilities led the group of engineers to recommend that the "bridge," which was really two distinct bridges separated by a tunnel through an island, would consist of: (1) a unique pair of double-deck suspension bridges, each with a main span of 2,310 feet, arranged in tandem and sharing a common central anchorage in the middle of the water; (2) a 540-foot tunnel through Yerba Buena Island, with a bore larger than any other tunnel in the world; and (3) a great truss bridge laid out in a sweeping curve, with a cantilever section fourteen hundred feet long, which made it the longest and heaviest cantilever span in the U.S. and the third-longest in the world, flanked by a number of other spans exceeding five hundred feet. (It was on this latter portion of the bridge that a section of the upper deck fell during the 1989 Loma Prieta Earthquake, and the traffic disruption during the month when the bridge was closed for repairs provided many opportunities to reflect on the importance of the communications link that the bridge provided between San Francisco and communities, like Oakland, on the east side of the Bay.) In June 1933, the start of construction was marked on the island by a ceremony that included the explosion set off from Washington by President Roosevelt, and the symbolic beginning of excavation with the use of the golden spade. Chief engineer Purcell expressed the hope that traffic would be using the bridge by January 1937.

The opening of the San Francisco–Oakland Bay Bridge actually took place late in 1936, ahead of Purcell's public hopes as well as of the completion of the Golden Gate. Like all such events, the opening provided an opportunity to look both backward and forward. Among the episodes in San Francisco Bay history that was recalled on the occasion was the story of "a shrewd and likable fellow" named Joshua A. Norton, who had come to California during the 1849 Gold Rush, "attained considerable importance and amassed a fortune," only to lose it and his mind. Returning to the area after years of absence, he declared himself "Emperor of the United States, Protector of Mexico and Sole Owner of the Guano Islands," and issued paper money, which was honored by the locals, who humored him. Among the many imperial proclamations issued by Emperor Norton was one ordering the Coast Guard to blockade Carquinez Strait, long before Steinman's cantilever crossed it, and one inviting Abraham Lincoln and Jefferson Davis to meet and arbitrate an end to the Civil War, an invitation they did not accept. But the document most on the minds of those celebrating the completion of the Bay Bridge was the following:

PROCLAMATION

We, Norton I, Emperor of the United States and Protector of Mexico, do order and direct . . . that a suspension bridge be constructed from the improvements lately ordered by our royal decree at Oakland Point to Yerba Buena, from thence to the mountain range of Saucilleto. . . . Whereof fail not under pain of death.

Given under our hand this 18th day of August, A.D. 1869, and in the 17th year of our reign, in our present Capitol [*sic*], the city of Oakland.

(signed)
Norton I.—Emperor

Though Norton's bridge might have been an even grander span than the one built, at the same time snubbing San Francisco and making the Golden Gate Bridge unnecessary for getting to Marin County, at least from Oakland, the order certainly leaves no doubt that dreams of bridges were grand during the emperor's reign. In 1936, the builders of the Bay Bridge saw the historical anomaly not as just an amusing footnote to the story of their own bridge, but as a testament to dreams of all kinds: "Who is bold enough to say that they will not some day be fulfilled?" Many decades after Norton flourished, the actual bridge inspired the Spanish-language poet Jorge Carrera Andrade to write *Canto al Puente de Oakland*, one verse of which reads, in translation:

Your length like a river or like buoyant hope
miles of iron and of sky interwoven—
can only be measured with the music
or the metres of dream.

Just as so many New York City bridges owe their existence and appearance to a group of engineers who worked for government bodies of one form or another, so did the San Francisco–Oakland Bay Bridge owe its final form to the talents and abilities of California state engineers like Purcell and Andrew. Consulting engineers play a crucial role whenever it comes to particular questions of detail, experience, and precedent, but the creative and political sympathy and savvy of career government employees around the nation have also played significant roles in shaping the built environment. Among such engineers was Conde McCullough.

Conde Balcom McCullough was born to a physician and his wife in 1887 in Redfield, South Dakota. As a young man, he attended Iowa State College, from which he received his bachelor's degree in civil engineering in

Artist's conception of how the San Francisco–Oakland Bay Bridge would look when completed

The completed San Francisco–Oakland Bay Bridge, showing its tandem suspension bridges, tunnel, and cantilever sections

1910. After a first engineering job in Des Moines, he joined the Iowa State Highway Department, beginning as a designing engineer in 1911 and rising to assistant state highway engineer by the time he left, in 1916, to join the Civil Engineering Department at Oregon State College. Within two years, he had risen to the rank of professor and was head of the department, but he left the college the following year to become state bridge engineer for the State Highway Department. In order better to understand and deal with the legal constraints on his job, McCullough also went to law school, at Willamette University, receiving the bachelor of laws degree and being admitted to the Oregon State Bar in 1928. He wrote a considerable number of articles and books on bridges, economics, and law, including—with his attorney son, John McCullough—a two-volume work, *The Engineer at Law*.

Conde McCullough's creations in steel and reinforced concrete are even more responsible for Oregon's overall reputation for beautiful bridges than are Lindenthal's and Steinman's efforts in the state. McCullough's Bridge of the Gods and his Caveman and Rogue River bridges, this last incorporating the innovative prestressing techniques developed by the French engineer Eugène Freyssinet, are as graceful and whimsical as their names. The Coos Bay cantilever, which in 1936 completed one of the last major links in the Oregon Coast Highway, was designed by McCullough and was dedicated to him after his death in 1946. The Conde B. McCullough Memorial Bridge thus joined the exclusive group that includes the Eads Bridge at St. Louis and the Roebling Bridge at Cincinnati in being named for its engineer.

Not every engineer who works for the government or a government-related agency gets an opportunity to be as broadly based in his work as McCullough did in the course of a career, but some certainly have and do. The shy Ammann, for example, who may have appeared on the surface to be apolitical and uninterested in law or public affairs, did engage in politics of a private nature. After all, it was he who wooed the future governor of New Jersey with plans for a Hudson River Bridge, which he in turn could advocate in his inaugural address, and for which he also could recommend Ammann himself as the project engineer. The independent Steinman, on the other hand, engaged more in a politics of a quite open and different kind—namely, the politics of his profession, for which he had begun to emerge, in the 1920s, as the most energetic and articulate spokesman.

4

In 1925, David Steinman, then president of the American Association of Engineers, wrote an article on "Outstanding Practice Problems of the Pro-

The Conde B. McCullough Memorial Bridge over Coos Bay on the Oregon coast, one of the few bridges in America named for their engineers

fession," which appeared in *Engineering News-Record*. The article reported on the results of a survey that asked "representative engineers of national reputation for their views on ethical conduct in negotiations for professional services." Steinman had not, of course, discovered this issue, which involved "how far the engineer may solicit an engagement without invitation; whether he should decline to do so competitively; whether a warning against competition should be included in the code of engineering ethics; and what can be done by the profession to combat the evil of inadequate fees." However, though Steinman had no doubt heard a lot relating to such topics in Lindenthal's office a decade earlier, the issues had become much more openly articulated in the meantime, going well beyond the pages of trade journals.

The American Society of Civil Engineers, begun in 1852 as the American Society of Civil Engineers and Architects, joined in 1916 the so-called founder societies, which then included the American Institute of Mining Engineers, dating from 1871; the American Society of Mechanical Engineers, from 1880; and the American Institute of Electrical Engineers, from 1884 and now known as the Institute of Electrical and Electronics Engineers. The proliferation of engineering societies in America followed only shortly after the same phenomenon in Britain, where the Institution of Civil Engineers, originally intended to encompass all of engineering that was not military, became only one among a plethora of specialized institutions, such as the Institution of Mechanical Engineers and the Institution of Electrical

Engineers. These were formed in large part because proponents of new areas of developing technology were not so easily integrated into the existing societies by their more traditional counterparts. By the early years of the twentieth century, there was such a diversity of engineering societies, differentiated largely by the technical specialty of their members, that engineers felt that there was no single voice for the profession itself.

Among the organizations formed "to address the social and economic interests of the engineer regardless of technical discipline," was the American Association of Engineers, founded in 1914. There were two schools of thought among its founders, one of which "favored establishing a labor union affiliated with the American Federation of Labor," and the other of which "visualized a professional society, avoiding the coercion normally inherent in labor organizations." The latter philosophy prevailed, and the American Association of Engineers was about twenty-five thousand strong when Steinman assumed its presidency. At the time, the question of professional ethics was a lively topic of debate, even though the technical societies had been talking about such matters for fifty years. In 1912, the American Institute of Electrical Engineers and the American Society of Mechanical Engineers had finally adopted codes of ethics, and so, in 1914, did the American Society of Civil Engineers. The codes of these founder societies, as well as those of the American Association of Engineers, were thought to be too general and too subject to interpretation by some of its members, however, and in 1923 a number of "practice cases," or case studies, had been issued to remove some of the ambiguities.

This did not solve everything, of course, and when Steinman's article on outstanding practice problems appeared in *Engineering News-Record*, it occasioned an editorial on the subject. According to the editors, it had to be admitted that the profession's ethics were "neither satisfactorily formulated nor universally followed," and existing codes were "made up of fine words which no one can controvert but of such embracing vagueness that interpretation is a matter of individual desire." The editorial continued: "Certain things are permissible, certain other things lie beyond the pale. In between is territory where the engineer may roam at will. There is no Supreme Court which interprets the law and publicly enforces it." Steinman's undertaking for the American Association of Engineers was held out as a step toward "disciplinary control." The "question of competition for business" was seen to be central to the problem, and engineers were thought to need "something comparable in solemnity to the doctor's Hippocratic oath." Unfortunately, it was a difficult time to be discussing such ideals, since business for engineers was no better than it was for the rest of the economy:

In plain words there are more practising engineers, acting as principals, than there are jobs. There is the constant struggle for the majority of those principals to get enough of the existing work to maintain themselves as principals and to keep from falling back into the ranks of the employed, ranks that are apparently not so crowded as are the upper strata of the employing class. Work, too, must be sought to a large extent not from old clients, as is the case with the doctor and the lawyer, but from new interests who too frequently know nothing about engineers or engineering. Competition becomes a dominant factor in independent engineering, and the selfish motive strong in those who are trying to practice it.

The problem of the code, therefore, is to set up an altruistic motive that will reinforce and justify the selfish one. It is not enough merely to assume that everybody agrees that certain things are not to be done; reasons must be given why they are not to be done. Because such reasons generally go back to the underlying necessities for truth and justice and honor, they lack force, for generalities on ethical conduct are always subject to individual interpretation. They must, therefore, be put on the more practical ground that self advancement lies also in the advancement of the profession one practices.

These were tough times in which to aspire to such ideals, but discussions of the kind initiated by Steinman were seen as the way to proceed to raise simultaneously the general status of the profession and the level of its practice. In the meantime, another engineering organization, the Federated American Engineering Societies, was formed, with Herbert Hoover as its first president. This organization of societies, not of individuals, was created "to further the public welfare whenever technical knowledge and engineering experience are involved." A third organization, the Engineers' Council for Professional Development, was formed in 1932 "primarily to increase the input of the practicing profession into the educational process." The issues remain to this day, having outlasted the organizations, which have been transformed several times since.

Another development in the 1920s and into the 1930s was the increasing number of states that had instituted registration laws, thus placing "engineering on a par with law and medicine as legally restricted and recognized learned professions," according to Steinman, who was among the most outspoken proponents of such registration laws. Between 1907, when Wyoming enacted the first such statute, and 1935, engineering registration was established in thirty-two states containing over 85 percent of the engineers in the country. Among the arguments he put forth in his many talks and articles on the subject were the following:

> The public needs to be protected against the quack, the incompetent, the unscrupulous, and the impostor, who do not belong in our profession but nevertheless practice in its name. . . .
>
> The public judges a profession by the examples it meets. When the public sees men who are unlettered and untrained holding themselves out as "engineers," respect for the engineering profession is weakened or destroyed. When the public sees the word "engineer" on the shop window of a plumber, an electrician, a radio dealer, or an automobile mechanic, a wrong picture of the engineering profession is implanted.
>
> For years, the engineering profession talked about this problem—the abuse and misuse of the term "engineer"—but nothing was done about it. Finally with the aid of registration laws, means for successfully protecting our designation became available.
>
> At first, public officials were slow to co-operate. They declared that we could not "copyright the dictionary." We pointed to the precedents of the other legally established professions which had successfully "copyrighted large chunks of the dictionary." Any unlicensed man hanging out his shingle as a "lawyer," a "physician," a "dentist," or an "architect" will be promptly arrested and subjected to the penalties of the law.

Steinman believed so strongly in registration that he thought it should be a requirement for membership in engineering societies, but established groups like the American Society of Civil Engineers were not receptive to such an idea. Indeed, as with college degrees and other non-society designations they would not include the letters "P.E.," indicating registration as a professional engineer, after the names of engineers appearing in society publications. Thus, in 1934, Steinman invited representatives of four relatively young state societies of professional engineers to join him at the Columbia University Club in New York for an organizational meeting of a new group, the National Society of Professional Engineers, whose membership would be restricted to registered professional engineers and whose activities would be limited to the "nontechnical concerns of all engineers." Not surprisingly, Steinman became the society's first president.

With the establishment of registration laws and the growing proliferation of engineering schools, entry into the profession via the self-taught route of an Eads, or even the semiformal educational route of a Lindenthal, became less and less common. Though state licensure regulations included grandfather clauses so established practitioners were not excluded no matter what their route to their practice, and allowed for responsible experience as a substitute for formal education, earning an engineering degree was increasingly the way to become an engineer.

As president of a national group, Steinman spoke and wrote frequently on matters relating to the profession, including engineering education. Whereas so many in his profession less than a century earlier had little if any formal education, Steinman expected the engineer of the twentieth century to be a lettered individual. Then, when identifying himself to strangers as an engineer, he would not hear "an involuntary exclamation," as Herbert Hoover once did from a woman he had met while traveling, followed by her admission, "Why, I thought you were a gentleman!" Hence, among the ways Steinman saw to advance the status of the profession was the manner in which engineers were educated. This had changed to a considerable degree since the nineteenth century, but he saw reasons for it to change further still:

> The four-year course may have been adequate two generations ago, but the increasing content of essential engineering knowledge and the growing recognition of the desirability of a background of liberal and cultural studies for a professional man have altered the picture. Those of us who took a complete college course before entering an engineering school have never regretted it. . . . Personally, I favor a pre-engineering college course of two years, ultimately of four. This is in line with the best standards achieved in other professions.

Although Steinman might have changed "men" to "men and women" if he were writing today, he would have to change little else, for the issue of what form an engineering education should take, and whether it should follow a general college degree, is still a matter of some discussion. A strong argument can also be made that engineering, which is "essentially a mode of thought based on a mastery of the laws of nature," should be a component of all liberal education in an age that must deal with problems not only of bridging ever-wider chasms, both literally and figuratively, but also of undoing some of the inherited neglect and environmental legacies of earlier times.

In addition to advocating the liberal education and registration of engineers, Steinman pushed for use of the professional title "Engineer" or "Engr." with personal names, which he likened to physicians' use of "Dr." Adopting the practice for himself, Steinman began to sign his letters "Engr. D. B. Steinman." On this issue, however, not even *Engineering News-Record* was in his corner. In an editorial commenting on Steinman's introduction of the proposal at the first annual meeting of his National Society of Professional Engineers, the magazine retorted: "Engineers above all are supposed to be logical; do they propose to follow the present plan to its logical conclusion with Physician Jones, Dentist Smith, Chiropractor Brown—or Barber Cavello, for that matter, for barbers too are licensed in the interest of public safety?"

Among the letters to the editor on the subject was one from Ing. Robert B. Brooks, Jr., who pointed out that the "Mexican engineer is a titled individual." Indeed, in many Spanish-speaking countries, the earned title "Ingeniero" is a mark of distinction, as is the title "Ingenieur" in Germany. But such traditions were not easily introduced in America, "where titles have been looked upon with disfavor." No matter the inconsistency of *Engineering News-Record* in forgetting that the young country did confer the titles of doctor, senator, general, captain, professor, and the like; the time was not propitious for Steinman to be suggesting the adoption of the title of engineer. Though his commitment to the issue never fully disappeared, it seems to have flagged a bit after he completed his two terms as president of the National Society of Professional Engineers. Among the things that competed for his time and attention were the new opportunities that had arisen for engineers generally to undertake bridge projects following such eminently successful and prominent models as the George Washington Bridge, in which the structure was paid for by the tolls levied on the traffic using it.

5

All the great bridge designers seemed to want to hold the record for the longest span, but there were only so many locations that needed or could justify a bridge of record size. Among the last of the great unbridged crossings in the United States that remained unspoken for in the mid-1930s was the entrance to New York Harbor known as the Narrows. Ammann was only one engineer working clandestinely at the time on plans for a bridge at that location. Steinman also saw the crossing not only as the opportunity to regain the span record for the East Coast, but also as the opportunity of a lifetime for an engineer who wished to be memorialized in his work. Though perhaps not quite so obsessive about what Steinman would call his "Liberty Bridge" as Lindenthal was with his North River Bridge, Steinman nevertheless worked on and off on the design for twenty-five years, possibly having an idea for the structure as early as 1926. It was planned to have a main span of 4,620 feet, "a thousand feet longer than the George Washington span" and over four hundred greater than the Golden Gate. Steinman would also point out that it would have a clearance of 235 feet above high water, "100 feet higher than the East River Bridge," of his great hero Roebling, and towers eight hundred feet high, "higher than the Woolworth Building."

David Steinman's unrealized Liberty Bridge

After the death of Holton Robinson, Steinman practiced under his own name for fifteen years. The cover of a brochure issued by D. B. Steinman in the late 1940s was dominated by a sketch of Liberty Bridge. A smaller reproduction of this same sketch had appeared without identification or comment on the inside title page of a Robinson & Steinman brochure dating from the early 1930s, but now a description of the cover declared that Steinman's dream would be "the world's greatest engineering achievement. Furthermore, spanning the gateway to America, it will be a symbol of our free, vital civilization, a portal of hope and courage—an inspiring symbol of the spirit of America." These postwar words may have been intended to rouse support for his dream bridge, and they may indeed have done that among his friends and associates, but Steinman apparently did not have the ear of Robert Moses, the person who, perhaps more than any other single individual, controlled whether a bridge would be built across the Narrows and, if it would, who would build it. Ironically, Steinman, the supreme politician of his profession, seems to have been much more naïve in the local politics of bridge building than Ammann or Strauss in their quests to erect a great bridge in a great municipality. Nevertheless, as late as 1948, in an interview that appeared in *The New York Times*, Steinman said, "I expect Liberty

Bridge to be built and hope to be identified with it." After that achievement, he would be ready to retire, he allowed, but an aging engineer had to do more than hope to win the competition for a great bridge.

Steinman continued to promote his Liberty Bridge, and himself, in his own way. The back cover of the same brochure that carried a sketch of the span contained a photograph of "the hands of Dr. Steinman at work on plans for the great span over the Narrows," taken by the photographer Frank H. Bauer for a book of studies of "the hands of outstanding representatives of the various arts and professions."

To accompany his hands using dividers and scale, Steinman took a quote from John Ruskin about building not for "present delight" but "forever" with stones that "will be held sacred because our hands have touched them," by descendants who will say, "See, this our fathers did for us." Steinman, who never showed any such admiration for his own father, evidently thought so much of the portrait of his hands that it formed the larger-than-life focus, surrounded by images of many of his already realized bridges, in a mural in an engineering-faculty lounge that he would donate to the University of Florida. Before he did that, however, Steinman thought there might be an opportunity for greater exposure of his engineer's hands immortalized with dividers and scale over drawings of his dream bridge. In fact, he thought the image would form the perfect basis for the design of a postage stamp that was to be issued to commemorate the centennial of the American Society of Civil Engineers, in 1952. As late as 1957, a biographical sketch of Steinman described such a stamp as having been issued, but the stamp that was actually released in 1952 showed not an engineer's hands but two bridges—a covered wooden bridge and a steel suspension bridge, which represented the century of engineering progress. Steinman must have been greatly disappointed that his stamp design was displaced in the final decision, but he may have been even more disappointed that it was Ammann's George Washington Bridge that represented the century of progress. That the hands did appear as part of the design of an official first-day cover envelope may have been but small consolation.

Though it could be said that the George Washington Bridge was indeed the most significant structure to mark the century of progress since the founding of the American Society of Civil Engineers, an equally strong argument might have been made for not including it, or for employing the image of any one of several other bridges. After all, the George Washington was over twenty years old in 1952, making it more a symbol of eight decades, rather than a century, of progress. Had nothing of significance happened in bridge engineering, if that was indeed to be the metaphor for progress, since 1931? The light suspension bridges with sleek girder-stiffened decks that culminated in the Tacoma Narrows Bridge were not suitable candidates, for obvious reasons, but it could also be argued that the George Washington itself made engineers do what they did to those bridges. And what of the Golden Gate Bridge? Did it not represent progress beyond the George Washington? In short, the George Washington was a curious choice for the stamp. To understand why such a choice was made, however, requires a detour onto some routes of engineering progress that remain incompletely mapped to this day.

After the George Washington demonstrated that a stiffening truss was not absolutely necessary for the success of a suspension bridge, roadways supported by shallow stiffening girders were a natural development. As we have seen, the thin, ribbonlike profile provided by such designs was in keeping with the aesthetic goals of the time, and so Ammann, Steinman, Moisseiff, and their contemporaries were designing bridges with more and more slender profiles. Problems had begun to appear in bridges built as early as 1937. The Fykesesund Bridge in Norway, which had a 750-foot span suspended by rolled I-beams, and the Golden Gate Bridge, which had a conventional truss, oscillated in the wind, but it was the eight-hundred-foot span of Steinman's own Thousand Islands Bridge over the St. Lawrence River, completed in 1938, and the 1,080-foot span of his Deer Isle Bridge in Maine, opened in 1939, along with Ammann's Bronx-Whitestone Bridge, finished that same year, that drew the greatest attention to the problem, es-

Official first-day cover and U.S. postage stamp commemorating the centennial of engineering in America, incorporating, respectively, David Steinman's hands working on plans for his not-to-be-realized Liberty Bridge and Othmar Ammann's George Washington Bridge

pecially with the collapse of the Tacoma Narrows, which had essentially the same plate-girder construction as these.

Even before that disaster, Steinman and Ammann disagreed as to how best to retrofit their wavy bridges. Both of Steinman's spans had been fitted with cable stays that were stretched between points on the tower near the roadway and the suspension cables. Thus installed, they were designed to stay, or steady, the main cables, and thereby check oscillations of them and the suspended roadway to an acceptable level. Ammann's Bronx-Whitestone Bridge, on the other hand, had cables stretched between the tops of the towers and the roadway, which proponents believed would check the motion of the roadway directly. Within a month of the collapse of the Tacoma Narrows Bridge, which had been fitted with cable stays of yet another kind, *Engineering News-Record* published separate articles on the alternatives endorsed by

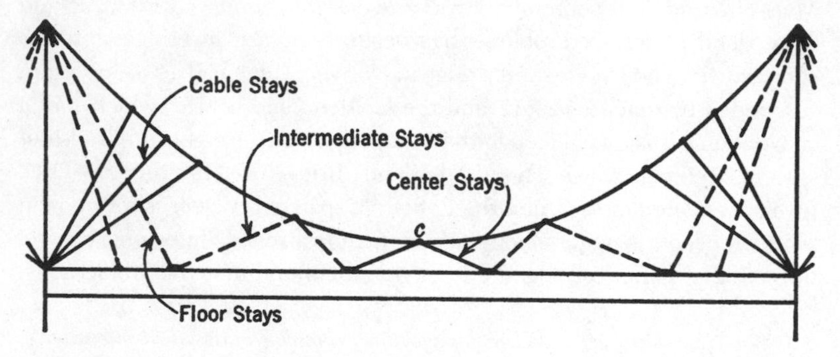

Terminology used for various means of attempting to suppress or reduce oscillations of suspension-bridge decks

Steinman and Ammann. Some time after these pieces appeared, Steinman brought the issue out in the open with a letter to the editor in which he challenged the implication that Ammann's solution was found to be preferable to his own after "elaborate tests on a model conducted at Princeton University." In fact, Steinman contended, his system of cable stays was not included in the tests, which were carried out for the Triborough Bridge Authority.

Steinman concluded with nine reasons why he believed that the system adopted to steady the Bronx-Whitestone was "less efficient and effective than the system previously successfully applied on the Thousand Islands and Deer Isle bridges." Among his reasons were factors relating to temperature changes, tower flexibility, side-span motion, torsional oscillations, and various technical details having to do with the nature of harmonic motion. The letter was followed, in the same issue, by a response from

Ammann, who labeled as "valueless" Steinman's "general unqualified assertions" that were "unsubstantiated" by analysis or experiment, and speculated that his criticism of the Bronx-Whitestone solution was motivated by Steinman's "unsuccessful attempts to sell to the Triborough Bridge Authority his services and the use of his patented stay ropes which he endeavors to advertise as being superior to anything else." In an attempt to refute some of the more technical of Steinman's points relating to the dynamic behavior of bridges, Ammann revealed some of his own prejudices: "They involve such a complex problem that no one, not even the most learned physicist, could make a reliable analysis without experimental investigation. Dr. Steinman's medley of arguments is pure guesswork expressed in impressive sounding scientific words." Extensive studies involving models were required to resolve the matter, according to Ammann, and all installations called for "constant watching" to be sure they did not slip the way those on the Tacoma Narrows Bridge had done. The report of the committee of Ammann, von Kármán, and Woodruff on the collapse of that bridge gave no acknowledgment of the disagreement with Steinman over the form that stays should take. Steinman would later tell *Engineering News-Record* that the committee was composed of his "competitors" and that he was left out.

Politics and personalities can most easily enter where there are no incontrovertible solutions to technical problems, for the analytical difficulties can be horrendous and may rest upon assumptions that can always be called into question. Model tests, including the computer-based ones that are possible today, are also subject to criticism for their assumptions, and even when these are agreed upon, there can never be an exhaustive study of all possible conditions under which the bridge and its cable systems can operate. As for the "most learned physicist" that Ammann referred to, the analysis of such an individual, who might be a onetime academic like Steinman, remains a sore point among engineers to this day, for physicists tend to deal with such idealized systems that many bridge engineers fail to see the analyses as representing real bridges in real winds. Indeed, the problem epitomized by the Tacoma Narrows Bridge continues to stir controversy and debate among theoreticial engineers, practical engineers, and physicists alike. Whatever explanations of that collapse may be claimed or proposed remain open to the accusation that they are pure theory, for the simple reason that the very phenomenon they are intended to explain—namely, the actual oscillation and collapse of a full-scale suspension bridge across the Tacoma Narrows—is not available for verifying the theory. As for the retrofitted bridges of Steinman and Ammann, they have been made even more difficult to analyze with the added complications of their stays and stiffening systems. Though Steinman's cable-stay solution was never admitted to be superior to Ammann's, the latter's Bronx-Whitestone Bridge was finally retrofitted with the stiffening trusses that essentially made the question of cable stays moot, and incidentally destroyed the bridge's sleek lines.

Thus, when it came time to decide what bridge to put on a stamp commemorating a century of engineering, the choice between an Ammann reality and a Steinman dream also became a choice between the two camps of engineering approaches and responses to the Tacoma Narrows collapse. Furthermore, since the centennial of the American Society of Civil Engineers was being viewed as an occasion to define the centennial of the profession of engineering itself in America, that organization no doubt had desired to have a say in whose bridge should be pictured on "their" stamp. They would naturally turn to the work of Ammann, who after his break with Lindenthal had become the consummate organization man. He was more identified with bridges than anyone in New York, where the headquarters of the American Society of Civil Engineers was located, and his George Washington Bridge was the topic of an entire volume of the society's *Transactions*. This must certainly have made a portrayal of that bridge preferable to a photograph of the hands of Steinman, who in his promotion of professional-engineering registration could actually have been seen as a threat to the oldest professional-engineering group in America. In 1953, the year after the stamp was issued, Ammann was made an Honorary Member of the society, thus achieving its coveted "Eminence Grade" of membership. Steinman, on the other hand, continued as an ordinary Member and was never recognized by the ASCE as having achieved "eminence in engineering."

Among the reasons for Steinman's lack of recognition by some segments of the engineering establishment must certainly have been his insistence on keeping the embarrassment of the Tacoma Narrows collapse more in the forefront of discussion than many engineers, such as Ammann, would have liked. The more it was talked about, the more attention it might call to the underlying influence of the George Washington Bridge, and to other spans built in the design climate of the 1930s. In the early-to-mid-1940s, Steinman's desire to understand and articulate theories on the stability of suspension bridges, not to mention to build still larger ones, had brought plenty of attention to the most ignominious event in engineering history. But his interest in bridges became nicely complemented by his desire to pursue literary endeavors.

The book that Steinman wrote with Sara Ruth Watson, *Bridges and Their Builders*, had its origins when Steinman, the very visible engineer and promoter of his profession, was approached by the publisher G. P. Putnam's Sons to write a book for the general reader on the history of bridges. After entering into a contract to do so, he had not been finding time to complete the ambitious project when, early in 1941, in Tampa, Florida, he met Watson, who taught at Fenn College in Cleveland, at a meeting of the American Toll Bridge Association, an organization Steinman had founded about a decade earlier. He was at the meeting to give his classic demonstration of bridge-deck instability, using (like von Kármán) a crude model and an electric fan, in his lecture, "Bridges and Aerodynamics," and Watson was there to lecture on "Bridges in Poetry and Legend."

When Steinman checked in at the meeting and met Watson, she "struck him as so charming" that he offered on the spot to turn the book contract over to her, according to his biographer Ratigan. However, Steinman and Watson agreed to write the book jointly, and it was immensely successful. The chapter on the Roeblings and the Brooklyn Bridge so captivated Irene Steinman that she suggested it be made into a movie. To his response, "I can't write a movie," Irene retorted, "David, you can do *anything*." Since this was no doubt what the egoist Steinman wanted to hear, he set out first to

David Steinman, with simple model and electric fan, giving one of his many lectures on the aerodynamics of suspension bridges

write an entire book on the Roeblings and their bridge, seeing this as a necessary first step toward writing a screenplay. This new book project was to take five years to complete, during which time the movie notion seems to have been forgotten, but not other ones.

In his early sixties, Steinman began to write poetry, some of it bordering on the devotional but much of it about bridges and bridge building, and the verse was eventually collected in several volumes, including *I Built a Bridge* and *Songs of a Bridgebuilder*. His poetry, like almost everything else he did, brought him recognition and awards, and he must have relished the attention that poetry societies gave an engineer who advocated the liberal education of his colleagues so that they might be more readily perceived also, as a group, to be citizens of culture and stature. In one of his poems, Steinman praised the life of the mind, as nurtured in college, where eager young students go

> *To spark the things of spirit that transcend*
> *The shibboleths of ancestry and creed.*

Such ideas must have consoled Steinman, who had repudiated his ethnic origins, even before he committed the thoughts to verse. His major prose-writing project of the time gave him, in addition, a surrogate family to research—the Roeblings and their Brooklyn Bridge. In the preface to his finished work, Steinman perpetuated the myth of his own life story, making himself a child of the bridge rather than of an immigrant family that lived in the squalor and hunger that surrounded it:

> A boy grew up in the shadows of the Bridge. He loved to walk over the span and to explore its marvels. He was awed by its vastness, by the majesty of the towers and by the power of the cables; and he was fascinated by all the details of the construction—the anchorages and the cables, the trussing and the beams, the slip-joint at mid-span, the machinery of the cable railway, the stone work of the towers, and the magic of the radiating stays. When he returned from these pilgrimages he would recount to his playmates and to his elders the wonders he had seen. To him it was truly a "miracle bridge"; and, as he wondered how so marvelous a work could have been created, he was fired with the ambition to become a builder of suspension bridges. In a background of poverty, this far-flung ambition seemed beyond the boy's reach; but the spirit of the Bridge, and later the story of its builders, had entered his heart—and the dream came true.

It was, Steinman continued, "in partial discharge of that debt of inspiration" that he undertook to write his book on the Roeblings, perhaps imagining them to be his professional progenitors. Gleaning information from "thousands of sources—original manuscripts, family letters, diaries, memoirs, notes, reports, periodicals, newspaper files, biographical works, scrapbooks, technical literature, records of historical societies, and correspondence," Steinman cobbled together a gripping story, if in a ponderous book. The first edition of *The Builders of the Bridge* appeared in 1945, the same year that his longtime partner, Holton Robinson, died. Soon afterward, as if released from some constraint by the event, Steinman dropped his elder's name from the firm's, as he had omitted his parents from his biography. In 1948, the firm of D. B. Steinman received a contract for modernizing the Brooklyn Bridge by eliminating the trolley tracks so that it could carry six lanes of vehicular traffic. Steinman assumed, as a further labor of love, this responsibility to modify yet preserve the bridge that had inspired him as a youth.

A second edition of Steinman's story of the Roeblings appeared in 1950, and it differed from the first mainly in its acknowledgment of a woman's contribution to the Brooklyn Bridge enterprise. When Colonel Washington Roebling was struck with caisson disease in 1872, at the age of thirty-five, and became bedridden in a room overlooking the construction site of the Brooklyn Bridge, where three years earlier his father had suffered the accident that was to claim his life, control of the bridge project might have passed on to another engineer had it not been for Washington's wife, Emily Warren Roebling. According to Steinman, writing elsewhere,

> She grasped her husband's ideas and she learned to speak the language of the engineers. She made daily visits to the bridge to inspect the work for the Colonel and to carry his instructions to the staff. She became his coworker and his principal assistant—his inspector, messenger, ambassador, and spokesman—his sole contact with the outside world.

Emily Roebling in fact functioned as assistant to the chief engineer. In a speech on the occasion of unveiling a tablet memorializing her, Steinman related how, upon the day the bridge was dedicated in 1883, Washington Roebling turned to Emily and told her, as a more generous Lindenthal might have told his assistant Ammann or Steinman upon the completion of the Hell Gate Bridge, "I want the world to know that you, too, are one of the Builders of the Bridge." In an epilogue added to the second edition of his book, Steinman claimed as one of its accomplishments the attention that the book had directed "to the heroic contribution of a woman in the building of the Bridge." He was, in part, atoning for the fact that he had forgotten her in giving his story of the builders of the bridge the subtitle *The Story of John Roebling and His Son*, and that he had seemed to dedicate it to the great men alone. Perhaps Steinman was coming to realize, no matter how subliminally, the debt he owed his own father and mother.

6

Steinman was able to devote time to literary pursuits in the 1940s in part because it had been a slow decade for new bridge building. This was caused not so much by the collapse of the Tacoma Narrows—that should only have affected the genre of suspension bridges, the way the fall of the Quebec had adversely affected only cantilever bridges a few decades earlier—as by World War II, which had focused so much attention on the destruction of existing bridges rather than on the erection of new ones, both literally and metaphorically. Writing after the war, Steinman noted that, "compared to the 1930s, when nearly every year witnessed a new bridge triumph, this slowing down of an accelerated tempo is an unusual situation." This article in *Engineering News-Record* presented a series of tables giving such information as the world's longest spans in various categories and recording "progress in bridge building as recorded in successive record span lengths." Suspension bridges had dominated that progress for the previous century, with only rare anomalies, such as the Firth of Forth and Quebec steel cantilevers or the Hell Gate and Bayonne arches. Steinman, and others who had dreamed of designing and constructing even greater suspension bridges, must have worried, especially when they looked at the historic record, that in the wake of the Tacoma Narrows disaster their bridges of choice might become as unpopular as cantilevers had earlier in the century.

Steinman still wanted to build the record-setting Liberty Bridge, which was on his drawing board, on the cover of his firm's brochure, and in the frontispiece of his book with Sara Watson. When *Bridges and Their Builders* was issued in a revised edition by Dover Publications in the mid-1950s, however, Liberty Bridge no longer occupied a position of honor; by then it was clear that his dream bridge was not to be in Steinman's trophy case. In the meantime, he had begun to dream of other great spans, such as those crossing the straits of Mackinac, in Michigan, and Messina, in Italy, which awaited the design of exceptional bridges. Yet, if suspension structures were to make credible bridge proposals for such crossings, the matter of aerodynamic stability would have to be addressed. One approach was to test bridge models in wind tunnels, the way airplane-wing designs had then been studied for some years. Such an approach was, however, open to the limitations of experimental work generally, which meant that it gave only specific information on a specific test of a specific model of a specific design. A cleverly selected array of experiments could provide rather conclusive evidence about the phenomena and design under consideration, but there would always remain uncertainty as to whether the critical conditions had been tested or whether the model gave a true representation of the behavior of the full-scale bridge.

Theoretical studies, on the other hand, could encompass general conditions and thereby deal, in principle, with every conceivable combination of wind and resistance, for example. Whereas Ammann was able to dismiss Steinman's "guesswork expressed in impressive sounding scientific words" during their letter exchange on cable stays, a more mathematically based description of the rigidity and aerodynamic stability of suspension bridges was more difficult to refute. Steinman, with his theoretical background and experience translating the mathematical Melan, was recognized to be capable of producing such a description, and he published it in the November 1943 issue of the *Transactions of the American Society of Civil Engineers*, exactly three years after the Tacoma Narrows collapse. He modeled with mathematical formulas of considerable generality the cables and stiffening girders of a suspension bridge, and proceeded to pursue their mathematical and physical implications for the engineering of such bridges. He was able to conclude from his formulas that by "increasing weight, depth, rigidity, and bracing," or adding stays and devices of various kinds, much as John Roebling had written about and done in the previous century, engineers could make suspension bridges stable in the wind. However, Steinman also pointed out that "these methods resist or check the effects, but do not eliminate the cause." He was also able to conclude from his theoretical analysis that modifications to the cross section of a bridge, such as "using open spaces in the floor or by adding horizontal fins or other wind-deflecting elements" could eliminate the cause of instability. He found it "more scientific to eliminate the cause than to build up the structure to resist the effect," a point of view with which von Kármán would no doubt have concurred. The idea of cutting slots in a bridge deck to obviate its oscillation was, in fact, one of the recommendations that emanated from the board of engineers appointed to investigate the Tacoma Narrows collapse, and the rebuilt bridge across the Narrows did incorporate the idea.

Finally, Steinman concluded his *Transactions* article with a more personal request, that readers "share with him his faith and conviction that suspension bridges of all span lengths can be designed economically to any desired degree of rigidity and with assured aerodynamic stability." Not surprisingly, given the interest in the subject in the wake of the Tacoma Narrows collapse, Steinman's work attracted discussions that occupied more pages than did his paper, as did his responses to these discussions. In general, however, the reactions of readers, especially to his conclusions, were favorable.

In the 1950s, after a decade in which literary and historical pursuits competed for his time as a theoretician and a designer, Steinman rededicated himself with rejuvenated interest to promoting bold new suspension bridges. In part because of theoretical work like his on aerodynamic stability—which provided guidance to wind-tunnel tests of new deck designs, which in turn confirmed theoretical predictions—there was renewed interest and confidence worldwide in building long-span suspension bridges. One project that had been shelved during the 1940s was the crossing of the Straits of Mackinac, which had so separated the Upper from the Lower Peninsula of Michigan that the Upper Peninsula was for all practical and economic purposes more a part of Wisconsin than of Michigan. Thousands of cars would wait sometimes almost a full day to get ferry service across the straits during summer-vacation time. At least as far back as 1888, when Cornelius Vanderbilt was attending a directors' meeting at the Grand Hotel on Mackinac Island and said, "What this area needs is a bridge across the Straits," an obvious advantage had been seen in such a structure. In one of his later poems, "The Bridge at Mackinac," Steinman would not only set the scene but also use rhyme to clarify the pronunciation of the place name. Whereas the island's name is pronounced as it is spelled, this is not so for the waters in which it stands:

In the land of Hiawatha,
Where the white man gazed with awe
At a paradise divided
By the straits of Mackinac—

Regardless, however, of how the place names were pronounced, it had not been until the 1930s that legislation encouraged the serious consideration of a bridge across the straits. Even then, although the technological climate was right, the financial promise of a self-supporting toll bridge for seasonal traffic in the upper Midwest was not so bright as it was in the traffic-growth areas on the East and West Coasts.

By 1950, the sight of thousands of cars waiting for ferries had renewed interest in a bridge. An Inter-Peninsula Communications Commission appointed by Governor G. Mennen Williams promoted a resurrection, with an influential membership, of the Bridge Authority that had been abolished during the war. Though there was some question as to whether the Authority could actually finance or build a bridge, they could certainly gather technical and financial information. The engineering questions were to be addressed first by a board of three consulting engineers, to be recommended by Dean Ivan C. Crawford of the University of Michigan.

A major suspension span would likely be among the bridges of choice for the Mackinac crossing, and so the appointment of a credible consulting board was saddled with the decade-old legacy of the Tacoma Narrows disaster. Whereas Ammann had been a member of the expert committee that reported on that accident, Steinman had subsequently been a much more visible theorist as to how such an event could have happened and be prevented in future bridge designs. The naming of consulting engineers was further complicated by the debate that had ensued between Ammann and Steinman as to how to deal with their own flexible bridges. In the end, Dean Crawford extricated himself from the dilemma by recommending *both* Ammann and Steinman for appointment as consulting engineers for the project, along with Glenn Woodruff, the San Francisco engineer who had sat on the Tacoma Narrows investigatory panel with Ammann and with the aerodynamicist von Kármán.

The board of engineers reported in January 1951, after six months of study, that a "perfectly safe suspension bridge" could be built across the straits, for a cost of approximately $75 million. An independent report on traffic and financing matters supported the economic feasibility of such a project. A final decision was delayed for various reasons, including: questions of steel availability during the Korean War; suggestions that there were unsuitable foundation conditions beneath the straits; and a stipulation that none of the preliminary consulting engineers could be picked for the actual construction project. This last was a frustrating obstacle, for Ammann and Steinman were the two most logical builders. In the end, the Michigan Legislature granted the Bridge Authority the right to engage the engineer of its choice.

In the meantime, the federal Advisory Board on the Investigation of Suspension Bridges, which had been appointed by the commissioner of public roads in 1942 to coordinate research relevant to suspension-bridge design, especially with regard to aerodynamic stability, had issued its preliminary report of tentative findings, which was published in 1952 by the American Society of Civil Engineers. The authors of the 1941 failure report—Ammann, von Kármán, and Woodruff—were members of the Advisory Board, but the "competitor" Steinman was not. Steinman's articles were, however, prominently referenced in the new report. The behavior of some particular bridges, including the Bronx-Whitestone, was discussed, and an unpublished report by Ammann to the Triborough Bridge Authority was quoted as admitting that the effect of the stay cables had "not been sufficient to prevent nor apparently reduce the exceptional oscillations of larger amplitude."

Such disappointing behavior had led to the addition of the present stiffening truss to the bridge, and it gave Steinman a victory of sorts.

Financing for a Mackinac Bridge was not easy to come by, and the Michigan Bridge Authority had no funds to engage an engineer to produce a design. Only Steinman would agree to undertake the job on speculation. Thus, in January 1953, he was selected as designing engineer of the Mackinac Bridge; Woodruff was later named as his associate. Preliminary plans and estimates were ready within two months, and construction contracts were negotiated, as required, before the bonds could be issued in late 1953. Steinman's design incorporated features endorsed by the Advisory Board, including space between the deep stiffening trusses and the outer edge of the roadway. This was to raise the critical wind velocity, at which oscillations of the deck could start, from the forty-two miles per hour that had become associated with the failed Tacoma Narrows Bridge to a calculated value of 642 miles per hour. An additional feature—namely, an open-grid roadway under the two center traffic lanes—raised the critical wind velocity to "infinity."

Steinman's design, when drawn to scale, showed the Mackinac Straits Bridge to be larger than the Golden Gate. Though the older structure still retained the record for the longest suspended span between towers, the Mackinac Bridge was actually longer in total suspended span, by almost a thousand feet. When measured from the end of one anchorage to the end of the other, the suspension bridge itself was over eighty-six hundred feet long, and thus surpassed by over two thousand feet the overall length of any suspension bridge extant. Steinman had, in a way, gotten to build the largest suspension bridge on earth. When he wrote, in collaboration with Michigan newspaperman John T. Nevill, the story of the design and construction of the enormous structure, the book was entitled *Miracle Bridge at Mackinac*. At least in his own mind, Steinman was no doubt likening his crowning achievement to the now dwarfed Brooklyn Bridge of his youth. In another work, the "official picture history" of the new bridge, Steinman wrote of the structure and himself:

> The Mackinac Bridge is my crowning achievement—the consummation of a lifetime dedicated to my chosen profession of bridge engineering. As far back as 1893, when I was a newsboy selling papers near the Brooklyn Bridge, I told the other newsboys that someday I was going to build bridges like the famous structure that towered majestically above us. They laughed at me. Now I can point to 400 bridges I have built around the world, and to my masterwork—the Mackinac Bridge—the greatest of all. The realization, one after another, of dreams that seemed hopeless leaves me reverent and humble.

The Mackinac Bridge

Though Steinman may have had a curious way of expressing his humility, he was no doubt humbled at this time, for it was also clear that the Mackinac Bridge would have to be his "Liberty Bridge," because the New York Narrows project had in the meantime been given to Ammann by Robert Moses, who was in effect his own banker.

There were still other bridge prizes to be pursued, of course, and Steinman had been pursuing them. Yet, even if he had not been growing old, and even if he had not always acknowledged the essential role of assistants in helping him reach his goals, including lesser ones than his "crowning achievement," Steinman the chief engineer knew that he could not have succeeded in his quests without a talented and broad-based staff. As Ammann had acknowledged his dependence upon his assistants, so did Steinman at the conclusion of the Mackinac project. Among those who were central to the success of the enterprise were R. M. Boynton, C. H. Gronquist, and J. London. Boynton, a 1920 civil-engineering graduate of the University of Maine, had been with Steinman since 1928 and was responsible for the substructure of the bridge. Carl Gronquist, who received B.S., M.S., and C.E. degrees from Rutgers University, joined Steinman after receiving the master's degree in 1927, and was in charge of the superstructure. London, who received both his B.S. and his C.E. degree from the City College of New York in the early 1920s, had joined Steinman in 1922 and had responsibility for the approaches, lighting, and equipment associated with the Mackinac Bridge. Together, they represented a new generation of engineer, one that came out of the many newer American public schools of engineering that in the early twentieth century overshadowed the once dominant position of the European tradition and private schools like Rensselaer Polytechnic Institute.

In 1960, Steinman added the names of three partners to his firm's name. Whereas he had practiced as D. B. Steinman since the death of Holton Robinson, now the consulting firm would be known as Steinman, Boynton, Gronquist & London. The new firm needed a new brochure, of course, and in it a brief background on the organization, with no false modesty, stated its credentials: "Since 1921, the members of the firm have been designers or consultants on over 400 bridges on five continents, many of them being among the most renowned bridges in the world." The "record cost" of the Mackinac Bridge, almost $100 million, was described as more than that of the George Washington and Golden Gate bridges combined. This "artistically and scientifically . . . outstanding" structure, the "longest suspension bridge in the world," was further described in more personal terms: "Here is Dr. Steinman's and his firm's crowning achievement. It represents the attainment of a new goal of perfect aerodynamic stability, never before attained or even approximated in any prior suspension bridge design."

Not only past achievements were pictured in the consulting firm's brochure. In a foreword signed by Steinman, he wrote of "the great spans of tomorrow," and it was one of these especially that had recaptured his imagination. As early as 1950, the Italian Steel Institute had retained Steinman to prepare plans for a crossing of the two-mile-wide Strait of Messina, between Sicily and the Italian mainland. The legendary passage through which Ulysses had to sail between Scylla and Charybdis, the strait is the site of the occasional mirage known as the *fata morgana*. How the poet Steinman must have longed for the commission, and the occasion to commemorate its achievement in verse. There was no time for poetry when courting engineering commissions, however, and the bridge sketched in the brochure was described as having a record five-thousand-foot main span, stiffened against railroad traffic, aerodynamic forces, and earthquakes. According to the consulting firm's brochure, commencement of construction awaited only the financing of the $150 million cost.

The Strait of Messina bridge design proposed by David Steinman

Perhaps it was the adrenaline that the Mackinac commission released that caused Steinman to produce a new outpouring of articles on bridges and aerodynamics in the early-to-mid-1950s, but it was the bridge across the Strait of Messina that became his new sought-after achievement. Steinman knew that, no matter how much he spoke of the total suspended span or the eighty-three hundred feet between abutments or the five-mile overall length of the roadway of the Mackinac Bridge, the main suspended span was the technological achievement by which records were really kept, and the Mackinac's was only thirty-eight hundred feet long, a full four hundred feet less than that of the Golden Gate, and less still than that of Ammann's Verrazano-Narrows Bridge would be. If Steinman really wanted to hold the record, he had to be identified with a bridge like the one he had proposed across the Strait of Messina.

Among the articles Steinman had written, more than incidentally promoting his new dream bridge, was one entitled "Suspension Bridges: The Aerodynamic Problem and Its Solution." This appeared in 1954 in *American Scientist*, the journal of Sigma Xi, the research honor society that had been founded early in the century as a scientific counterpart to Phi Beta Kappa. In this comprehensive piece, renderings of the Mackinac and Strait of Messina bridges, drawn from the same perspective, appear on facing pages. There is a strong physical resemblance between the two bridges' towers, and the clear implication had to be, if the one, why not the other. There were certainly no technical impediments in Steinman's mind, as his article clearly argued. He showed how he had physically checked with stays the aerodynamic motion of his Deer Isle Bridge, without having to resort to a retrofitted truss, and he pointed out how he had solved the mathematical problem of understanding what it took to control aerodynamic motion in bridges on the drawing board. Forty years after its appearance, the paper is remembered by engineers and scientists alike as having been a definitive resolution of the problem of suspension-bridge oscillations, both practically and theoretically, in spite of a renewed interest in 1990 in revisiting and reanalyzing the Tacoma Narrows collapse on the occasion of its fiftieth anniversary. Another article by Steinman, a historical perspective on bridges generally but with a special emphasis on suspension bridges and the aerodynamic problem, had appeared in *Scientific American*. It concluded with a discussion of bridges of the future, of which the Strait of Messina span was the clear successor to the one across the Straits of Mackinac.

7

Perhaps those whose dreams of bridges go to the lengths that Steinman's did cannot ever stop dreaming of bettering themselves. The Messina bridge project was to be left on Steinman's drawing board, however, when he died in 1960. He had become ill barely six months after establishing the partnership that would associate his name with projects well beyond his death. His obituary in *The New York Times* remembered him as the designer of the

Henry Hudson Bridge, the work he had effectively completed as a student at Columbia, as well as of "more than 400 others spanning rivers and harbors in many parts of the world." An editorial in that paper called his "greatest success" the bridge in Michigan that had come to be recognized as the "world's longest suspension bridge" and to be called affectionately "Big Mack." Ironically, the hometown paper incorrectly spoke of Steinman as having been "born on the Lower East Side four years before Brooklyn Bridge was opened on May 24, 1883," which would have made his year of birth the same as Ammann's. The paper did not misspeak, however, when it referred to Steinman's belief that a bridge could be "a poem stretched across a river" and that "bridges are an index to civilization." Though the editorial recognized Steinman to have been a poet who wrote in steel, it by no means remembered him only as a dreamer: "He helped in the negotiations and the rivalries that must proceed—sometimes it seems endlessly—before a great bridge is built." It should not diminish Steinman's accomplishments to say that this prosaic praise might also have been written of any of his few significant rivals and peers.

But if the popular press remembered Steinman affectionately, only allowing that "rivalries" were a part of bridge building, the engineering press did not recall him so warmly. *Civil Engineering*, the magazine of the American Society of Civil Engineers, treated his death as but another bit of society news, albeit with a picture of an aged Steinman holding a drawing of his last dream, the Messina Strait bridge. He was acknowledged to be "regarded as [*sic*] one of the great engineers of the twentieth century," but the reserved tone of the notice of the death of the "famous bridge builder," who had been a member of the society for half a century, only hinted at the legacy Steinman had left behind in the profession he so loved. Unlike Ammann, who had received the society's highest formal recognition, Steinman seems to have been thought of as just another dues-paying member, albeit one of some notable accomplishment who had been active for fifty years. In fact, he had been the promoter of what was seen as a competing organization, the National Society of Professional Engineers. Given his aspirations toward honors and awards, he may have been disappointed at not being made an honorary member, or at least a fellow, of the civil-engineering society. He would be further shunned by having not so much as an abstract of a memoir of him published in the society's *Transactions*. But such mean-spiritedness had been foreshadowed.

A year before he died, *Engineering News-Record* had profiled Steinman in the same "Men and Jobs" series in which its editors had profiled Ammann a year earlier. The contrast of the two treatments is striking. Ammann's was titled "An Artist in Steel Design," and it portrayed the "unobtrusive looking man" as one who disliked attention and preferred to stay at home rather than go to parties that served "no particular purpose." Instead of being a loner, however, he was a firm "believer in the conference table and in amalgamation of talents to do a job." When asked by the interviewer to describe the "typical" engineer's personality, Ammann replied:

> We may lack glamor and sparkle. We might even be considered dull by many people, but I don't believe it. I think that the fact that we are dealing so intensely with concepts outside the layman's ken makes us often not understood by them.
>
> This is actually the engineer's number one problem today. He must learn somehow to communicate more easily—both with his colleagues and with the public. Most of us [engineers], when we have something to say, will qualify our statement to death until we're bogged down and the point is lost to all but those who have the patience to dig to see what we really mean. Even then, one is not often sure.

Ammann suggested teaching more communication skills to student engineers as a way of correcting the problem, but neither he nor the editor who was interviewing him seemed to want to pursue directly what role fundamental traits of personality may have to do with it all. Rather, the interview continued with a digression into Ammann's keen sense of office detail, which was reported to be exemplified by his knowing "most of his employees by name and personality," and by the fact that he "scanned carefully" everything that left the office.

Among the things reported to ruffle Ammann's feathers was any expression of admiration for the "rugged individualist," because "a man like that is nothing but an egotist." Ammann believed that "People are meant to work together. Nobody wants to see a one-man show." There can be little doubt that Ammann's rival, the rugged individualist and egotist David Steinman, was a target of these remarks. It was Steinman, more than Ammann, who had reached out to and communicated with the public with an effective glamour and sparkle.

Steinman's profile, perhaps at his instigation and after an appeal for equal time, appeared under the title of the magazine's cover story, "What Measure for This Man?" He was described as having had a full life, "full of disappointments and frustrations as well as of recognition and financial rewards." By his own admission, his great disappointment was being denied "the focus of my life ambition," the Liberty Bridge he had spent thirty years promoting. *Engineering News-Record,* which had been so much an interpreter of the profession, surmised, moreover, that the "loss of affection among contemporaries may be the greatest of his sacrifices," for he was a man who wanted "very much to make friends" but whose personality had not been one "to take or leave."

To its own rhetorical question of what Steinman's life added up to, the magazine responded with more questions. Would it be "the mighty Mackinac Bridge"; the book about the Roeblings, *The Builders of the Bridge*; the book for juveniles, *Famous Bridges of the World*; the poems; the National Society of Professional Engineers, which he founded; his tireless efforts for passage of registration laws for engineers; or his later "speech-making campaign when he barnstormed the country" explaining the collapse of the Tacoma Narrows Bridge, which he believed he could have saved? It was this last effort especially that did "not endear him to his contemporaries who had a part in the investigation" of the failure that was so embarrassing to the profession.

According to Steinman, "All my competitors were on a committee to investigate the collapse and I was scrupulously left out." He also allowed that "channels of information" were closed to him, as were "channels of publication," but the anonymous reporter did not feel the engineer said these things with bitterness, only "somewhat sadly." Indeed, one of Steinman's distinguishing traits may have been his willingness to discuss openly engineering embarrassments for the good of the entire profession. In 1929, for example, when heat-treated wire was showing signs of weakness in the cables of his Mount Hope Bridge in Rhode Island and the Ambassador Bridge in Detroit, both then under construction, the cables were dismantled and replaced with conventional cold-drawn wire. Rather than helping the incidents to be forgotten, as some thought human nature and professional pride might dictate, Steinman "distinguished himself by helping to record fully and promptly the findings of this unfortunate experience."

But it was the strictly personal qualities of the man, more separated from professional practice than issues of the cables or instabilities of bridges, that had finally to be addressed in a profile. In the year since Ammann had blasted the rugged individualist in the same department of *Engineering News-Record*, Steinman had added the names of other engineers to his own, and his firm was running easily without "The Doctor," whose identity had sometimes seemed to be one with it. His relationship to his employees was nonetheless reported to be perhaps "an outstanding facet" of his personality: "They call him generous, thoughtful, receptive, ethical, quixotic, brilliant, warm, human, a team man and character-builder." He, for his part, considered all of them his "brother engineers."

Steinman's methods of doing business were perhaps affected by his "quixotic" quality. He admitted to having "put off earnings to fight causes," and thus it was not surprising that the practice "barely broke even most years until the Fifties," when the Mackinac Bridge project was realized. His methods of "promoting professional engagements" were considered one likely legacy of his career, for "he would do considerable engineering on a proposed bridge in hopes of some day getting to design it in detail and see it built." His Liberty Bridge was among a list of forty other such proposed bridges. During the 1920s, Steinman had traveled everywhere around the country looking for prospective sites for toll bridges to design, but he later speculated on why he found it difficult to get very far with state highway departments: "I didn't know and don't want to know the political ropes."

Though Steinman may not have known or even wanted to know the politics of bridge building, he did seem to have an instinct for the politics of self-promotion. Perhaps the gentle-looking man who was, like so many builders of large bridges, slight in physical stature, worked hard at promoting his own accomplishments because he had repudiated the more modest but nevertheless essential ones of his parents. Or perhaps he felt that competing for publicity and recognition of a less tangible kind was not quite the same as fighting tooth and nail for a bridge commission. Whatever his motivation, however, Steinman was a notorious self-promoter, leading at least one reporter to state that "perhaps his greatest contribution will some day be judged to be his public relations effort."

Steinman seemed to seek and need the limelight as a flower does the sunlight, and no one knew this more objectively than the press: "Editorial offices for years have been on the receiving end of the Steinman mail—poems, itineraries, news releases, pictures." Although little of the material was usable to editors, it did call constant attention to the profession of engineering. It was estimated that no other engineer since Herbert Hoover or Charles Kettering, the crusty inventor of the electric starter for automobiles, had done more in his time to make engineering known to the public, and in Steinman's case, "he identified himself, indeed integrated himself, with his profession so thoroughly that it would be difficult to say what effort he puts forth for self and which for his profession." Though this perhaps self-generated confusion of himself with engineering did not win the approval of some of Steinman's contemporary engineers, such as Ammann, it was in the final analysis a true measure of the man.

When Steinman died, a little more than a year after the assessment of him appeared in *Engineering News-Record*, the journal editorialized on "Dave Steinman" with the same ambivalence, noting that, "unfortunately, his great accomplishments were sometimes clouded by his personality, which frequently made him the center of controversy." He was likened in ego and outspokenness to Frank Lloyd Wright, and was said to have done

David Steinman the self-promoter, shown here posing among the floor-stay and suspender cables of the Brooklyn Bridge

for engineering in his lifetime what Wright did for architecture in his. The question of Steinman's "real contribution" still irked the editors, however. They allowed that he "personified civil engineering" and that he was a "nearly unique" interpreter of his profession to the public, leaving behind too few to fill this important role, but in the end they would not grant him the accolade he no doubt would have most wanted to hear. Not one of his bridges was named in the editorial, not even the great Mackinac or his dream Liberty or his proposed Messina Strait. Instead, the magazine whose predecessor forty-seven years earlier had run a picture of Steinman's first bridge, the modest timber cantilever he built with a troop of Boy Scouts in Idaho, grouped all of his structural achievements anonymously into a single sentence that at the same time negated them: "His bridges, which will remain as great monuments to him, would probably have been designed by others if he had not come along."

Though this kick at the casket might be in one sense true, it diminishes more the editor's credibility as a student of engineering than the greatness of Steinman's accomplishments. No doubt, had Steinman and his colleagues not designed and built the Mackinac Bridge in Michigan, the Deer Isle Bridge in Maine, the St. Johns Bridge in Oregon, the Carquinez Straits Bridge in California, the Florianópolis Bridge in Brazil, or even the timber cantilever in Idaho, those places might have had their bridges sooner or later, by others if Steinman had not come along. But can anyone, after knowing how the personality of the engineer informs his designs, believe that any of Steinman's bridges would be quite the same span if designed by another? The Liberty Bridge remained a paper bridge, because the Verrazano-Narrows was built—clearly an Ammann bridge, rooted in the same aesthetic as his George Washington and Bronx-Whitestone bridges. Had Ammann not come along and filled the role of chief bridge engineer for the fledgling Port Authority, who knows what the George Washington Bridge and all its descendants would look like today? There is little doubt that bridges would stand where they do now, but they would be different bridges, embodying the personal style and ideas of whoever's bridges they were, and they would affect our present sense of bridgeness differently than do those that actually exist. Eads, Cooper, Lindenthal, Ammann, and Steinman each built his own kind of bridge in his own time, and each of them has left a legacy that has influenced the bridges and bridge builders that have followed. This is, and will always be, the essence of the endeavor.

REALIZE

The great bridges of the great engineers remain as spectacular today as they were when they were dedicated, but even the greatest bridges may be least appreciated by those who benefit most from them. A search for a more efficient means than ferries to move railroad trains and then motor vehicles impelled the engineers of the century of bridge building that extended from the 1840s to the 1930s to design more and more ambitious spans. However, unless these bridges are approached with a proper perspective, whether by the armchair traveler from behind a book or the actual traveler from behind a steering wheel, their greatness and achievement can hardly be appreciated. The right of way that approaches a massive cantilever head-on affords no view of the bridge to speak of, and the way across can appear from a train window to be little more than a long series of slanted steel obstructions to the view of a majestic river. The *whoosh-whoosh-whoosh* of relative motion gives no hint of the human dreams and drama that began over a blank and silent drawing board.

The view from the rear seat of an automobile can give us no better a perspective of a highway bridge, especially if it is one on a straight and heavily trafficked road. Who can really appreciate the structural-engineering achievement of a spectacular bridge when traveling in one car among hundreds in one of several lanes and at the same time trying to help the driver pick out the one relevant sign for the next connecting road on the interstate-highway route? Sometimes we can ride over bridges of immense technical

The Bixby Creek Bridge, a reinforced concrete arch, on California's coastal highway

achievement without even sensing their magnitude or grandeur or knowing that we are on a bridge at all. This is especially true of arch bridges whose structural muscle lies wholly below the roadway and so is invisible from the road. One such bridge is the Rio Grande High Bridge, which carries U.S. Route 64 over the gorge of the Rio Grande River in northern New Mexico. Traveling west on this route from Taos, one encounters what appears to be a remarkably flat plain extending for miles, essentially unobstructed by vertical vegetation or human artifacts. From the level of the road approaching it, the gorge itself is an invisible cut in the plain, and its great depth can only be appreciated by parking the car and leaning over the parapet on one of the slight outward extensions of the bridge so thoughtfully provided for sightseers, from which the great steel arch below may seem to make the gorge appear even deeper than it is. Such virtually invisible and largely anonymous bridges are numberless on the numbered roads across America, on the winding Pacific Coast road, in the gouged land of the Southwest, in the hilly land of the East, and even in the flat Midwest, where bridge approaches rise like Indian mounds to carry local traffic over the interstate.

Ironically, the greatest bridges and the ones that we tend to approach from the most favorable prospects are those in large and crowded cities, where the buildings push the roadways almost into the water, so that they must spiral back toward the bridge as if drawn to its grandeur. In New York City, for example, one has a spectacular view of the George Washington Bridge for what seems like miles when approaching it from the south on the West Side Highway and the Henry Hudson Parkway. In rush hour, the traffic on this road is often stop-and-go, and when it is stopped one can admire Othmar Ammann's masterpiece and reflect upon how much more convenient than ferries it nonetheless is for crossing the Hudson. At night, looking out the window of a terrace restaurant just east of Columbia University, one can see, over the rooftops of Manhattan to the north, the overwhelming scale of the bridge outlined in lights. Over the rooftops to the east, the lighted outline of the Triborough Bridge looms, and, beyond it, that of the Bronx-Whitestone. As one drives south on the FDR Drive, New York City's oldest suspended structures—the Williamsburg, Manhattan, and Brooklyn bridges—appear in sequence; driving under their approach spans and beside their towers provides a sense of scale that is missing in a drive over them. In San Francisco, the towers of the Golden Gate and Bay bridges dominate so many views that they have become defining landmarks of the city by the bay.

Although perhaps less appreciated visually, the importance of the Bay Bridge for communication between San Francisco and Oakland was demonstrated when a section of its upper deck fell onto its lower one during the 1989 Loma Prieta Earthquake, closing both roadways for about a month. The Northridge Earthquake, which struck Los Angeles in 1994, demonstrated further the vital link that a major bridge provides in a highway system. With key bridges out, Los Angeles commuters found themselves stuck in day-long traffic jams on detours in the early days following the disaster. Similar frustrations occurred in Connecticut in 1983, when a section of a bridge over the Mianus River fell without warning, leaving a gap in the heavily traveled Interstate 95. Though traffic was rerouted through neighboring towns, drivers were frustrated and residents annoyed until the bridge section was replaced.

Barring accidents, bridges, like health, are most appreciated when they begin to deteriorate and fail. Thus politicians seemed to become interested in bridges when they found that so many of them were structurally deficient. In 1992, for example, this included about one out of every five of the half-million or so bridges in the United States—an improvement over some previous years. Among the most dramatic stories is of a bridge whose condition began to deteriorate almost from its beginning. The Williamsburg Bridge is said to have been "born of a dare" when Leffert Buck was challenged to "build a bridge longer than the Brooklyn Bridge, in half the time and with less money," and the span appeared to have fulfilled the requirements when it opened in 1903. Buck achieved the time and cost savings in part by coating the wire strands in the bridge's four cables with graphite and linseed oil rather than galvanizing them with a molten-zinc mixture, as had been done with the Brooklyn. At the time, the procedure was considered radically different, and the wisdom of the decision came under serious question when broken wires and corrosion were discovered in the cables before the bridge was a decade old. Before it was two decades old, the bridge's cables were wrapped in a galvanized steel cover, but severe rusting continued inside. When the bridge was forty years old, hundreds of gallons of linseed oil were poured onto the cables at the tops of the towers, with the expectation that it would seep down the cables and into their interstices and slow the corrosion. Two decades later, the procedure was repeated, with fish oil and mineral spirits. In the meantime, painting of the steel in the towers and roadway had been neglected, and so they had also developed severe rusting. Other East River bridges—the Brooklyn, Manhattan, and Queensboro—had also been the victims of deferred maintenance, blamed on a period of fiscal crisis that had struck New York, and by the mid-1980s they were undergoing repair and rehabilitation work costing on the order of half a billion dollars.

The serious deterioration of the cables of the Williamsburg Bridge put it in a category of its own, however, and a major decision presented itself: should the bridge be fixed, or should it be torn down and replaced by a new structure? Replacing the cables without closing the bridge was likened to "restringing a pearl necklace while it is around someone's neck," but building an entirely new bridge "in this era of environmental impact statements" was thought to invite "legal challenges that could result in substantial delays." Several proposals that might be described as "radical" were considered. One involved erecting new towers over the old ones and hanging a new deck under the old, which traffic would continue to use during construction. When the new deck was finished, the old one could be closed and demolished, after which the new deck could be moved into position to receive traffic. A second proposal involved first building narrow new bridges on each side of the old, then tearing down the deteriorated bridge and building a third bridge in its place, and finally joining all three to provide a wide new roadway. Still another proposal called for building two larger bridges on either side of the old one, tearing it down, and then moving the two new bridges entire toward each other to be joined together as a unit. While such proposals were being considered, the engineering firm of Steinman, Boynton, Gronquist & Birdsall was monitoring strain gauges that it had installed on the eyebars connecting the cables to the anchorages, in order to have advance warning of any accelerated deterioration of rusted wires. The consulting engineers were effectively using a technique introduced by their progenitor, David Steinman, seventy years earlier, when he wanted to check theory with reality on the Hell Gate Bridge. But just as the Williamsburg situation was described as "a case study of how not to treat a bridge as well as how not to build one," the Hell Gate itself was at the same time the subject of a different kind of scrutiny.

In the late 1980s, New York Senator Daniel Patrick Moynihan was made aware that the Hell Gate Bridge had not been painted in fifty-odd years, except for the daring work of graffiti artists who had left their marks high atop the stone parapets and steel arches. Since Moynihan had lived for a while as a child in Astoria, known mostly as the site of the Steinway piano factory but also as the eastern approach to the Hell Gate, he took a special interest in the bridge, which he called "a great engineering miracle." Moynihan, who was chairman of the Subcommittee on Water Resources, Transportation, and Infrastructure of the Committee on Environment and Public Works, was disappointed that no one at either Amtrak or the Department of Transportation appeared to have an interest in the bridge. He was further annoyed when his inquiring letter to the Department of Transportation brought no response, and so he held a special hearing on Capitol Hill to discuss the matter. The nearby Triborough Bridge, he pointed out, was constantly being painted by a crew that did nothing else, and that was the way to take care of a bridge so that it did not rust away. Not all bridges, unfortunately, were under the watchful care of an agency like New York's Triborough Bridge and Tunnel Authority, which collected sufficient tolls to maintain its works.

Moynihan pointed out that generally we were "disinvesting in the American plant" and that "the national roof is leaking." The estimate for painting the Hell Gate Bridge was $43 million, however, about a third of which would have to be spent just in removing the accumulated rust and dealing in an environmentally sound way with the lead-based paint that still covered the bridge here and there. Since the heavy steel bridge was deemed structurally sound by W. Graham Claytor, Jr., president and chairman of Amtrak, he had little sympathy for spending so many millions of dollars for "cosmetic purposes." So reported *The New Yorker* in a story on the city's "eighth bridge" in early 1991. Whether or not he needed such a story to push him to persist, before the year was out Senator Moynihan had worked the halls of Congress, and $55 million had been appropriated to repair and repaint the Hell Gate Bridge, "the least famous of the eight across the East River," but the one dearest to the senator's heart. Not every bridge has such influential friends.

The appropriation of the money was only the first step in getting the Hell Gate Bridge painted, however. In the spring of the following year, a special train carried Moynihan, Mayor David Dinkins, and other local politicians to the bridge for a ceremonial first brushstroke, as *The New Yorker* reported. The ceremony had a special twist to it, for the key bridge link in the northeast corridor rail line between Boston and Washington was to be painted "an entirely new color—Hell Gate Red!" The color had been chosen by a "committee of color experts," which included architects, a representative of the Municipal Art Society, the "minimalist painter" Robert Ryman, and the "color consultants" Taffy Dahl and Donald Kaufman from the firm Donald Kaufman Color. A press kit described the new color as "deep cool red" and placed it in "the family of red colors historically associated with railroads," so that the Hell Gate would be readily distinguished from the city's automobile bridges. No mention was made of the distinctive red that had covered the Forth Bridge for a century. The consultants' color was to "complement the greens and blues of the landscape, adding to the richness of the scene," which made it sound as if the bridge were in a pristine natural setting rather than the litter-strewn, graffiti-pocked center of New York. Proponents of the color appeared to consider as a plus that it also would "disguise any rust that developed on the bridge"; no one seems to have mentioned that this should not be the point of a paint job. Paint should serve a prophylactic as well as a cosmetic purpose, but, rather than disguise rust, it might be better to highlight any that might begin to develop, so that it could be attended to before it spread too far.

When the train arrived at the bridge, dignitaries and reporters alike noted the considerable number of holes in the walkway of the aging bridge. However, the order of business was not to repair the structure but to apply the ceremonial first strokes of paint. Shortly after the mayor and the senator began wielding their rollers, observers saw that the color was not quite what they had been led to believe was Hell Gate Red. To some it looked like "sort of a vermilion," to others a pink. The representative of the Municipal Art Society assured Senator Moynihan that after the appropriate number of coats, the paint would dry to the appropriate color, that in fact the ceremonial paint was a benign latex substitute for the real thing, whose fumes, it was feared, would have "knocked out half the Democratic leadership of the city."

The story of painting the Hell Gate is full of art and artifice, of politics and chicanery. The engineers—Lindenthal and his assistants, Ammann and Steinman—and their engineering of the structure seventy-five years earlier were not a part of the story or the event in the early 1990s. They had by and large been forgotten in the hoopla of color consultants and committees, who will no doubt ignore the bridge for another several decades at least, especially if its rust is disguised by its color. To the engineer, painting a bridge is as necessary as changing the oil in a car; it is neglected at the peril of the machine, which at least one architect, Le Corbusier, understood did not have to have grossly moving parts. Every bridge is a machine of sorts, moving ever so slightly under the action of the traffic, the push of the wind, the heat of the sun, or the growth of the rust that should not be allowed. Arresting rust and other deleterious movements of a bridge are matters for sound engineering, not for sound-byte politics. It has been estimated that as much as 2 percent of new-construction cost should be earmarked each year for maintenance, including painting, for the life of a major bridge structure. Neglect, euphemistically called "deferment," of maintenance is only postponement of the inevitable, as the cases of the Williamsburg and Hell Gate bridges so forcefully demonstrate.

The Intermodal Surface Transportation Efficiency Act of 1991, behind which Senator Moynihan was the major force, has encouraged aesthetic improvements to infrastructure generally, and these can double as protection against deterioration. In the Baltimore area, for example, Stan Edmister, who calls himself the nation's first "bridge-maintenance artist," has used multiple layers of high-gloss paint to provide a protective coating that he claims will last fifteen years, which is about twice the time that conventional bridge paint lasts. In 1992, Edmister began painting bridges on Interstate 83 with "eye-catching, rich colors that break from the light blues and greens traditionally used to make bridges disappear into the landscape." His idea of painting different structural elements of the spans different colors evokes the practice of Victorian times, when the Crystal Palace, for example, was decorated with such a polychromic scheme, as Alhambra Jones applied his "science of color" to it. London's light-blue, white, and red Blackfriars Bridge, the blue-and-white suspended approaches of stone-clad Tower Bridge, and the red Forth Bridge near Edinburgh, are fine extant examples of the late-Victorian sense of color. Indeed, from 1890 until recently, the Forth Bridge was conscientiously kept painted "Forth Bridge red," and the constant job occupied twenty-four painters, who worked steadily in a twelve-year cycle to keep the entire structure covered with five coats of paint. The vastness of the endeavor was so well known in Britain that "painting the Forth Bridge" is still a metaphor there for an endless task.

For some time in America, even before there were modern color consultants and bridge artists, engineers have been receptive to using paint for the decoration as well as for the protection of steel. As early as 1902, Chicago's Loop elevated structure was painted "a pale buff color" at the request of merchants who wished to brighten the street below. That effect was achieved, "but soot and dirt which collected on the upper portions and mud which spattered the columns soon began to mar the appearance of the structure, so that it was eventually repainted the original dark gray." In 1920, at the annual meeting of the American Railway Bridge and Building Association, J. R. Shean, of the Pacific Electric Railway, argued that "canary yellow, pearl gray or light olive green" finishing coats on steel bridges would make them "more in harmony with the surroundings." Though there would be some discoloration from dirt and smoke, he reasoned that the lighter colors would last longer because of their greater "resistance to heat rays."

David Steinman was a strong advocate of the creative painting of bridges. He introduced the use of color in structures he designed because he had grown tired of seeing bridges painted funereal black and battleship gray and he wanted "to get away from these sad, somber, cold colors and into something warm and bright, to harmonize with and be a part of the landscape." His Mount Hope Bridge, his first departure from tradition, was painted "a light-greenish tint." Steinman "progressively became bolder, using verde green, jade green, apple green, foliage green, and forest green" in later bridges. His St. Johns Bridge in Portland, Oregon, for example, whose high roadway provides over two hundred feet of navigable clearance, in 1931 was painted "a pleasing shade of verde green," to blend in with the trees, rather than the yellow-and-black stripes that had been suggested to warn airplane pilots of its presence. The major suspension span of Steinman's Thousand Islands International Bridge, dedicated in 1938, had its steelwork painted a "patina green." His most daring use of color, perhaps, was in the Mackinac Bridge, for which he chose "a two-color combination—foliage green for the spans and cables, and ivory for the towers, to express the difference of function"—namely, tension and compression, something Waddell had suggested in the nineteenth century. The onetime academic Steinman knew he might be "joshed about the *ivory towers*," but he felt they were appropriate for the structure. Furthermore, he reflected, "I may be called a dreamer. But the Mackinac Bridge is a wonderful dream come true!"

Among the most distinctively painted of American bridges is, of course, the Golden Gate, and its consulting architect, Irving Morrow, first proposed "using an orange-red color for the towers," with deeper shades for the suspenders, cables, and approaches. He also thought that the scale of the structure should be "emphasized, rather than played down," and he thought that a "red, earthy color" would be appropriate for contrast with the alternately foggy-gray and blue skies over the Golden Gate. In the end, a single color was chosen, to tie the bridge into the red-orange rock of the Marin hills. The color finally settled upon has been variously described as "red lead" and "iron-oxide red," but it is officially known as "International Orange," and it is in fact strikingly similar to that of Forth Bridge red. Morrow, who was also responsible for the sculptural detailing of the towers, helped make the bridge what has been called "the world's largest Art Deco sculpture." However, the Golden Gate Bridge remains a healthy work of structural art largely because it has been properly cared for, being painted continuously since its completion in 1937. The job of covering the ten million square feet of steel surfaces takes a full-time force of painters forty-eight months to complete, after which they, like their onetime counterparts on the Forth Bridge, begin all over again.

The continued collection of tolls on a bridge like the Golden Gate assures its continued maintenance. In his final report, Chief Engineer Strauss countered the call for "free bridges" with the observation that, like free lunches, "there are no free bridges" and "all bridges must be paid for in taxes of some sort." He called tolls a "users' tax" and described them as "the only method by which the age-old isolation of San Francisco could be ended." However, Strauss also noted that the "modest and fair" toll rate of fifty cents set by the Bridge District in 1937 was less than half what he had based his financial calculations on, thus jeopardizing its future. Fortunately, his estimates of use were overly conservative. The first year, the Golden Gate was crossed by about nine thousand vehicles a day; a half-century later, the bridge was being crossed by more than six times that amount, reaching an aggregate of over one billion cars. In 1968, the Golden Gate became the first bridge to institute one-way tolls, thus relieving congestion for about half the traffic. Toll revenue now exceeds $50 million annually, which is enough for "maintenance, repairs, modernization, equipment, supplies and salaries" to operate the bridge, plus some left over to subsidize public transportation. This is a far cry from New York's toll-free East River bridges, whose engineers have had to fight for money to paint and repair an infrastructure neglected and long forgotten by the vast majority of politicians.

Rust and corrosion may attack a bridge slowly over many years, but an earthquake can do its harm in a few seconds. Besides the damage it did to the San Francisco–Oakland Bay Bridge, the 1989 Loma Prieta Earthquake caused a mile or so of the theretofore undistinguished, if not downright ugly, Nimitz Freeway in Oakland, also known as the Cypress Structure, to collapse on scores of cars and trucks, crushing, trapping, and killing forty-two people. The Golden Gate Bridge survived that quake unscathed, but it has subsequently been ordered to undergo retrofitting to prepare it to withstand the "big one" that continues to threaten California. Earthquakes were not unknown to San Francisco when the Golden Gate was being planned and designed, of course, and there was considerable controversy between geologists and engineers about the nature of the foundation on which the bridge would rest. What most complicates any design that must take earthquakes into account is that there is no single earthquake to design against. As the 1994 quake in Los Angeles and the 1995 one in Kobe, Japan, demonstrated, each time the earth shakes, it may move in a different direction, with a different amplitude, and with a different frequency. The wind is almost more predictable.

In order to design bridges against earthquakes, engineers must make judgments as to the range of directions, amplitudes, and frequencies that are most likely to occur in the vicinity of the structure. Designing against an earthquake that might measure upward of an 8.5 on the Richter scale, for example, would result in a very, very conservative design structurally, the building of which would require an enormous financial investment. In fact, engineers today face almost the same degree of uncertainty and ignorance in designing against earthquakes as mid-nineteenth-century engineers did in designing against the wind. Only the experience gained on real bridges in real circumstances can confirm or refute the soundness of the judgments that have been made. Many of the San Francisco–area bridges and viaducts built from the 1930s through the 1950s were created in a design climate that accounted for earthquake forces in a particular way. For example, Charles H. Purcell, chief engineer of the San Francisco–Oakland Bay Bridge, described in 1934 how earthquakes would be taken into account in its design:

> Since the structure is in a region which has suffered, and may again suffer, more or less violent earth shocks, unusual precautions have been taken to safeguard against the hazard. All elements of the bridge are designed for an acceleration of the supporting material [the earth] of 10 per cent that of gravity. It was readily recognized that the usual criteria for earthquake design would not be satisfactory in dealing with this [extraordinary] structure. In view of this, an exhaustive study was made and design methods were evolved which took into consideration the various peculiarities of the problem.
>
> In the case of the channel piers, the horizontal force created by the acceleration of the mass will be augmented by forces due to the movement of the pier through the water and the soft mud immediately below. In fact it is conceivable that the soft mud may have an acceleration of its own in a direction opposite to that of the pier. These forces were incorporated in the analysis. In dealing with the superstructure, and particularly the suspension spans, the elastic and mechanical flexibility of the elements were fully considered. For earthquake design a 40 per cent increase in basic unit stress was permitted.

Though Purcell's "unusual precautions" to accommodate a ground acceleration of 10 percent that of gravity may have seemed conservative in the 1930s, they proved to be less so as earthquake experience accumulated. The 6.6-magnitude Northridge Earthquake that struck near Los Angeles in early 1994 had both horizontal and vertical ground accelerations far in excess of 10 percent. Even before that earthquake, however, experience subsequent to the construction of the Bay Bridge revealed gaps in contemporary understanding and the resulting weaknesses in designs, and corrective measures were prescribed for it and other older bridges. Unfortunately, identifying engineering needs and raising funds to implement them are not necessarily commensurate. For example, had engineers argued in 1988 for spending tens of millions of dollars to encase the steel piers of the East Bay Crossing in concrete to stiffen them against the sideward swaying of an earthquake, further studies might have been called for to establish why a bridge that had appeared perfectly sound for half a century suddenly needed to have thickened the slender legs that had contributed to the structure's grace. However, after the 1989 earthquake revealed that the flexibility of the Oakland span caused enough swaying to dislodge one of its deck sections, the money was available fast enough so the work was completed in early 1992.

Over a billion dollars was earmarked for the strengthening of highway structures throughout the state in the wake of the Loma Prieta Earthquake, but that work was incomplete when the Northridge Earthquake struck. The California Department of Transportation, known as Caltrans, had necessarily to engage in a form of triage for bridges, and when the Los Angeles area shook in January 1994, at least one of the major highway spans scheduled for strengthening the next month collapsed. For some time afterward, even geologists were a bit frustrated that they "had not yet pinpointed the source of the powerful earthquake that appears to have emanated from a previously unmapped thrust fault." There were no great fissures opened up by this quake described as "the costliest natural disaster in U.S. history," with associated costs estimated as high as $30 billion. Among the highway bridges that collapsed and caused the greatest inconvenience for commuters were necessarily ones that involved the longer spans required at the junctions of major routes.

Some of the bridge failures during the Northridge Earthquake were unusual in that bridge decks appeared to have been bounced vertically as well as slid horizontally. Whereas the San Francisco earthquake of 1989 was characterized by a slow horizontal shaking that shifted the East Bay span off its supports and caused the elevated freeway to collapse like a house of cards on a shaky table, the Los Angeles earthquake involved large vertical motions that dominated the structural response in some locations. All but two of the bridge structures that collapsed in the 1994 quake were built before the 1971 San Fernando earthquake, and Caltrans came in for some criticism that it had given strengthening these low priority. The two newer structures that collapsed were said to have been negligently designed by highway engineers, but a Caltrans spokesman pointed out, "You can't design a bridge to resist every possible earthquake—and this one came from an unknown fault." Such charges and defenses will no doubt continue to be made as long as bridges are built—and collapse. But if no bridges ever collapsed, engineers would then come in for criticism because they were designing structures to resist incredibly large earthquakes, storms, and even terrorist attacks that might never happen. And if engineers were demanding and getting, perhaps with the aid of stories and pictures of fallen bridges, all the funds they needed to design against everything, what other needs of society would be neglected? Priorities for health and safety are never easily determined, whether they relate to bridges or to the people who use them.

Must we thus expect, if not allow, a bridge failure to occur now and then? The history and promise of bridges suggest that we must, for reasons that have to do with neglect of the past and its relevance for the future. Neglect of the past is often embodied in a short-term historical memory, thinking, with hubris, that one's own generation's engineering science and technology have progressed so far beyond what they were a generation or two earlier that the bridges of one's professional progenitors, and even one's mentors, make pretty pictures but not examples or models for modern engineering. A historical perspective on bridges and their engineers reveals not only that such shortsightedness is nothing new, but also that it has led to disaster time and again.

A close reading of the history of major bridge failures is contained in a remarkable piece of scholarship by Paul Sibly and his adviser, then at University College London, Alastair C. Walker. Among the conclusions of their work, published in 1977, was the strong temporal pattern that bridge failures had followed from the middle of the nineteenth century. What Sibly and Walker noted was that the collapses of the Tay, Quebec, and Tacoma Narrows bridges, which occurred in 1879, 1907, and 1940, respectively, were very nearly thirty years apart. A less commonly remembered incident, but one that was equally dramatic and in its own time the subject of investigation by a royal commission, was the collapse of Robert Stephenson's Dee Bridge in 1847—further reinforcing the observation that a thirty-year cycle was associated with bridge failures. To test their hypothesis, which pointed to a major bridge failure about the year 1970, Sibly and Walker looked at incidents around that time and found that, indeed, in 1970 there were two significant failures of a new type of steel bridge, known as a box girder, then under construction in Milford Haven, Wales, and in Melbourne, Australia.

The pattern of bridge failures laid out by the historical studies of Sibly and Walker revealed several characteristics besides a thirty-year cycle. For example, each of the bridge types involved was of a different kind (trussed girder, truss, cantilever, suspension, and box girder), and each had been evolving within a design climate of confidence and daring when the accident occurred. The stories of the Quebec cantilever and the Tacoma Narrows suspension bridge epitomize this aspect of the thesis. Indeed, the different types of bridges that failed had often been introduced or developed with renewed vigor two or three decades earlier in response to a failure that drove a different kind of bridge out of favor. Thus the cantilever-bridge type, made most famous by the spans over the Firth of Forth, was introduced in the wake of the disastrous fall of the high girders of the Tay Bridge. When the Quebec Bridge collapsed, so did the reputation of the cantilever as a prime competitor of the suspension bridge for long spans.

Among the speculations Sibly and Walker offered to explain the thirty-year cycle was the nature of engineering practice, in which there developed "a communication gap between one generation of engineer and the next." This can certainly be true when aging engineers like Theodore Cooper remain aloof from their projects, as he did in the case of the Quebec Bridge, and when less experienced engineers associated with a project defer to the presumed infallible experience and judgment of a more eminent engineer, as happened with the Tacoma Narrows Bridge. In the case of Gustav Lindenthal and the succeeding generation embodied in his assistants Ammann and Steinman, the inflexibility of the mentor in failing to modify the plans for his dream bridge, to accommodate the changing nature of transportation across the Hudson River in the twentieth century, not only cast his judgment in doubt but also opened up rifts that prevented substantial communication between professional generations.

But perhaps the most significant factor that widens simultaneously the communication and generation gaps between engineers of different times is the ever-present development of engineering science and the tools of analysis. No twentieth-century suspension-bridge engineer seems ever to have lost a reverence for the works of John Roebling, as epitomized in his Brooklyn Bridge. However, with the growing development of analytical tools, as embodied in the deflection theory that Leon Moisseiff seemed so effectively to have developed and applied, the methods of Roebling, which relied upon physical more than mathematical argument, appeared to have been superseded. Unfortunately, with the relegation to dusty archives of Roebling's verbal reasoning about his concerns over stiffness and wind, the natural forces and response of bridges to them that so concerned him ceased to be a primary concern to more mathematically minded engineers, who remembered the old master's bridges primarily as aesthetic models. The limitations of this shortsighted view of engineering history became immediately apparent in the wake of the collapse of the Tacoma Narrows Bridge, and the subsequent revitalization of the suspension-bridge form took place only in light of newly embraced aerodynamic theories and wind-tunnel testing. This new perspective led to such innovations as the winglike decks and inclined suspender cables of the Severn and Humber spans in England, the latter the longest in the world until the completion of a bridge in Denmark and the Akashi-Kaikyo Bridge across Japan's Akashi Straits. The Severn span, however, has not been without its own problems, and it has had to be "strengthened" to carry the heavier lorries that have been allowed to use Britain's motorways since the bridge's original design and construction. No matter how strong the bridges are, users of the United Kingdom's greatest spans across its wide estuaries have been buffeted by winds so forceful at times that the largest lorries have been instructed to cross in pairs, thus reducing somewhat their chances of being blown over.

The pattern of bridge development established by Sibly and Walker suggests that in the late twentieth century there should be not only another radically new bridge type evolving toward more and more daring lengths and slenderness but also that a major failure can be expected in that type sometime around the turn of the millennium. The genre that seems so eerily poised to continue the thirty-year cycle of major bridge failures has developed from an old type that was rediscovered in Europe in response to the exigencies of rebuilding an infrastructure destroyed by the war. Though the superstructure of many bridges in Germany had been damaged, their foundations and piers were often reusable. The challenge to engineers was to design for these prewar foundations lighter bridge decks that could then carry the heavier postwar traffic. Cable-stayed bridges had been conceived centuries earlier, but they had never before been built on the scale or in the numbers that they began to be in Germany in the 1950s. For some time after that period, such bridges were thought to be the most economical and suitable choice for spans no longer than about twelve hundred feet, or somewhat shorter than the main span of the Brooklyn Bridge. By the 1980s, however, cable-stayed designs were being proposed with span lengths that had previously been thought to be in the exclusive realm of the now more conventional suspension bridge.

The Sunshine Skyway Bridge across Tampa Bay, among the longest and most photographed of the cable-stayed bridges in the United States, has a main span of twelve hundred feet. Completed in 1987, the Florida crossing remained among the top dozen or so longest cable-stayed bridges in the world into the early 1990s, when the genre really took off to new lengths. Many long spans were completed in Japan and Canada in the late 1980s, and in 1991 the longest cable-stayed bridge finished in Europe was the Queen Elizabeth II Bridge across the Thames at Dartford, with a main span of almost fifteen hundred feet. The French, however, had already at that time under construction over the mouth of the Seine a cable-stayed bridge with a main span in excess of twenty-eight hundred feet—the magnificent Pont de Normandie—and Danish engineers let it be known that they were considering a cable-stayed span approaching four thousand feet in length to complete what is known as the Great Belt link between Denmark's two largest islands. Though this design was eventually rejected in favor of the suspension bridge, itself remarkable with its main span of more than a mile in length, the daringness of the Danish cable-stayed proposal became the topic of some discussion among engineers.

Cable-stayed bridge proposed over the Mississippi at Cape Girardeau, Missouri

The Sunshine Skyway Bridge across Tampa Bay, shown on the cover of a brochure

British engineers questioned whether it was wise for the French even to attempt a cable-stayed bridge with a main span almost twice the existing record. Questions of how the incomplete structure at the mouth of the Seine would behave in the wind were central to such endeavors, and there were warnings that scaling up in so large a leap from existing bridges was a prescription for disaster. Engineers proposing the doubling or even tripling of existing spans were confident, however, claiming that the larger bridges were "perfectly" possible because of modern computer modeling and construction techniques. Special devices were fitted to the incomplete spans to stabilize them in the wind during construction, and when the deck was finally completed, in the summer of 1994, many an engineer breathed a sigh of relief. After six years of design and construction, the span was opened to traffic early in 1995.

Regardless of how sophisticated the computer models or construction techniques, whether the cable-stayed Pont de Normandie, with its record main span, would successfully cross the mouth of the Seine depended at least in part on luck. Excessively high or unusual winds that were not factored into the computer model could hold as much of a surprise for the engineer of the Pont de Normandie as the excessive weight of steel that was inadvertently omitted from the calculations for the Quebec Bridge held for its engineer. Any model, whether a simple equation on the back of an envelope or an elaborate numerical one in the gigantic memory of a supercomputer, is only as good as its fundamental assumptions. The Tacoma Narrows Bridge fell because the most sophisticated deflection theory used to design it did not take into account the dynamic effects of the wind. In sum, the undoing of a project will derive not so much from its size or scale as such, as from an imperfect understanding. What is an insignificant detail in a cable-stayed bridge of relatively modest size can grow to surprising importance as the size of a bridge grows. The sage advice to increase the size of bridges slowly reflects an awareness of this scale effect among more experienced engineers, but younger engineers, full of confidence in their computers, often think such caution to be a mark of excessive conservatism.

Making great leaps in size does not, of course, doom a bridge to failure, and daring young engineers can use the historical examples of the Forth and the George Washington bridges to defend their ambitious designs. The cable-stayed bridge that might fulfill the inexorable prophecy of Sibly and Walker's pattern of failures will not necessarily be the longest of the genre. The Tacoma Narrows Bridge had, after all, only the third longest main suspended span in 1940. But though mere size may not bring bridges down, it may often be the focus of attention with regard to warnings about their behavior. Being only the third-greatest span, with a modest traffic load and in a relatively remote location, the Tacoma Narrows was not a structure that called attention to itself—even though it was the most slender of bridges—until it began to oscillate in the wind and collapsed. It may be similar with cable-stayed bridges. The Pont de Normandie and others in the vanguard of design will be more carefully planned and watched than those that will be almost but not quite as large—those that will be, ironically, of little more than local significance or remarkableness during their design and construction. It may be from the great but not necessarily the greatest that we can expect the most unpleasant surprises.

If a major collapse of a cable-stayed bridge is to be prevented, there must be as much attention paid to the maintenance of the engineering-design infrastructure as to that of the physical infrastructure. This means that engineers should be as sensitive to the historical cycles of success and failure that have plagued the design enterprise as they are to the cycles of freezing and thawing that can plague their physical roadways and bridges. Care of the design infrastructure requires the maintenance of lines of communication between engineering generations, so that new tools and models are not used in ignorance of past experience. The surest way to break the vicious cycle of bridge failures identified by Sibly and Walker must certainly begin with the recognition that neglected patterns from the past become unconscious patterns for the future. Only by bringing these patterns into sharp focus and by seeing the modern engineer as reinventing, albeit with faster and more powerful tools, the bridges of the past and of different cultures, can we hope to realize dreams that do not spiral into nightmares. The bridge-and-structural engineer Henry Tyrrell articulated this almost a century ago, when he wrote the opening lines of the preface to his 1911 history of bridge engineering:

> Proficiency in any art or science is not attained until its history is known. Many a student and a designer finds, after weary hours of thought, that the problems over which he studied were considered and mastered by others, years or centuries before, perhaps with better results than his own.

The earlier years of the cable-stayed bridge genre brought to the fore such individual engineering personalities as the German Fritz Leonhardt, who practiced in Stuttgart; the greatest spans today are being designed by firms that carry the names, but not necessarily the personalities, of the older generation. As the presence of Ammann and Steinman continues to be felt through the firms of Ammann & Whitney and of Steinman, Boynton, Gronquist & Birdsall, commonly referred to simply as Steinman, so does that of Leonhardt in the firm of Leonhardt, Andrä und Partner, to which in the early 1990s the patriarch still contributed his philosophy, if not his daily presence. However, not all of the most recent record spans of cable-stayed structures have been designed by firms that tie themselves in name to great engineers of the past. The Danish firm with the trendy name CowiConsult, "one of the world's leading bridge designers," was the one designing, in its bridge department in Copenhagen, the world's longest cable-stayed span. Other great bridge-designing firms, such as Britain's Acer-Freeman-Fox, and the American Sverdrup Corporation, still tie themselves explicitly to the names of their entrepreneurial ancestors, but the increasingly used anonymous collective designations of "partners" and "corporation" indicate the trend away from the small partnership team or the dominant personality of an individual consulting engineer.

Whether their designs are attributed to forceful individuals or to anonymous corporations, cable-stayed bridges are not the only uniquely twentieth-century bridge type, nor is steel the only twentieth-century bridge material. Among other notable categories that have been very successful, not only as great structures but also as works of art, are the concrete bridges of the Swiss engineers Robert Maillart, in the first half of the century, and Christian Menn, after midcentury. Both have concealed steel, as reinforcement and as cables, in their bridges, which are primarily concrete structures. David Billington has described both Maillart and Menn as structural artists whose works are monumental pieces of sculpture as well as utilitarian works, and he has written in illuminating detail especially about Maillart's great concrete bridges.

As concrete has challenged steel, so have new materials challenged them both. Advanced composite materials made of glass, carbon, and polymer fibers, originally developed for the aerospace industry, are being introduced into experimental bridges, enabling them to weigh as little as one-tenth as much as conventional steel or concrete designs. One such bridge, a 450-foot-long cable-stayed road bridge over Interstate 5 in San Diego, has been aided in the materials-testing-and-design phase by a grant from the Federal Highway Administration. Because the cost of the new stuff is as much as twenty times that of conventional materials, such bridges can generally be expected to need this kind of financial support, and to remain in the experimental category until the materials become economically competitive. That will not prevent engineers from dreaming of using the newer materials to span such continuing challenges as the Strait of Messina and the Strait of Gibraltar.

In the meantime, others continue to work with conventional materials but in unconventional forms. Among the most talked-about individual bridge designers of the late twentieth century, in any material or form, is Santiago Calatrava, whose training as both an engineer and an architect has given him and his work a special cachet. Calatrava was born in the second half of the twentieth century, in 1951, in Valencia, Spain; he studied architecture there before going to the Swiss Federal Institute of Technology in Zurich, where he became also a civil engineer. Since opening a practice in Zurich in 1981, he has been responsible for structures of dramatic space and volume, mostly connected with transportation, throughout Europe. His best-known structures tend to be his bridges, however, including the canted-towered, cable-stayed Alamillo Bridge, built for the 1992 Expo in Seville, and the Bach de Roda Bridge in Barcelona, whose inclined arches and suspender cables enclose pedestrian paths that broaden midspan to plazas to create a secure yet open space that is both protective and inviting. The engineer-architect Calatrava has been accused of being more the latter than the former, however, for he has said that he wants "to win back engineering objects like bridges for architecture." Never mind that such talk can reopen old wounds and spark debate between the professions; in the final analysis, Calatrava's work will be judged by the standards of both, and there are indications that he forgets some of the fundamental principles of structural-engineering art in his pursuit of appearances.

Santiago Calatrava's Alamillo Bridge in Seville, Spain

In his Alamillo Bridge, for example, Calatrava employed a massive counterweight under the roadway to add enough tension to the longest cable so that it would be taut and not sag under its own significant weight. This added considerably to the cost of the bridge, and at the same time sacrificed structural honesty. Calatrava may see such compromises as necessary to win back bridges for architecture, yet the great engineers have never felt they were deliberately wresting bridges from architects. Even if modest spans like the ones Calatrava has designed may fairly be viewed as pieces of sculpture as well as utilitarian crossings by those who commission them, it is not necessarily obvious that bridges of record span, which will always remain first and foremost engineering problems requiring engineering solutions, should be saddled with significant extra weight, be it physical or metaphorical, for the sake of appearance alone.

The city of St. Paul, Minnesota, recently commissioned not an engineer but a sculptor, James Carpenter, "to conceive the form for a bridge" across the Mississippi River. Among the forms developed by the New York artist, who worked in conjunction with a German engineer, was a skewed-decked, cable-stayed bridge supported by a V-shaped pylon located on an island six

Computer-generated image of Calatrava's unrealized East London crossing of the River Thames

hundred feet from either shore. Though a computer-generated image of the bridge inserted electronically into photos of the site showed the structure to be a striking design, with elements clearly intended to make the bridge distinctive, the cost would have been more than twice that of a conventional span. It fell upon the City Council to decide whether to spend an extra $15 million dollars, even though the bulk of it might have to be requested from federal bridge replacement funds, for a structure whose principal characteristic might be said to be difference for the sake of being different. The city's Public Works Department was instructed to work out estimates for a cable-stayed bridge with its two spans in alignment, but the cost for such a structure was still considered prohibitive, because it was learned that "federal funds would not be available for anything beyond the least expensive design."

Though the mayor clearly preferred the artist's signature design, political and economic realities led him to lean toward a less expensive double-arch option, also produced by the artist, which was not unlike an unrealized Calatrava proposal for a single-arch crossing of the Thames in East London. In fact, this medium-priced design was the public's choice, according to opinion surveys, but in the end the mayor recommended a third type of bridge to cross the river at Wabasha Street. This was the least dramatic, least distinctive, and least expensive of the original options—a box-girder bridge that could be built for $20 million and could include "pedestrian amenities such as ornamental lighting, windshields, pedestrian outlooks, and stair and elevator towers" to the island in the river. As in so many other cases regarding bridges and their appearance, in the final analysis the politicians and citizens of St. Paul had to settle for what they could afford. The dream of an artist, neither more nor less than that of an engineer, was alone insufficient to dictate reality.

A public tension between means and wants often only highlights a more constant tension between function and form. Though it ebbs and flows as surely as do the waters over which many a monumental bridge is built, the ongoing push and pull between designer and financier, between engineer and architect, between engineering and art, come to our attention mainly when a question of design bobs to the surface. Disagreements over form will no doubt remain as long as there are engineers and artists who see their objectives as different. Bridge design is among the most visible and vulnerable arenas in which such competition has taken and will continue to take place, and it is wise to recall that the greatest bridge engineers have always taken an equal interest in the structural and artistic values of their designs, the mediators more often than not being safety and economics. Not that engineers are more willing than architects to sacrifice beauty for brawn, or looks for lucre; the greatest bridges that engineers have built are clearly the ones that unite and achieve both structural and aesthetic goals, and often with striking strength and economy in their context. Above all, however, engineers know that, first and foremost, their bridges must stand into the future against weight and wind and want. The most beautiful bridge, when negelected in structural design and maintenance, can become, fallen, the most ugly pile of concrete and steel. That is not bridge building.

As new materials, computational techniques, and generations of engineers come to dominate the world of bridge building, as they will especially the projects involving the greatest technical challenges, there will necessarily come to the fore competitions and disagreements among designs and their designers. This is to be expected in any creative endeavor; we should not be surprised that it is heightened in bridge building, which is among the most visible, symbolic, and evocative of all interactions between engineer and engineer, and between engineers and society. Artists and architects may challenge the engineer, but ultimately only the engineer will be able to cantilever out over technically uncharted waters to build bridges greater than any before. Though knowledge of structural principles is of course essential in such an endeavor, a sense of history provides the judgment for engineers to dream effectively beyond the present. Even if dreams come easily to dreamers, bringing a dream to reality takes a view that is firmly founded on experience of what can and cannot be done technically, plus a confidence in what is humanly and economically possible at a given time. Both of these qualities may be necessary to achieve greatness in bridge building, yet they alone seem not to be sufficient to bring a particular dream to reality. There appears to be, as Ammann saw it, a certain element of luck involved in the enterprise.

Whether designed by engineer or architect, artist or Boy Scout, every bridge is a legacy to its environs and to its users. The environment itself, especially when it is cruel by geography or polluted by society, cannot be expected to be any more respectful of a bridge than it is of an automobile or an endangered species. The society of users, who are in fact willy-nilly the stewards of the world's bridges and of the greater infrastructure, must recognize that every artifact that has been or ever will be created, whether in now traditional steel and concrete or in the composites of the future, must be maintained as well as used. By understanding this and the origins of our bridges and other artifacts of civilization, and the humanness of those who once dreamed of what we now so often take for granted, we not only engage ourselves in the technosocial endeavor that involves engineers at its core, but we also understand how their human natures and their dreams affect the way we experience our cities and towns, our borders, and our open spaces.

NOTES

Reference notes are keyed to phrases and quotes on the text pages indicated. Where successive quotes and information have come from the same source, only the first or the most prominent occurrence is referenced. Full bibliographical citations for articles and books identified only by author, or, where multiple works by an author are cited, by author and year, are given in the bibliography following these notes.

Biographical information, especially for less well-known engineers, is not always readily available or conveniently indexed. Among the most extensive sources of personal information on engineers are the biographical dictionaries of the American Society of Civil Engineers and the American Society of Mechanical Engineers. The most readily available information on the lives and careers of deceased members of the ASCE, to which virtually all bridge engineers discussed in this book belonged, is often in the memoirs published in that society's *Transactions*. These memoirs were generally written by associates and published several years after obituaries. So that the memoirs most relevant to this book can be more readily identified, they have all been grouped in the bibliography under the entry "Memoirs of Deceased Members," and there alphabetically by subject. References to such entries are indicated by the notation "Memoir" in the notes. Where biographical material is not referenced, its source is the two volumes of *A Biographical Dictionary of American Civil Engineers*.

The following abbreviations are used in the notes:

ASCE = American Society of Civil Engineers
ASME = American Society of Mechanical Engineers
BDACE = *A Biographical Dictionary of American Civil Engineers*
DAB = *Dictionary of American Biography*
EN = *Engineering News*

ENR = *Engineering News-Record*
NYT = *The New York Times*
TASCE = *Transactions of the American Society of Civil Engineers*

Chapter 1 Imagine

page

8 The earliest bridges: see, e.g., Steinman and Watson, ch. 2; cf. Tyrrell (1912), ch. 1; Waddell (1916), ch. 1.

9 Homer: Waddell (1916), p. 11.
Persian kings: ibid.
Herodotus: ibid., p. 5.
China, northern India: Steinman and Watson (rev. ed., 1957), p. 17.
Middle Ages: see ibid., ch. 4, esp. pp. 53–54.

10 Altopascio Order: ibid., p. 54.
"were nicked": ibid.
Frères Pontiffes: ibid.

11 Palladio: Palladio, third bk., chs. V–IX; cf. Waddell (1916), pp. 11–12.
brothers Grubenmann: Waddell (1916), p. 12.
Squire Whipple: BDACE, vol. II; "Memoir."
"father of American bridge building": BDACE, vol. II.
President Eliphalet Nott: Reynolds, p. 475.
Rensselaer: ibid., p. 466.
Whipple patented: U.S. Patent No. 2,064.

15 Statue of Liberty: Hawkes, p. 76.

21 one out of every five: Secretary of Transportation, p. 5.

Chapter 2 Eads

22 James Buchanan Eads: see esp. Dorsey; Steinman and Watson, ch. 10; Vollmar; Yager.

23 Young James: S. R. Watson and Watson, p. 107; Morgan, p. 92; *Popular Science Monthly*, Feb. 1886, p. 545.
"This is going to be": S. R. Watson and Watson, p. 108.
a fire broke out: Jacobs and Neville, p. 58.
Barrett Williams: Yager, p. 17.
read at will: *Popular Science Monthly*, Feb. 1886, p. 545.
West Point: Grayson, pp. 18, 22.

24 Franklin Institute: ibid., p. 24.
engineering schools: Grayson, pp. 24–30.
"the last patroon": *National Cyclopaedia of American Biography*, vol. II, p. 397.
Rensselaer Institute: Grayson, p. 28; see also Reynolds, pp. 466–67.

25 "many of the fledgling": Schodek, p. 13.
Loammi Baldwin: BDACE, vol. I; see also Ford, pp. 278–80; Schodek, pp. 340–2.
"learned engineering": Ford, p. 278, with the name spelled Laommi.

"who would become": Ford, p. 279.

26 insurance underwriters: Scott and Miller, p. 71.
five hundred explorations: ibid.

27 Future snag boats: *Popular Science Monthly*, Feb. 1886, pp. 545–46.
"I had occasion": Eads (1868), p. 21; see also Woodward, p. 3.

29 "most powerful": Baxter, p. 244.
"the *St. Louis*": quoted in S. R. Watson and Watson, p. 111; see also Baxter, pp. 242–44; *Popular Science Monthly*, Feb. 1886, pp. 547–48.
outlining a proposal: Woodward, p. 7.

30 Charles Ellet, Jr.: see Lewis.
"Mr. Ellet promises": quoted in Woodward, p. 7.

31 "The time is inauspicious": ibid., p. 9.
Britannia Bridge: see Clark, vol. II, p. 815.

32 "comparative merits": J. A. Roebling (1841), p. 193.
"become indispensably": quoted in Woodward, p. 10.
"much thought": Homer, p. 10.
a tunnel under the river: Woodward, pp. 10–11.

33 "charging that bridges": Scott and Miller, p. 77; see also Kutler.
"it cost nearly half": Scott and Miller, p. 77.
geography had been undone: ibid., p. 78.
"the future Great City": ibid., p. 78.

34 "the datum plane": Kouwenhoven (1982), p. 542.
Lucius boomer: Scott and Miller, p. 79.

35 Simeon S. Post: see BDACE, vol. I; U.S. Patent No. 38,910.
"most beautiful contrivance": Palladio, p. 66.

36 "mathematical bridges": see, e.g., Labrum, ed., pp. 100–101.

37 "not produce injurious effects": U.S. Patent No. 38,190.
"measured in the center": Kouwenhoven (1982), p. 542.

38 A convention of civil engineers: see Eads (1868), p. 37.
"some 'engineering precedent' ": ibid., p. 41.
"In view of the great importance": ibid., p. 3.

39 "Mr. Boomer's bridge": ibid., p. 5.
"needlessly extravagant": ibid., p. 10.

40 "If the upper member": ibid., p. 11.

41 "every known method": ibid., p. 17.
"bow-string girder": ibid., p. 12.
"catenary or suspended arch": ibid., p. 17.

42 upright versus suspended arch: ibid., pp. 48–57.
cast steel: ibid., p. 57.
several patents: see U.S. Patents No. 83,942; 89,745; 95,784; 132,271; 142,378; 142,379; 142,380; 142,381; 144,519.

43 Henry Flad: BDACE, vol. I.
Charles Pfeiffer: Scott and Miller, p. 82; cf. Eads (1868), p. 4. Though Scott and Miller use the spelling "Pfeifer," as does Eads in this particular source, the predominant spelling is followed here. See, e.g., Eads (1884), p. 44.
"based his first": Scott and Miller, p. 90.

"improvement in arch bridges": U.S. Patent No. 95,784.

44 Flad was issued a patent: U.S. Patent No. 132,271.
"After careful revisions": Eads (1868), p. 4.

45 "I cannot consent": quoted in Scott and Miller, p. 67.

46 pneumatic caisson: ibid., p. 189; see also Steinman and Watson, pp. 185–86.
"a visit": Reavis, p. 10.
"For a while": ibid., p. 11.

48 sixty-six feet: Eads (1884), p. 557.
"where all things": Reavis, p. 10.
In March 1870: see, e.g., Scott and Miller, pp. 106–7.

49 Brooklyn Bridge: D. McCullough (1972), p. 298.
John Roebling's bridge: ibid., pp. 90–92.
Roebling's son: ibid., p. 157.

50 went to Europe: ibid., pp. 165–67.
"cassoons": ibid., p. 167.
"correct some statements": *Engineering*, May 16, 1873, p. 337; see also *Engineering*, June 27, 1873, p. 458; Sept. 5, 1873, pp. 195–96; D. McCullough (1972), pp. 344–47.
"the first practical": quoted by Eads in *Engineering*, May 16, 1873, p. 337.

51 "I trust I shall not": *Engineering*, May 16, 1873, p. 337; see also Eads (1884), p. 68.

52 Francis Collingwood, Jr.: see, e.g., D. McCullough (1972), pp. 145, 374.
Collingwood Prize: see ASCE *Official Register*.
"The arches": Eads (1868), p. 33.

53 Piper & Schiffler: Keystone Bridge Company, p. 7; quotations from Carnegie, p. 116, which puts the date at 1862.
"proud of having": Carnegie, p. 117.
"did not leave": ibid., p. 45.

54 "unusual character": ibid., p. 119.
"not stand up": quoted in ibid., p. 120.
"was seemingly one": ibid., p. 120.
"Must we admit": Eads (1868), p. 44.
"first large": Carnegie, p. 155.
John Piper: ibid., p. 120.
meet the specifications: see, e.g., Scott and Miller, pp. 95–96, 109–10.
First gentleman: Steinman and Watson, pp. 181–82.

55 Linville proposed: Keystone Bridge Company, p. 16.

56 By the end of the summer: Steinman and Watson, pp. 201–5; cf. Scott and Miller, pp. 124–28.
fourteen heavy locomotives: Kouwenhoven (1974), p. 175.
discovered by Marquette: Yager, p. 81.

57 "Yon graceful forms": Eads (1884), p. 42.

58 "Everything which prudence": ibid., p. 43.
Among those individuals: ibid., p. 44.

59 fireworks: Kouwenhoven (1974), pp. 178–80, figs. 5, 8; see also Scott and Miller, p. 130.

went bankrupt: Kouwenhoven (1974), p. 180.
official name: ibid., pp. 159–60.
"conducted principally": Morgan, p. 120.
Arthur E. Morgan: see, e.g., *Current Biography*, 1956.

60 before there was a law: Morgan, p. 96.
"flamboyant in their gaudy paint": Scott and Miller, p. 125.
"drop the case": quoted in Morgan, p. 113; see also Scott and Miller, pp. 125–26.
"bible": Morgan, p. 78. system of jetties: ibid., pp. 131–32.
retrospective scrutiny: ibid., p. 151.

61 Board of Army Engineers: ibid., pp. 133–37.
"in a narrow executive capacity": quoted in ibid., pp. 140–41.
lengthy review: Eads (1884), pp. 304–29.
total bill: Morgan, p. 142.
final legislation: ibid., pp. 142–49.

62 "If the profession": Eads (1884), p. 48.
"the most difficult piece": quoted in Vollmar, p. 20.
thirty feet: Morgan, p. 167; cf. Vollmar, p. 21, which puts the depth at thirty-one feet when the jetties were completed in July 1879.
"the savings on transportation": Vollmar, p. 21.

63 "The key-note": Eads (1884), p. 53.
"directing the great sources": quoted in, e.g., J. G. Watson, p. 9.
"*Le Grand Français*": D. McCullough, p. 56.

64 "The question": Eads (1884), pp. 411–12.

Chapter 3 Cooper

67 Theodore Cooper: "Memoir."

68 Hoosac Tunnel: see Jacobs and Neville, pp. 36–51.
in the navy: "Memoir," p. 828.

69 "tripped on an unbalanced plank": Woodward, p. 184.
"He was conscious": ibid., p. 185.

70 found another tube broken: ibid., p. 190.
Cooper moved about: "Memoir," p. 829; see also D. McCullough (1972), p. 341.
Tay Bridge: see, e.g., Koerte; Paxton, ed.; Prebble.

71 east coast of Scotland: Koerte, p. 21.
Scottish firths: ibid., p. 18; Paxton, ed., pp. 25–26.
Stockton & Darlington Railway: Straub, pp. 167–68.

72 Bouch went to Scotland: Prebble, p. 20.
"the most insane idea": quoted in Koerte, p. 21.

73 "The tremendous impetus": Prebble, p. 24.
"[T]he simple reason": ibid., p. 20.
"after some twenty years": ibid., p. 36.
At Dundee: Shipway (1989), p. 1089.
foundations: ibid., p. 1092.

75 "Q: Sir Thomas": quoted in Koerte, pp. 103–4.

"for very limited surfaces": Prebble, p. 202.
"the fall of the bridge": quoted in Shipway (1989), p. 1096.
"beeswax, fiddler's rosin": Prebble, p. 193.
76 "no absolute knowledge": quoted in Prebble, p. 212.
"We find that": quoted in Koerte, p. 108.
suspension bridge: Shipway, in Paxton, ed., p. 44.
77 new Tay Bridge: Chrimes, p. 135.
Barlow, Son & Baker: Shipway (1989), p. 1097.
"The massive character": reproduced in ibid., p. 1100.
The stumps: Koerte, pp. 108–9.
consulting engineers: see, e.g., Koerte, p. 134.
John Fowler: see Westhofen, pp. 276–81.
78 "major schemes": Chrimes, p. 135.
Benjamin Baker: McBeth, p. 95; see also biographical sketch in Westhofen, p. 281.
79 "should be well enough": B. Baker (1887), p. 142.
Gerber bridge: see Shipway, in Paxton, ed., and Shipway (1990), for illustrations of Gerber and related bridges.
80 Cincinnati Southern Railway: Jackson, p. 174.
Charles Shaler Smith: BDACE, vol. II.
Charles Conrad Schneider: ibid.
Octave Chanute: ibid., vol. I; see also "Memoir," p. 1665.
81 "Canti-lever Bridge": *Spanning Niagara*, p. 17.
82 "This is a question": EN, Nov. 5, 1887, p. 335.
" 'How are you getting on' ": B. Baker (1887), p. 116.
"between England and Scotland": *Scientific American*, Feb. 4, 1888, p. 70.
82 "To get an idea": B. Baker (1887), p. 116.
83 The image of the bridge: Mackay (1990b), pp. 8–9.
84 "Indeed, I have evidence": B. Baker (1887), p. 116.
85 "The best evidence": ibid.
87 "was invited": Mackay (1990b), p. 16.
"ingenious illustration": EN, June 11, 1887, p. 385.
88 "*Each semi-arc*": quoted in B. Baker (1887), p. 238.
"no deaths": ibid., pp. 170–71.
"Happily there is": ibid., p. 238.
89 Construction of the Forth Bridge: Birse, in Paxton, ed., pp. 128–29.
"may by its freedom": Westhofen, p. 218.
steel in British bridges: Birse, in Paxton, ed., pp. 126–28.
Clyde Rivet Company: ibid., p. 128.
90 "You can fold": B. Baker (1887), p. 210.
certainly was stiff: Shipway, in Paxton, ed., p. 62.
"straddle legged": Shipway (1990), p. 1097.
91 "Holbein straddle": ibid.
"the 'cantilever fever' ": EN, April 7, 1888, p. 270.
"it was useless to criticise": EN, Dec. 28, 1889, p. 616.
"there would never be": quoted in ibid.

"You all know about": quoted in a letter to *Engineering News,* Oct. 10, 1907, p. 391.

92 Two Eiffel Towers: *Engineering*, May 3, 1889, p. 501.
formal opening: Cox, in Paxton, ed., p. 90.

93 "Aberdeen to New York": Mackay (1990b), p. 112.
Channel tunnel: see, e.g., Hunt; cf. *Engineering*, Oct. 30, 1868, pp. 389–92.
"given reasonable care": Grant, in Paxton, ed., p. 91.
trans-Siberian Railway: see NYT, Aug. 15, 1994, p. A4.
bridge connecting Siberia: see, e.g., G. T. Pope.
Joseph Strauss: Golden, p. 5.
Tung-Yen Lin: see G. T. Pope; cf. ENR, June 7, 1962, pp. 53–54; ENR, July 25, 1994, pp. 38–40.
Kinzua Viaduct: Jackson, p. 145.

95 "the first authoritative specifications": ENR, Aug. 28, 1919, p. 443.

96 Ashtabula Bridge: Jacobs and Neville, p. 56; see also Macdonald.

97 exact cause of the failure: Jacobs and Neville, p. 57; see also Macdonald.
"not only alarmed": Cooper (1889), p. 21.
He documented: ibid., p. 25.
"worked out independently": ibid., p. 27.
"the first paper": ibid., p. 22; see Cooper (1878).
"must provide for": Cooper (1889), p. 51.

98 "the American system": ibid., p. 49.
absent from British practice: ibid., p. 51.
"If an engineer": quoted in ibid., p. 50.
commission of five: EN, Sept. 6, 1894, p. 187.

99 "cantaliver": see EN, Dec. 27, 1894, p. 534.

100 "the first practical solution": Cooper (1889), p. 21.

101 Edward Wellman Serrell: *Spanning Niagara*, pp. 23, 25.
Quebec Bridge Company: Royal Commission, pp. 12–15.

102 Cooper preferred the cantilever: ibid., pp. 16–17.
pace of design work: ibid., p. 37.

103 "employ a competent": ibid., p. 42.
"This puts me": ibid., p. 43.
"provided the efficiency": ibid., p. 46.
"*de facto*, chief engineer": ibid., p. 75.
hypercritical: ibid., pp. 50–52; cf. EN, Oct. 31, 1907, p. 474.

104 "a technical man": Royal Commission, p. 50.
rejecting the proposed procedure: ibid., p. 79.
Over the next three weeks: ibid., pp. 79–85.
"it looked like a serious matter": ibid., p. 88.
"a grinding sound": NYT, Aug. 30, 1907, p. 1.

106 "for not having visited": NYT, Aug. 31, 1907, p. 1.
qualified earlier reports: NYT, Sept. 1, 1907, p. 1.
"the Nestor": EN, Oct. 31, 1907, p. 473.
"The Canadian Commission": ibid.

107 "I should have been glad": ibid., p. 474.

"maintain a judicial attitude": ibid., p. 469.
"vigorous language": NYT, Nov. 21, 1907, p. 4.
108 "These errors of judgment": Royal Commission, p. 9.
109 "Why, if you condemn": EN, Oct. 3, 1907, p. 364.
"Mr. Cooper states": Royal Commission, pp. 49–50.
110 "The Quebec Bridge collapse": EN, Oct. 3, 1907, p. 365.
"commonplace in appearance": *Scientific American*, Feb. 12, 1910, p. 148.
111 "for the sake of": Royal Commission, p. 56.
"Twice the hopes": ENR, Nov. 27, 1917, p. 579.
112 "the entire responsibility": quoted by Lindenthal in ENR, Nov. 16, 1911, p. 583.
114 a lengthy tract: EN, Nov. 16, 1911, pp. 581–86; Nov. 23, 1911, pp. 613–19.
115 "the most important": EN, Nov. 16, 1911, p. 599.
116 "If five or more": ibid., p. 583.
"Causes of the Disaster": ibid., p. 582.
"While the Quebec Bridge Co.": ibid.
118 "foresaw": NYT, Aug. 25, 1919, p. 11.
119 "consulting work": ENR, Aug. 28, 1919, p. 443.
memoir of Cooper: see "Memoir."
121 His total assets: NYT, March 25, 1919.

Chapter 4. Lindenthal

122 "the Nestor": EN, Dec. 21, 1916, p. 1188.
"dean" of American bridge engineers: ENR, Aug. 8, 1935, p. 208.
Lindenthal was born: see "Memoir"; see also Buckley, p. 40.
"educated at": DAB, suppl. 1, pp. 498–99; *National Cyclopaedia of American Biography*, vol. XVI, p. 117.
The issue of Lindenthal's education: see Buckley, pp. 56–57.
123 According to a memoir: "Memoir," p. 1790.
"received practical training": ibid.
"was put to work": Buckley, p. 56.
"to start a life": quoted in ibid.
"incline plane and railroad": *National Cyclopaedia of American Biography*, vol. XVI, p. 117.
"stood a little over six feet": Buckley, p. 57.
124 "Lindenthal was neither": ibid., p. 56.
After the Centennial Exhibition: BDACE, vol. I; Smith et al., p. 244.
125 Monongahela River: *Scientific American*, Sept. 22, 1883.
Lewis Wernwag: see Nelson, pp. 59–60.
"an American engineering superlative": see Nelson.
"the most stunning": Jackson, p. 321.
"in the course of time": *Scientific American*, Sept. 22, 1883, p. 180.
riverboat captains could arrange: Gangewere, p. 29.
126 "would not be subject to": Schodek, p. 129.
"the triumph of architectural skill": *Scientific American*, Sept. 22, 1883, p. 180.

127 Smithfield Street Bridge carried: Lindenthal (1883); Jackson, pp. 151–52; Schodek, pp. 129–31; Billington (1983), pp. 123–24.
portal motif: see Pennsylvania Historical and Museum Commission, p. 129.
128 "new bridge at Pittsburg": *Scientific American*, Sept. 22, 1883, p. 180.
"from an excellent photograph": ibid.
130 "practicability of a railroad bridge": TASCE, vol. 97 (1933), p. 422.
"There was keen competition": ibid.
"given thought": ibid., p. 423.
131 "annoyance and even danger": Lindenthal (1887), [p. 1].
Arthur Mellen Wellington: BDACE, vol. II.
with Calvert Vaux: D. McCullough (1972), p. 146.
132 "1874–'78": quoted in EN, May 23, 1895, p. 337.
"great work": ibid.
"It would be well": quoted in *Engineering Education*, July–Aug. 1990, p. 524.
"the influence of his energy": EN, May 23, 1895, p. 338.
"devoted his leisure": ibid.
133 "from a man": EN, July 9, 1887, p. 24.
"Are the proposed tunnels": quoted in ibid.
134 "not as a publication": Lindenthal (1887), copyright page.
136 "Prof. Lindenthal": NYT, Jan. 5, 1888, p. 5.
137 "public move": EN, Nov. 12, 1887, p. 348.
"wagon-ways": EN, Nov. 19, 1887, p. 359.
Fort Lee: NYT, Jan. 27, 1888, p. 8; Jan. 28, 1888, p. 3.
138 "certainly not so formidable": EN, Jan. 7, 1888, p. 1.
A profile diagram: EN, Jan. 14, 1888, p. 22.
"protect them absolutely": ibid., p. 30.
"the most prominent feature": EN, Jan. 28, 1888, p. 57.
139 "the first definite description": ibid., p. 62.
"architectural excellence": EN, Feb. 4, 1888, p. 78.
140 "The graceful suspension": ibid., pp. 78–79.
141 "It is certainly true": ibid., p. 79.
143 four types of bridges: EN, March 3, 1888, pp. 153–54.
"cantilever fever": April 7, 1888, p. 270.
"as they were then": EN, March 3, 1888, p. 154.
144 so committed to the suspension concept: ibid., p. 155; cf. EN, March 24, 1888, p. 226.
144 New-York and New-Jersey Bridge Company: Ammann (1933b), p. 5.
opposition on the New York side: NYT, Feb. 5, 1888, p. 3.
Early in 1888: EN, April 14, 1888, p. 283.
"It plainly contemplates": ibid., p. 294.
145 Henry Flad: NYT, July 3, 1888, p. 4.
"we shall have a bridge": NYT, Dec. 25, 1888, p. 4.
"much-talked-of bridge": *The American Architect and Building News*, Dec. 8, 1888, p. 267.
critical appraisal by Max Am Emde: *The Engineer*, vol. 67 (1889), p. 411.
146 "Ignorance of it": EN, July 20, 1889, p. 58.

"the bridge is not intended": ibid., p. 59.
"If English and Scotch railways": EN, March 8, 1890, p. 228.
147 "take hold of the project": NYT, Oct. 12, 1889, p. 1.
American Association for the Advancement of Science: EN, Nov. 9, 1889, pp. 435–37; Nov. 16, 1889, pp. 464–65; Nov. 23, 1889, pp. 486–87.
"Zoölogists tell us": EN, Nov. 9, 1889, p. 436.
149 "If well maintained": EN, Nov. 23, 1889, p. 487.
"Man is more destructive": ibid.
150 "*Like half a rainbow*": ibid.
In the early spring: EN, April 5, 1890, p. 313; July 5, 1890, pp. 12–13.
151 "a few hackfuls": EN, Jan. 2, 1892, p. 15.
two bridge companies: see EN, March 16, 1893, p. 258.
"The North River Bridge Co.": EN, Jan. 2, 1892, p. 15.
152 location of the bridge: cf. Billington (1977), table 1.
"near Desbrosses St.": TASCE, vol. 97 (1933), p. 423.
"somewhere between Seventieth and Eightieth": NYT, Feb. 5, 1888, p. 3.
"at about Sixtieth St.": NYT, March 24, 1888, p. 5.
"between 10th and 181st Sts.": EN, April 14, 1888, p. 283.
"between Washington Heights and Spuyten Duyvil": NYT, July 3, 1888, p. 4.
"at Fourteenth-Street": NYT, July 7, 1888, p. 5.
"at Fort Washington": ibid.
"at any point": NYT, Oct. 16, 1888, p. 3.
"near 13th St.": EN, April 6, 1889, p. 299.
"about Forty-second St.": EN, May 10, 1890, p. 434.
"to recommend": see EN, Sept. 6, 1894, p. 187.
Bouscaren: BDACE, vol. I.
153 Burr: ibid.
Morison: ibid.; E. E. Morison.
Charles Walker Raymond: see "Memoir."
"of the unanimous opinion": EN, Sept. 6, 1894, p. 187.
154 in favor of a suspension bridge: ibid., p. 192.
"the maximum length": EN, Nov. 22, 1894, p. 423.
"for information": EN, Dec. 6, 1894, p. 465.
"one of the most valuable": EN, Nov. 22, 1894, p. 428; cf. EN, Nov. 1, 1894, p. 364.
"attract a traffic": EN, Nov. 22, 1894, p. 428.
Consolidated: EN, Dec. 13, 1894, p. 479.
secretary of war: EN, Dec. 20, 1894, p. 503.
Traffic on Brooklyn Bridge: see, e.g., EN, Nov. 19, 1887, p. 359.
155 Niagara Gorge Bridge: EN, Dec. 27, 1894, p. 534.
"To Mr. Roebling": ibid.
"engineers are only now": ibid.
first prize: EN, June 28, 1894, p. 546.
"One design": ibid., p. 547.
156 "there is no knowing": EN, Dec. 27, 1894, p. 534.
"immense rigid trusses": EN, June 6, 1895, p. 361.

Charles MacDonald: Shanor, pp. 139–40.
"unless something were done": EN, July 11, 1895, p. 25.
a cornerstone: Shanor, p. 141.

157 a tunnel: EN, Feb. 13, 1896, p. 97.
"to appreciate the fact": EN, Nov. 25, 1897, p. 346; cf. Dec. 9, 1897, p. 378.
Hudson Tunnel Railroad Company: EN, June 16, 1892, p. 609.
John Fowler: EN, Sept. 15, 1892, p. 245.
British money: EN, June 16, 1892, p. 609.

158 calls for additional bridges: NYT, May 2, 1883, p. 5; May 14, 1887, p. 8.
Frederick Uhlmann: EN, Jan. 30, 1886.
Leffert Lefferts Buck: see "Memoir."

159 plans for the Williamsburg Bridge: EN, July 30, 1896, p. 76.
"utterly opposed": EN, Aug. 20, 1896, p. 126.
final price tag: see, e.g., D. McCullough (1972), pp. 506, 509.
"judgment, skill": EN, Jan. 27, 1898, p. 60.

160 "ignorance of the true value": ibid., p. 60.

161 "An engineer may not": EN, March 3, 1898, p. 144.
"Roughly speaking": EN, Dec. 17, 1903, p. 535.

162 "the heaviest suspension bridge": Hungerford, p. 26.
"So far as engineering science": ibid., pp. 26–27.

163 "slipping to Brooklyn": NYT, Sept. 2, 1906, p. 1.
Two additional supports: NYT, Nov. 10, 1909.
additional steel: EN, May 14, 1914, p. 1082.

164 "Mr. Buck designed": NYT, June 9, 1911, p. 7.
"perfect condition": ENR, Dec. 8, 1921, p. 939.
Engineering News-Record: see ENR, April 5, 1917; cf. McGraw; Mehren; C. W. Baker.
"Such bridges": ENR, Dec. 8, 1921, p. 924.

165 "dragged woefully": Hungerford, p. 117.
Lindenthal deducted: ibid., p. 118.
The Roebling firm: NYT, Sept. 7, 1905, p. 7.
excluded by New York politics: see D. McCullough (1972), p. 374.
R. S. Buck: EN, Feb. 19, 1903, p. 183; see also TASCE, vol. 40 (1898), p. 160.

166 first semiannual report: EN, Aug. 21, 1902, p. 124.
The new plans: EN, Feb. 19, 1903, p. 184.
board of five engineers: ibid.
"small suspension bridges": EN, March 12, 1903, p. 229.

167 "one of the most experienced": ibid., p. 234.
"they are to be preferred": ibid., p. 243.
The final report: EN, July 2, 1903, p. 24.

168 "even more disappointing": EN, July 9, 1903, p. 38.
"a chain-bridge": EN, July 23, 1903, p. 79.

169 The debate over the Manhattan Bridge: see, e.g., EN, July 30, 1903, p. 102; Aug. 6, 1903, p. 124; Aug. 13, 1903, p. 144.
editorial stance: EN, Aug. 13, 1903, p. 142.
"no greater than": EN, Oct. 1, 1903, p. 296.

"yet much to learn": EN, Oct. 29, 1903, p. 392.
"the best way": EN, Dec. 31, 1903, p. 590.
"personal spite": NYT, Dec. 9, 1905, p. 15; cf. p. 8.
still advocating a chain: NYT, March 31, 1906, p. 8.
Mayor McClellan: NYT, June 16, 1908, p. 5.

170 formally opened: EN, Jan. 6, 1910, p. 27; see also NYT, Dec. 12, 1908, p. 1.
"to watch the construction": NYT, April 12, 1909, p. 5.
Ralph Modjeski: see Durand; *Current Biography*, 1940; DAB, suppl. 2; "Memoir."

171 "The première tragedienne": "Memoir" of Modjeski, p. 1624.
"someday he would build": Modjeska, pp. 245–46.
student with Ignace Paderewski: Durand, p. 246.
"as an honor conferred": *Civil Engineering*, March 1931, p. 568; see also *Civil Engineering*, April 1931, p. 667.

172 "When I was four": *Journal of the Western Society of Engineers*, vol. 36, no. 2 (April 1931), p. 73; see also Durand, p. 255.
"father of bridge building": "Memoir" of Modjeski, p. 1624.
"It is that": *Journal of the Western Society of Engineers*, vol. 36, no. 2 (April 1931), p. 79.

173 "prolonged applause": ibid., p. 72.
His posing: Government Board of Engineers.
Delaware River Bridge: Carswell, pp. 37, 39.

175 "a clean bill of health": NYT, Sept. 14, 1909; see also ENR, Oct. 14, 1909, pp. 401–9.
The Manhattan Bridge: see, e.g., Billington (1983), p. 136.
Lindenthal actually raised: ENR, April 27, 1911, p. 517.
"scheme was rescued": NYT, July 23, 1880, p. 2.

176 "by enormous passenger elevators": NYT, March 16, 1887, p. 5.
legality of running railroad tracks: NYT, Nov. 8, 1889, p. 3.

177 only one pier: Reier, p. 44.
The new location: EN, Nov. 29, 1894, pp. 439, 448.
Supreme Court: EN, Nov. 21, 1895, p. 350.
In his first report: EN, Aug. 21, 1902, p. 125.

178 Lindenthal's specifications: EN, Sept. 3, 1903, p. 206.
"largest cantilever bridge": NYT, July 4, 1905, p. 3; see also July 8, 1905, p. 14.
"and so weakened": NYT, Sept. 22, 1906, p. 6.
Work gangs: NYT, Feb. 14, 1908, p. 1.
dynamite was found: NYT, March 9, 1908, p. 11.
"seemed to be defying": NYT, March 13, 1908, p. 4.
Two independent consultants: Burr and Boller & Hodge, p. 3.

179 Burr did recommend: ibid., p. 26.
"must not be opened": NYT, July 30, 1908, p. 12.
fearless pedestrians: NYT, Aug. 18, 1908, p. 7.
Rudyard Kipling: see "The Bridge-Builders," in Kipling.
"birds in large flocks": NYT, Dec. 10, 1908, p. 3.
"architects and structural engineers": NYT, Dec. 27, 1908, pt. 5, p. 4.

"unpleasantly suggestive": NYT, Sept. 6, 1908, pt. 3, p. 4.
Miss Elinor Dolbert: NYT, Jan. 6, 1909, p. 3.
180 perhaps Wilbur Wright: NYT, Feb. 1, 1909, p. 1.
"veracious press agent": NYT, Feb. 4, 1909, p. 6.
181 Questions of safety: see, e.g., NYT, March 31, 1909, p. 2.
"Dr. Rainey": NYT, May 13, 1909, p. 1.
182 "new bridge ablaze": NYT, June 13, 1909, p. 1.
Hell Gate: EN, May 30, 1907, p. 583.
183 Henry F. Hornbostel: Buckley, pp. 44, 48.
refused to submit new plans: NYT, March 29, 1904, p. 10.
"a pair of immense pylons": EN, May 30, 1907, p. 583.
"in Westchester County": Ammann (1918), p. 1000.
leaning toward a scheme: ibid., p. 860.
184 three comparative designs: ibid., p. 865.
"a utilitarian structure": ibid., p. 868.
An arch design: ibid., p. 865.
Ammann's list of factors: ibid., pp. 869–70.
185 "more expressive of rigidity": ibid., p. 871.
"massiveness over lightness": Billington (1983), p. 126.
"a slight reversal": Ammann (1918), pp. 872, 874.
"rather a massive frill": Billington (1983), p. 128.
186 "although not objecting": Ammann (1918), p. 872.
original tower design left a gap: see ibid., fig. 8, facing p. 874.
"Lindenthal feared": Buckley, p. 48.
187 "Mr. Lindenthal conceived": Ammann (1918), p. 865.
188 "A great work of art": ibid., p. 863.
twice as heavy: EN, Jan. 8, 1914, p. 59.
189 staff of ninety-five: Ammann (1918), p. 985.
"with minor architectural changes": EN, Jan. 8, 1914, p. 59.
"The arches could": EN, Jan. 22, 1914, p. 203.
Lindenthal responded: EN, Feb. 5, 1914, p. 316.
190 "objection was made": Ammann (1918), p. 1032.
"Hell Gate Arch Bridge": EN, Dec. 31, 1914, p. 1311.
Federal Express: NYT, March 11, 1917, sect. III, p. 4; ENR, March 15, 1917, p. 453.
191 "chief memorial": ENR, Aug. 8, 1935, p. 208.
The Sciotoville: McClintic-Marshall Company, p. 3.
"perhaps the boldest": quoted in Billington (1983), p. 125
"the ultimate expression": Carl Condit, quoted in Jackson, p. 173.
"daring and handsome structure": in discussion to Lindenthal (1922), p. 963.
"genius that originates": quoted in ibid.; see also p. 912.
"the boldest bridge plan": Waddell (1916), p. 608.
192 "The Sciotoville Bridge": in discussion to Lindenthal (1922), p. 962.
"detailed, although somewhat belated": Lindenthal (1922), p. 910.
Rowland Prize: *Civil Engineering*, Sept. 1935, p. 594.
"in this unusual work": Lindenthal (1922), p. 953.

193 Portland, Oregon: Wood, p. 25; ENR, June 26, 1924, p. 1115; cf. Ratigan, pp. 194–95.
largest bridges on the West Coast: ENR, Aug. 8, 1935, p. 208.
"he never built two": "Memoir," p. 1793.
"his habit of looking": ibid.; cf. Leonhardt (1984b), pp. 32–34.
"one of the masters": EN, Dec. 21, 1916, p. 1188.
J. A. L. Waddell: see, e.g., *Who's Who in Engineering*, 1922–23; BDACE, vol. II; *National Cyclopaedia of American Biography* vol. XXVII; ENR, March 10, 1938, p. 354; Waddell (1928).
194 "of finding someone": EN, Dec. 21, 1916, p. 1188.
"the fact that parts": Lindenthal (1916), p. 1175.
"too active to find": TASCE, vol. 105 (1940), p. 1794.
Waddell kept his staff employed: Hardesty & Hanover (1987a), no. 1, p. 4.
"breezy and often gossipy": Lindenthal (1916), p. 1175.
195 Hell Gate: Waddell (1916), p. 27.
"the noted bridge engineer": ibid., p. 625.
"certainly of aesthetic appearance": ibid., p. 626.
"Gustav Lindenthal, Esq., C.E.": ibid., p. 126; cf. p. 482.
"A New York engineer": ibid., p. 586.
196 without one's name: ibid., p. 27.
"much valuable information": ibid., p. 126.
"one of the most prominent": ibid., p. 471.
"exceedingly favorable": ibid., p. 482.
"Messrs. Geo. S. Morison": ibid., p. 660.
"made in the late eighties": ibid., p. 662.
Hodge's plan: ibid., pp. 661–62.
197 Halstead Street Lift-Bridge: Waddell (1895); see also Waddell & Son.
198 series of partnerships: see Hardesty & Hanover (1987b).
199 Waddell & Son catalogue: Waddell & Son.
thoughtfully cropped photographs: Strauss Bascule Bridge Company.
"possession of a constitution": *Engineering*, April 15, 1938, p. 413.
"in longhand": in Waddell (1928), p. 17.
200 "The book": Lindenthal (1916), p. 1175.
"there are few structures": ibid., p. 1177.
201 "The author's repeated reference": ibid., p. 1178.
"The great bridges of New York": from letter quoted in ENR, Jan. 29, 1920, p. 250.
202 "resolutions of remonstrance": ENR, Feb. 26, 1920, p. 435.
Ralph Modjeski was chosen: Carswell, p. 6.
203 "the engineer who thinks": ENR, Nov. 24, 1921, p. 862.
"the finest example": ibid., p. 861.
Joseph Pennell: see, e.g., Fredrich (1993).
204 as early as 1818: *Delaware River Bridge*, p. 19.
In 1843: ibid., p. 20.
Waddell & Son: ENR, June 23, 1921, p. 1086.
Havana, Cuba: Waddell (1916), pp. 1156–57; fig. 52a, p. 1158.

Warren P. Laird: *Delaware River Bridge*, p. 23.
206 board of engineers: ibid., p. 6.
Suspension and cantilever: ENR, June 23, 1921, p. 1087.
"in lieu of": *Delaware River Bridge*, p. 24.
tolls: see, e.g., ENR, Sept. 3, 1925, p. 399.
"under the pressure": ENR, July 30, 1925, p. 167.
"if the outcome": ENR, Dec. 3, 1925, p. 900; see also July 30, 1925, p. 167.
"it probably ranks": ENR, Sept. 30, 1926, p. 530.
"There will be": *Delaware River Bridge*, p. 7.
208 "Bridge Engineering": Lindenthal (1924b).
"Engineers are sometimes": ibid., p. 657.
210 "Bridge construction": ibid.
211 "exceptionally severe winter": Lindenthal (1918a), p. 3.
"the slowing up and congestion": ibid., p. 10.
"it would be folly": ibid., p. 14.
"Dr. Engr. Gustav Lindenthal": see Waddell (1916), p. 1626.
"*Lower Deck*": Lindenthal (1919), p. 6.
212 his earlier pamphlet: ibid., p. 16. The pamphlet referred to is Lindenthal (1918a).
"the most backward": Lindenthal (1933), p. 7.
"deduce and predict": ibid., p. 5.
213 "He waved the inquiry": NYT, May 22, 1930, p. 29.
eighty-first birthday: NYT, May 21, 1931, p. 24.
"pigeonholed": ENR, Aug. 8, 1935, p. 208.
"grand old man": quoted in NYT, Dec. 16, 1932, p. 15.
214 two other engineers were honored: ibid.
216 eighty-fifth birthday: NYT, May 22, 1935, p. 4.
he died: see Lindenthal obituaries in *Civil Engineering*, Sept. 1935, p. 594; ENR, Aug. 8, 1935, p. 208; NYT, Aug. 1, 1935, p. 23.
"his dream of forty years": NYT, Aug. 1, 1935, p. 23.
only David Steinman: NYT, Aug. 3, 1935, p. 13.

Chapter 5 Ammann

217 Othmar Hermann Ammann: see, e.g., Durrer; Stüssi; Widmer; Wisehart; see also DAB, suppl. 7; *Current Biography*, 1963.
The river: Durrer, p. 26.
descendant of: *Consulting Engineer*, Feb. 1964, p. 10.
218 sketching pad: MacKaye, p. 23.
Swiss Federal Institute of Technology: Wisehart, p. 183.
Wilhelm Ritter: Billington (1980), p. 1109.
The final span illustrated: Ritter, p. 63.
worked one summer: MacKaye, p. 23.
Wartmann & Valette: *Current Biography*, 1963, p. 8.
"the engineer has greater": quoted in Wisehart, p. 183.

219 "My first serious interest": quoted in Widmer, pp. 5–6.
"Get all the experience": quoted in Wisehart, p. 183.
220 "the first door": Durrer, p. 27.
Union Bridge Company: Wisehart, p. 184; Widmer, pp. 4–5.
"designed twenty-five": quoted in Wisehart, p. 184.
"boss was simply": letter, Ammann to his parents, Dec. 3, 1904, trans. Margot Ammann Durrer. Courtesy of Margot Ammann Durrer.
"very modern": ibid.
221 Frederic C. Kunz: Widmer, p. 6.
222 "for their able assistance": Kunz, p. vi.
Among the many plates: Kunz, plate XLVIII A, facing p. 373.
"the actual boss": Widmer, p. 6.
new Quebec Bridge: see Kunz, plate XLVIII A, facing p. 373.
recommended him to Gustav Lindenthal: Wisehart, p. 185.
223 "OHA started position": reproduced in Widmer, p. 7; see Ammann (1918), p. 985; Widmer, p. 6; Wisehart, p. 185.
ninety-five engineers: Ammann (1918), p. 985.
left for Switzerland: Widmer, p. 7; MacKaye, p. 24.
the last rivet: Widmer, p. 7.
Ammann presented: Ammann (1918); see also Buckley, p. 45.
224 "obligation, for permission": Ammann (1918), p. 854.
reports of uncommon clarity: see Miller and Saidla, pp. 235–53.
"The paper is": Quimby discussion to Ammann (1918), p. 1020.
"and work on them": Miller and Saidla, p. 252.
"bent over his desk": letter, Margot Ammann Durrer to Henry Petroski, Nov. 8, 1993.
225 "Whenever I looked over": quoted in Durrer, p. 29.
"peculiar construction": Lindenthal (1922), p. 912.
226 "in this unusual work": ibid., p. 953.
"A great engineering work": Ammann (1918), p. 986.
227 "the position was not attractive": quoted in Widmer, p. 9.
turned the situation around: ibid.
Standard biographical sketches: see, e.g., *Current Biography*, 1963; DAB, suppl. 7.
North River Bridge Corporation: NYT, Jan. 19, 1921, p. 17; Jan. 22, 1921, p. 14.
"to obtain public support": NYT, April 17, 1921, sect. X, p. 15.
vehicular tunnels: see, e.g., ENR, June 20, 1918, p. 1202.
228 "coal famine": NYT, Jan. 27, 1918, sect. III, p. 8.
George Washington Goethals: see "Memoir"; S. R. Watson and Watson, pp. 121–28.
Six feet tall: S. R. Watson and Watson, p. 124.
229 "was due almost entirely": NYT, March 18, 1918, p. 18.
Goethals estimated: ibid.; NYT, June 29, 1918, p. 15.
The cost of $12 million: NYT, Feb. 25, 1919, p. 11.
Clifford M. Holland: NYT, June 5, 1919, p. 28; see also "Memoir."
"the youngest": NYT, June 15, 1919, sect. IV, p. 13.

230 "he spent more time underground": ENR, Oct. 30, 1924, p. 723.
"The duties of": NYT, June 5, 1919, p. 28.
231 salary of $10,000: ibid.
Several key appointments: NYT, July 2, 1919, p. 25.
"After very careful investigation": quoted in NYT, Feb. 15, 1920, sect. IV, p. 17.
232 Within days of the release: see ENR, March 25, 1920, pp. 624–26.
Goethals wrote: see ENR, April 8, 1920, pp. 729–32.
Holland suggested: NYT, Feb. 20, 1920, p. 26.
233 John F. O'Rourke: ENR, Aug. 2, 1934, p. 156.
O'Rourke would realize: NYT, Feb. 21, 1920, p. 13.
"The proposal to build": NYT, March 2, 1920, p. 21.
devote no more time: NYT, March 10, 1920, p. 16.
234 American Association of Engineers: see *Who's Who in Engineering*, 1937, p. xiv.
"answers to questions": NYT, July 16, 1921, p. 13.
"There can be no question": quoted in ENR, July 21, 1921, p. 125.
"successful enough to underbid": NYT, July 31, 1921, sect. II, p. 1.
235 "in the opinion": NYT, July 27, 1921, p. 5.
public debate: NYT, July 29, 1921, p. 17.
send Holland to Europe: ENR, July 28, 1921, p. 164.
cold-storage plant: see NYT, June 1, 1922, p. 1.
delays were estimated: NYT, Nov. 16, 1921, p. 10.
capacity of fifteen million: NYT, Dec. 28, 1921, p. 4.
experimental circular tunnel: NYT, Oct. 30, 1921, sect. II, p. 1.
236 twenty-foot granite shaft: NYT, Dec. 15, 1921, p. 5.
"Mr. Holland took a pick": NYT, April 1, 1922, p. 16.
"Mr. Falconer": NYT, June 1, 1922, p. 1; cf. NYT, June 7, 1922, p. 5.
a private affair: ENR, June 8, 1922, p. 971.
237 "purely commercial": NYT, Feb. 21, 1923, p. 16.
Bergen County: NYT, March 4, 1923, sect. II, p. 1.
"the construction of one": NYT, April 10, 1923, p. 1.
238 Port of New York Authority: NYT, March 6, 1923, p. 16.
vetoed two tunnel bills: NYT, May 31, 1923, p. 2.
"more wages": NYT, April 13, 1923, p. 36.
Costs were also rising: NYT, July 1, 1923, sect. VIII, p. 1.
"who did not come": NYT, Nov. 4, 1923, sect. 2, p. 8.
final cost was estimated: NYT, Jan. 15, 1924, p. 23.
Liberty Tunnels: NYT, Jan. 23, 1924, p. 16.
accumulated fumes: NYT, May 11, 1924, p. 1.
Yandell Henderson: NYT, March 3, 1924, p. 16; cf. NYT, May 18, 1924, p. 23.
tunnel sprang a leak: NYT, April 4, 1924, p. 15.
239 another sand-hog strike: NYT, April 10, 1924, p. 38.
"the only spot": NYT, Sept. 27, 1924, sect. I, p. 22.
"some of the engineers": NYT, Oct. 12, 1924, p. 14.
President Coolidge: NYT, Oct. 19, 1924, p. 1.
"in the depths": NYT, Oct. 26, 1924, sect. IX, p. 14.
The blast occurred: NYT, Oct. 30, 1924, p. 1.

"started working on tunnels": NYT, Oct. 28, 1924, p. 23.

240 "his continuous devotion": ibid.
"had expected": NYT, Oct. 30, 1924, p. 1.
the name Holland Tunnel: NYT, Nov. 14, 1924, p. 18.
"Throughout time": ENR, April 2, 1925, p. 545.

241 "able lieutenant": ibid., p. 572; cf. p. 575.
"nice controversy": ENR, Feb. 21, 1924, p. 309; see also March 16, 1922, p. 28.
In the 1924 election: NYT, Dec. 3, 1924, p. 13.

242 "the next of the great viaducts": NYT, Oct. 11, 1908, p. 7.
Pennsylvania Station: see, e.g., Goldberger, p. 31.
interstate bridge commission: NYT, June 12, 1909, p. 1.

243 "would seriously injure": NYT, Nov. 1, 1908, pt. 2, p. 8.

244 "the most feasible sites": NYT, Dec. 6, 1908, sect. II, p. 6.
"the steamship and tobacco man": NYT, June 12, 1909, p. 5.
"The problem": quoted in ibid.

245 relative riverbed conditions: see NYT, Feb. 9, 1909, p. 9; see also NYT, Oct. 5, 1910, p. 20.
George F. Kunz: NYT, Oct. 5, 1910, p. 20.

246 McDougall Hawkes: NYT, Oct. 6, 1910, p. 5.
Lindenthal wrote a letter: NYT, Oct. 17, 1910, p. 8.
charter was running out: NYT, March 16, 1912, p. 16.
long letter: NYT, Dec. 9, 1912, p. 20.

247 John F. Stevens: NYT, Dec. 13, 1912, p. 13.
Plans prepared by Boller & Hodge: NYT, Dec. 22, 1912, sect. 8, p. 2.
Henry Hodge: NYT, Dec. 22, 1919, p. 15; Dec. 24, 1919, p. 13; Jan. 1, 1920, p. 8.
"one of the foremost": NYT, Dec. 22, 1919, p. 15.
"The Hudson River": ENR, Dec. 23, 1920, p. 1246.

248 moving passenger platform: see, e.g., *Scientific American*, March 26, 1910, p. 257.
"in the vicinity": NYT, Jan. 19, 1921, p. 17.

249 "it was possible to bridge": NYT, Aug. 22, 1923, p. 15.
Ammann had his own stationery: reproduced in Doig (1990), p. 187.
considerable detail: see ibid.; see also Doig and Billington.

250 Joseph B. Strauss: see, e.g., van der Zee.
"March 22 / 1923": reproduced in Widmer, p. 12.
"think piece": Ammann (1923).

251 "In vain": letter, Ammann to his mother, Dec. 14, 1923, trans. Margot Ammann Durrer. Courtesy of Margot Ammann Durrer.
"It is over a year ago": quoted in Widmer, pp. 12–13.

252 "At the opportune moment": quoted in Widmer, p. 13.
public hearings: Doig (1990), pp. 169–70.

253 "A bridge at Fort Lee": reproduced in ibid., p. 188.
sketch of his suspension bridge: ENR, Jan. 3, 1924, p. 34.
"almost unnoticed": ENR, April 9, 1925, p. 588.
"such a man": quoted in Doig (1990), p. 171.

254 Benjamin F. Cresson, Jr.: ENR, Feb. 1, 1923, p. 223.

died suddenly: ibid.
William W. Drinker: ENR, Sept. 27, 1923, p. 531.
Since colonial days: City of Bayonne.
"the Port Authority ought": quoted in Doig (1990), p. 172.
255 "anxiously awaiting their decision": quoted in ibid., p. 174.
"We desire to secure": quoted in ENR, Dec. 18, 1924, p. 991; cf. editorial, p. 981.
"It calls for bids": ENR, Dec. 18, 1924, p. 981.
256 William Burr and George Goethals: see Reier, pp. 112–14.
"encouraging interview": Doig (1990), pp. 174–75.
Outerbridge Crossing: Reier, p. 114; Outerbridge and Outerbridge, p. 6.
Goethals Bridge: Reier, p. 114; ENR, March 22, 1928, p. 492.
group of consulting engineers: ENR, Dec. 13, 1928, pp. 873–77.
257 "the coat-hanger": Buchanan (1992), caption to Fig. 25.
"Lindenthal's departure": Reier, p. 61; cf. Freeman, pp. 217–18.
modeled the Sydney Harbour Bridge: see Freeman.
redundant diagonal member: ibid., p. 220.
Kill van Kull arch: see ENR, Dec. 13, 1928, pp. 873–77.
"the high regard": ENR, Dec. 16, 1926, p. 982.
258 five papers: see ENR, Feb. 24, 1927, pp. 336–37.
259 *De Witt Clinton*: ENR, Aug. 25, 1927, p. 321.
"there are compensations": "Addresses Delivered on the Occasion of Breaking Ground for the Hudson River Bridge," Port of New York Authority, Sept. 21, 1927, [p. 2].
submitted a wire-cable bid: ENR, Oct. 6, 1927, p. 563; Oct. 13, 1927, p. 592; Oct. 20, 1927, p. 648.
260 "esthetic considerations": ENR, Aug. 11, 1927, p. 215.
an early modification: ENR, May 24, 1928, p. 819.
model tests: ibid., pp. 819–22.
"He fully realizes": ENR, Nov. 15, 1928, p. 741.
261 The basic idea for the towers: see Ammann (1924).
"another important part": ibid., p. 25.
"The opposite extreme": ibid.
262 "the combination and utilization": ibid., p. 26.
"the architecture of the towers": ibid.
"Not since December 1911": *Civil Engineering*, Oct. 1933, pp. 583–84.
264 the Gilbert stonework: see, e.g., Billington (1977), p. 1663; see also Ammann (1933b), fig. 22.
"There is no part": Ammann (1933b), pp. 46–47, 51.
265 "the fact that": Moisseiff (1933), p. 165.
"When your car": Le Corbusier quoted in a press kit prepared by the Port of New York Authority for the dedication of the opening of the lower deck of the George Washington Bridge, 1962.
six months earlier: NYT, Oct. 18, 1931, p. 16.
266 "George Washington Memorial Bridge": NYT, Jan. 14, 1931, p. 1.
"ridiculous name": NYT, Jan. 20, 1931, p. 22.

267 would continue to call: NYT, Jan. 22, 1931, p. 22.
alternatives suggested: see, e.g., NYT, Feb. 12, 1931, p. 14.
"Cleveland Bridge": NYT, ibid., p. 27.
"named to suit": NYT, Jan. 22, 1931, p. 41.
public was now invited: NYT, Jan. 23, 1931, p. 1.
"Verrazano Bridge": NYT, Feb. 9, 1931, p. 39.
"volume of letters": NYT, Feb. 12, 1931, p. 14.

268 decided on "George Washington Bridge": NYT, April 24, 1931, p. 25.
"what the millions": NYT, April 25, 1931, p. 18.
"get its workaday name": ENR, May 7, 1931, p. 753.
"But probably the greatest": NYT, Oct. 25, 1931, p. 30.
"designer of the great": ibid.
"rendered special advice": Ammann (1933b), p. 65.
"Defying the age-old rule": NYT, Oct. 25, 1931, pp. 1, 30.

269 all the technical ingenuity: Doig and Billington, pp. 12 ff.
1923 article: Ammann (1923).
slightly skewed view: see, e.g., ibid., p. 1074.

270 unlikely extreme traffic condition: see Doig and Billington, pp. 15–16.
"every dollar spent": quoted in ibid., p. 10.
economic attractiveness: ibid., pp. 17–18.

271 "the stiffening system": quoted in ibid., p. 21.
"The setting": quoted in Widmer, p. 19.
"great monument to Mr. Ammann": see Ammann (1931b).
date of the first tunnels: ibid., p. 9.
"In my preliminary studies": ibid., p. 17.

272 "a man of vast": ibid., p. 25.
"There was one": ibid., p. 27; cf. Waddell (1916), p. 5, for another version of the story of squaws used in proof testing.
engineers who were not in attendance: see Ammann (1931b), p. 3.

273 James Wilkins: Golden, p. 11; Golden Gate Bridge (1994), p. 9.
Hetch Hetchy Valley: van der Zee, p. 25; see also NYT, Aug. 5, 1987, pp. 1, 10.
Aeroscope: see Anderson, pp. 128–31; cf. U.S. Patent No. 1,235,506.
"Alcatraz and the Angel Islands": Anderson, p. 130.

274 Joseph Baermann Strauss: *Who's Who in Engineering*, 1937.
"to build the biggest thing": Golden, p. 3.

275 "$100 and told him": ibid., p. 5.
New Jersey Steel and Iron Company: see Darnell (1984).
"element of movement": Strauss Bascule Bridge Company, p. 4.

276 O'Shaughnessy shared the data: van der Zee, p. 39.
It was an ungraceful hybrid: see O'Shaughnessey and Strauss, pp. 3, 5; cf. van der Zee, p. 42.
"to help stimulate": van der Zee, p. 41.
An attractive booklet: O'Shaughnessy and Strauss.
"new cantilever-suspension type": ibid., p. 6.
"so reasonable": see ibid., p. 12.

277 Golden Gate Bridge and Highway District: ibid., p. 43.

Charles Alton Ellis: ibid., pp. 48–50; cf. van der Zee and Cone.
278 Moisseiff prepared plans: van der Zee, p. 57.
Among the engineers: ibid., p. 78.
Ammann had accepted: ibid., pp. 78–84.
279 O'Shaughnessy: see, e.g., ibid., p. 86.
newly printed letterhead: see ibid., p. 72.
"the lowest ever written": ibid., p. 85.
Strauss sued: ENR, Nov. 19, 1936, p. 732.
Early plans: van der Zee, p. 94.
280 It was Ellis who assured: ibid., p. 100.
"the stepped-off type": ibid., p. 114.
John Eberson: ibid., pp. 114–15.
He produced a design: see ibid., p. 124.
Irving F. Morrow: ibid., p. 115.
281 "the Golden Gate Bridge": ENR, May 28, 1931, p. 877.
The university's president: van der Zee, p. 125.
"Mr. Strauss gave me": ibid., p. 126.
"who stands high": ibid., p. 128.
"even if Mr. Strauss:" ibid.
"a problem of this nature": ibid., p. 110; cf. van der Zee and Cone.
282 "the structure was nothing unusual": van der Zee, p. 142.
"vexed" even Moisseiff: ibid., p. 155.
"without diagonals": ENR, Jan. 25, 1934, p. 125.
"in protest over": van der Zee, p. 156.
"principal engineer": Golden Gate Bridge (c. 1987).
a plaque: van der Zee, p. 295.
283 Strauss's final report: Strauss.
Pedestrian Day: see van der Zee, pp. 296–98.
fiftieth anniversary: see, e.g., Kuesel, pp. 58, 59.
284 known among engineers: see, e.g., Ketchum and Heldermon.
$128-million project: ENR, Jan. 6, 1992, p. 12.
disagreements over liability: ENR, Jan. 4, 1993, p. 19.
"*At last the mighty task*": *San Francisco News*, May 26, 1937, Golden Gate Bridge and Fiesta Section, p. 4; cf. Golden Gate Bridge (c. 1987).
285 It is known that Ellis: van der Zee and Cone.
"When Telford planned": Ammann (1933a), p. 429.
286 Triborough Bridge Authority: see, e.g., Caro, pp. 657–59.
Tammany engineers: Caro, p. 391.
The proposed design: see Shanor, pp. 149–56.
288 "so that the structure": Embury (1938c), p. 265.
"It is now well established": Ammann (1939), p. 218.
289 "in no small degree": ibid., p. 217.
architect to the elite: Caro, p. 365.
"had the good fortune": Embury (1938b), p. 85.
290 worked with Dana *de novo*: ibid.
"wanted the anchorages": ibid., p. 86.

"until traffic conditions": ibid., p. 87.
"an anachronism": ibid., p. 88.
"how often do engineers": Embury (1938a), p. 4.
291 "a reversal of function": Embury (1938b), p. 89.
"it is always easier": ibid.
"pendulum action": ENR, Dec. 5, 1940, pp. 54–55.
Golden Gate moved sideways: Vincent, pp. 1817-1–1817-2.
292 Two much shorter suspension bridges: see Billington (1977), pp. 1667–68.
stranded on Deer Isle: NYT, May 31, 1978, p. 16.
"We have had to deal with": ENR, Dec. 5, 1940, p. 56.
293 Leon Solomon Moisseiff: see "Memoir"; DAB, suppl. 3, pp. 530–31.
294 "Although he did not": DAB, suppl. 3, p. 531.
Tacoma Narrows Bridge: see Ammann et al.
295 traffic over the bridge: ENR, Aug. 1, 1940, p. 139.
296 "Unless there are": Moisseiff, in Ammann et al., p. II-1.
"raise the west end": Moisseiff, in ibid., p. II-2.
"result in a neat": Moisseiff, in ibid., p. II-4.
"be about one cent per lb. less": Moisseiff, in ibid., p. II-6.
In a second part: ibid.
"height of the towers": Moisseiff, in ibid., p. II-7.
297 Theodore L. Condron: NYT, Apr. 14, 1955, p. 29.
identified the board: Condron, in Ammann et al., p. IV-1.
"full confidence in Mr. Moisseiff": quoted in ibid., p. IV-6.
Condron's tabulation: ibid., p. IV-5.
298 "certain tests had been made": Condron, in ibid., p. IV-3.
"Moisseiff and Lienhard": quoted in ibid., p. IV-3.
"there seems to be some question": Condron, in ibid., p. IV-4.
"In view of": Condron, in ibid., p. IV-3.
Freudian slipping: cf. Moisseiff and Lienhard.
299 "gratifying": Derleth, in discussion of Moisseiff and Lienhard, p. 1123.
"had its inception," ibid., p. 1122.
"With regard to the super-structure": Condron, in Ammann et al., p. IV-5.
300 "the Bridge was undulating": Cone, in ibid., p. IX-1.
"It might seem": quoted in Ammann et al., p. IV-6.
Construction bids: see Farquharson, pt. 1, p. 17.
301 Even before the bridge was completed: see, e.g., Ammann et al.
on November 7, 1940: see ibid.; cf. NYT, Nov. 8, 1940, pp. 1, 5.
302 "felt an obligation": Pagon (1934a), p. 348.
a series of eight articles: Pagon (1934a through 1935d).
"must reading": Farquharson, pt. 1, p. 14.
The letter: ENR, Nov. 21, 1940, p. 40.
Theodore von Kármán: *Current Biography*, 1955; NYT, May 18, 1963, p. 39.
303 "an eccentric study": NYT, May 18, 1963, p. 39.
The committee's report: Ammann et al.
"the Tacoma Narrows Bridge": ibid., "Summary of Conclusions," n.p.
bombastic autobiography: von Kármán.

304 "the bridge was built correctly": ibid., p. 212.
"took home from Cal Tech": ibid., p. 213.
"long standing of the prejudices": ibid., p. 215.
"difference in thinking": ibid.

305 "more complete theory": Woodruff, p. 214.
"The most perfect system": ibid., p. 215.
"analyze all the assumptions made": ibid., p. 214.
"probably induced": Ammann et al., "Summary of Conclusions," n.p.

306 Professor Farquharson: see Farquharson.
consulting board: see ibid., introduction.
"came as such a shock": ibid., p. 13.
"bridges severely damaged": ibid., p. 14.
J. Kip Finch: see Finch (1941).

307 "This time the problem": ibid., p. 407.
"the modern bridge engineer": ENR, March 27, 1941, p. 459.
John Roebling: see J. A. Roebling (1841).
J. Scott Russell: see Russell.
"It is also a mistake": Finch (1941), p. 459.

308 "assisting in the solution": ENR, Sept. 23, 1943, p. 451.
obituary: ENR, Sept. 9, 1943, p. 373.
unusual number of letters: ENR, Sept. 23, 1943, pp. 74–75.
"The one great disappointment": ibid., p. 75.
memoir of Moisseiff: see "Memoir."
"one of the best informed": "Memoir," p. 1509.

309 Charles S. Whitney: Cohen, p. 734.
Ammann & Whitney: Widmer, p. 20.
"the unprecedented increase": Port of New York Authority and Triborough Bridge and Tunnel Authority, p. 6.
The consulting engineer: ibid., p. 62.
"to help him": Dunham, pp. 90–91.

310 "While the truss members": Ammann (1946).
"inadequate vertical stiffness": ibid., p. 102.
motor-vehicle registration: Port of New York Authority and Triborough Bridge and Tunnel Authority, pp. 11, 13.
"masterpiece of traffic relief": NYT, Aug. 29, 1962, p. 31.

311 "small man": Durrer, p. 32. (Note that a typographical error has Ammann as only "4 feet, 6 inches" tall.)
"it took a few minutes": NYT, Aug. 30, 1962, p. 59.
bust of Ammann: see ENR, Sept. 6, 1962, p. 24.

312 "reflect favorably": NYT, Aug. 29, 1962, p. 59.
"no one man designed": NYT, Aug. 30, 1962, p. 28.
"one of America's outstanding engineers": NYT, Sept. 11, 1962, p. 32.
"anyone who would take": NYT, Aug. 29, 1962, p. 59.
statue of chief engineer Strauss: ENR, Dec. 7, 1939, p. 735; Feb. 1, 1940, p. 143; June 5, 1941, p. 862.
Clarence Dunham: see Dunham, p. 90.

313 Charles Worthington: NYT, Nov. 13, 1910, pt. 7, p. 2.
"mercantile development": ibid.
"Every one of them": *Rensselaer*, Dec. 1992, p. 7.
Milton Brumer: Rensselaer Polytechnic Institute, Distinguished Service Award, 1972, program.
"and it was on Milton Brumer": Cohen, p. 736.
"his simply styled office": NYT, Aug. 29, 1962, p. 59.
314 also kept an apartment: see NYT, March 26, 1964, p. 37.
"lucky": NYT, March 26, 1964, p. 38; cf. Katz, p. 36.
"at a thinly disguised": NYT, Oct. 23, 1964, p. 26.
Italy also issued a stamp: NYT, Nov. 29, 1964, sect. II, p. 35.
315 "a brave vagrant": NYT, Nov. 3, 1964, p. 30.
ever-popular numbers: Cohen, p. 739.
"the success of": Joseph Gies, quoted in Talese, p. 38.
Cornelius Vanderbilt: Réthi, pp. 8, 10.
"the most important link": ibid., foreword.
long-established neighborhoods: see Talese, ch. 2.
"essentially the application": Ammann, in preface to Réthi.
316 upper and lower decks: Talese, p. 45.
The opening ceremonies: Triborough Bridge and Tunnel Authority, Verrazano-Narrows Bridge Dedication, Nov. 21, 1964, program.
Robert Moses rode: see NYT, Nov. 22, 1964, p. 30.
317 "the careers of two men": ibid., sect. IV, p. 8.
"*Through the bound cable strands*": quoted in ibid.; see *The Bridge*, sect. viii, "Atlantis," in Crane, p. 55.
Talese had written: Talese.
"generals, admirals": NYT, Nov. 22, 1964, p. 30.
"I now ask": ibid.
318 "It's Ed Sullivan": Durrer, p. 32.
"reshaping the skyline": Blum, p. B1.
"The rain was coming down": Trump, quoted in ibid.
319 "dreamer in steel": "Remarks of Robert Moses on the Occasion of the Dedication of the Othmar Ammann College," Triborough Bridge and Tunnel Authority, February 18, 1968, program.

Chapter 6 Steinman

321 David Bernard Steinman: see Ratigan; see also *Current Biography*, 1957.
322 "his father lashed him": Ratigan, p. 23.
"softly weeping": ibid., p. 17.
"He could rattle off": ibid., pp. 21–22.
A boxed charlotte russe: ibid., p. 22.
Talk about building a new bridge: NYT, May 2, 1883, p. 5.
"steel wire suspension bridge": EN, March 7, 1895, p. 145.
Leffert Buck had had: EN, Feb. 13, 1896, p. 103; June 18, 1896, p. 393.

bare steel towers: EN, Aug. 20, 1896, p. 126; Aug. 4, 1898, p. 66.
He climbed upon the steelwork: S. R. Watson and Watson, p. 145.
323 Morrill Land Grant Act: see Grayson, p. 43.
"The most deserving case": quoted in *Current Biography*, 1957, p. 527.
324 "built by a troop": EN, Sept. 25, 1913, p. 614.
Steinman personally calculated: Ratigan, p. 101.
325 Steinman reported the results: Steinman (1918).
"special acknowledgement": ibid., p. 1042.
"able paper": Lindenthal, in discussion to ibid., pp. 1089–90.
326 unspecified "suggestions": Steinman (1918), p. 1072.
"The analysis": Ammann, in discussion to ibid., p. 1106.
"the expense": ibid., p. 1108.
In his closure: Steinman (1918), p. 1131.
deliberately introduced titles: ibid., p. 1132.
327 "wedding present": Ratigan, p. 101.
"commissioned him to write": ibid., p. 103.
biographer Ratigan: see Ratigan, [p. 460].
"Steinman, bridge engineering is easy": Ratigan, pp. 103–5.
328 service under Lindenthal: see, e.g., *Who's Who in Engineering*, 1959.
329 "in the slums": Daley, p. 33.
Irene Hoffmann: Ratigan, pp. 100–101.
he would be identified: see M. Davis.
"active in Presbyterian affairs": Daley, p. 33.
"a man who modestly": "Memoir" of Robinson, p. 1532.
He went into private practice: S. R. Watson and Watson, p. 147; cf. Daley, p. 33.
330 "a drafting table": "Memoir," p. 1533.
Holton Duncan Robinson: see "Memoir."
332 "he suffered excruciatingly": ibid., p. 1533.
"Even in his last years": ibid., p. 1535.
333 Steinman's article: Steinman (1924).
"most pleasing outline": ibid., p. 779.
"for better appearance": ENR, Nov. 27, 1924, p. 883.
334 The review itself: ENR, Dec. 21, 1922, pp. 1080–81.
Carquinez Strait Bridge: see Steinman (1927).
"to take the Island": see Robinson & Steinman.
335 "the desire to secure": ENR, Feb. 13, 1930, p. 272.
The towers were designed: S. R. Watson and Watson, p. 148.
"on a gusty, rainy day": "Memoir" of Robinson, p. 1536.
"professionally connected": ibid., p. 1535.
336 "helped to design": see, e.g., *Current Biography*, 1957, p. 528.
bridge to Oakland: see, e.g., Purcell et al.; United States Steel (1936).
"the greatest bridge": ENR, July 20, 1933, p. 89.
337 Charles H. Purcell: Smith et al., p. 242.
338 "among the lightest": Jackson, p. 264.
"studies and investigations": United States Steel (1936), p. 8.

The board of consulting engineers: Purcell, pp. 183, 187.

339 In an article: Purcell et al.
"consider the large number": ibid., p. 376.

340 two distinct bridges: United States Steel (1936), pp. 11–12.
Joshua A. Norton: ibid., p. 86.

341 "We, Norton I": reproduced in ibid., p. 87.
"Who is bold enough": ibid., p. 86.
"*Your length*": Andrade (unpaged).
Conde Balcom McCullough: Smith et al., pp. 242–43; see also Jackson, p. 304.

344 "Outstanding Practice Problems": Steinman (1925), p 851.

345 American Society of Civil Engineers and Architects: see C. W. Hunt, p. 17.
founder societies: see, e.g., Wisely, p. 308. The American Institute of Chemical Engineers became the fifth "founder society" in 1958.

346 "to address the social": Robbins, p. 3.
professional ethics: see, e.g., Wisely, pp. 128 ff.
codes of ethics: ibid. p. 127.
"neither satisfactorily formulated": ENR, May 21, 1925, p. 839.

347 "In plain words": ibid.
"to further the public welfare": Robbins, p. 5.
"engineering on a par": Steinman (1935b), p. 877.

348 "The public needs": ibid. pp. 877–78.
"nontechnical concerns": Robbins, pp. 5–6.

349 "Why, I thought": Hoover, pp. 131–32.
"The four-year course": ENR, Jan. 14, 1932, p. 65.
issue of what form: see Florman.
"Engr. D. B. Steinman": ENR, Feb. 13, 1936, p. 257.
"Engineers above all": ENR, Jan. 2, 1936, p. 25.

350 "Mexican engineer": ENR, April 23, 1936, p. 607; cf. Jan. 2, 1936, p. 25; Feb. 13, pp. 256–57; May 21, p. 749.
"Liberty Bridge": Daley, p. 33; cf. Steinman (c. 1929).

351 Robinson & Steinman brochure: Robinson & Steinman.
"I expect Liberty Bridge": NYT, Sept. 20, 1948, p. 27, as quoted in *Current Biography*, 1957.

352 late 1940s bridges brochure: Steinman (c. 1947).
"the hands of Dr. Steinman": see also *The Bent of Tau Beta Pi*, Dec. 1957, p. 23.
As late as 1957: *Current Biography*, 1957.

353 After the George Washington Bridge: see Billington (1977).

354 separate articles: Steinman (1941a); Ammann (1941).
"elaborate tests": Steinman (1941a).
response from Ammann: Ammann (1941).

355 "competitors": ENR, June 25, 1959, p. 58.

356 "Eminence Grade": Wisely, p. 105; see also pp. 106–109.
Sara Ruth Watson: Ratigan, p. 262.

357 American Toll Bridge Association: ibid., p. 219.
"I can't write": ibid.

358 This new book project: Steinman (1950).

"*To spark the things*": Steinman (1959), p. 55.
"A boy grew up": Steinman (1950), p. vii.
359 "She grasped her husband's ideas": Steinman (1954b), p. 26.
"to the heroic contribution": Steinman (1950), p. 420.
360 Writing after the war: Steinman (1948).
361 "guesswork expressed": ENR, Feb. 27, 1941, p. 317.
He modeled with mathematical formulas: Steinman (1943).
362 "What this area needs": Ratigan, p. 278.
"*In the land of Hiawatha*": Steinman (1959), p. 16.
Inter-Peninsula Communications Commission: Steinman (1957), pp. 23–24.
363 recommending *both*: ibid., p. 25.
"perfectly safe suspension bridge": ibid., p. 27.
"not been sufficient": quoted in Advisory Board on the Investigation of Suspension Bridges, p. 777.
364 Bankers evidently let it be known: Ratigan, p. 283.
critical wind velocity: see, e.g., Steinman, in Rubin, p. 17.
open-grid roadway: see Steinman (1943), p. 472; von Kármán, p. 215.
when drawn to scale: Steinman (1957), p. 166, fig. 17; cf. Rubin, p. 20.
"The Mackinac Bridge": Steinman, in Rubin, p. 18.
365 Among those who were central: Steinman (1957), p. 188.
366 "Since 1921": Steinman, Boynton, Gronquist & London, p. 3.
"Here is Dr. Steinman's": ibid., p. 4.
Italian Steel Institute: *NYT Magazine*, Oct. 11, 1953, p. 62.
According to the consulting firm's: Steinman, Boynton, Gronquist & London, p. 26.
368 "Suspension Bridges": Steinman (1954c).
Another article: Steinman (1954e).
His obituary: NYT, Aug. 23, 1960, p. 29.
369 An editorial: NYT, Aug. 25, 1960, p. 28.
"regarded as": *Civil Engineering*, Sept. 1960, p. 75.
"Men and Jobs": ENR, June 25, 1959, pp. 57–59.
"An Artist in Steel Design": ENR, May 15, 1958, pp. 136, 139–40.
370 "We may lack": quoted in ibid., p. 139.
"rugged individualist": ibid., p. 140.
"What Measure for This Man?": ENR, June 25, 1959, pp. 57–59.
371 "The Doctor": ibid, p. 59.
372 "Editorial offices": ibid.
"Dave Steinman": ENR, Sept. 1, 1960, p. 84.

Chapter 7 Realize

377 one out of every five: Secretary of Transportation, p. 5.
"born of a dare": NYT, Aug. 19, 1987, p. 12.
378 radically different: see Bruschi and Koglin, p. 122.
"restringing a pearl necklace": NYT, Aug. 19, 1987, p. 12.

379 "a case study": ibid.
"a great engineering miracle": Buckley, p. 59.
"disinvesting in the American plant": ibid.
"the least famous": NYT, Nov. 30, 1991, pp. 1, 31.
380 "an entirely new color": *The New Yorker,* April 27, 1992, pp. 30–32.
"a sort of vermilion": ibid., p. 32.
381 as much as 2 percent: Munich Reinsurance Company, p. 85.
"bridge-maintenance artist": ENR, Nov. 16, 1992, p. 23.
"Forth Bridge red": Grant, in Paxton, ed., p. 95.
twenty-four painters: ibid., p. 105.
"a pale buff color": ENR, April 22, 1920, p. 807.
382 "a light-greenish tint": Ratigan, p. 191.
"a pleasing shade of verde green": Steinman (c. 1947), p. 11.
"patina green": ibid., p. 12.
"a two-color combination": Ratigan, p. 300.
something Waddell: see ASCE calendar, 1991, caption for February.
"joshed about": Ratigan, p. 300.
"using an orange-red": van der Zee, p. 206.
"emphasized": ibid., p. 219.
red-orange rock: ibid., p. 265.
"red lead": Brown, p. 105.
"iron-oxide red": DeLony (1993), p. 143.
"International Orange": Golden Gate Bridge (c. 1987).
383 "the world's largest Art Deco": Brown, p. 105.
forty-eight months to complete: Golden Gate Bridge (c. 1987).
"free bridges": Strauss, p. 71.
toll rate: see van der Zee, p. 306; cf. O'Shaughnessy and Strauss, p. 12; Golden Gate Bridge (1994), p. 36.
one billion cars: Golden Gate Bridge (1994), p. 65.
one-way tolls: ibid., p. 64.
"maintenance, repairs": Golden Gate Bridge (c. 1987).
Nimitz Freeway: see, e.g., Levy and Salvadori, pp. 95, 105.
384 "Since the structure": Purcell, p. 187.
385 encase the steel piers: *Civil Engineering,* May 4, 1992, p. C-84.
"had not yet pinpointed": ENR, Jan. 31, 1994, p. 16.
"the costliest": *New Civil Engineer,* Feb. 3, 1994, p. 7.
386 large vertical motions: *New Civil Engineer,* Jan. 20, 1994, p. 1; ENR, Jan. 31, 1994, p. 16.
"You can't design": *New Civil Engineer,* Jan. 20, 1994, p. 3.
piece of scholarship: Sibly; Sibly and Walker; see also Petroski (1994), Ch. 10.
387 "a communication gap": Sibly and Walker, p. 208.
388 Cable-stayed bridges: see, e.g., Ito et al., eds.
390 "perfectly" possible: *New Civil Engineer,* Aug. 1, 1991, p. 8.
392 "Proficiency in any art": Tyrrell (1911), p. 3.
"one of the world's": *New Civil Engineer,* Aug. 1, 1991, p. 8; cf. O'Neill.
393 structural artists: see Billington (1983).

Maillart's great concrete bridges: see Billington (1979); Billington (1990).
Interstate 5: see *Science News,* May 15, 1993, p. 319.
Santiago Calatrava: see, e.g., Harbison.
Alamillo Bridge: Webster, p. 74.
"to win back": Metz, p. 60.

394 "to conceive the form": ENR, Jan. 11, 1993, p. 15.
computer-generated image: see Austin, pp. 41–42.
"federal funds": Minneapolis *Star-Tribune,* Nov. 17, 1993.
unrealized Calatrava proposal: Webster, pp. 72–73.

396 "pedestrian amenities": Minneapolis *Star-Tribune,* Dec. 23, 1993.

BIBLIOGRAPHY

Addis, William. 1994. "Design Revolutions in the History of Tension Structures." *Structural Engineering Review* 6: 1–10.

Advisory Board on the Investigation of Suspension Bridges. 1952. "Aerodynamic Stability of Suspension Bridges." *Proceedings—Separate No. 144A–144B,* American Society of Civil Engineers, pp. 721–81.

American Association of Engineers. 1933. *Vocational Guidance in Engineering Lines*. Easton, Pa.: Mack Printing Company.

American Institute of Steel Construction. [1948.] *Prize Bridges: 1928–1947*. New York: American Institute of Steel Construction.

American Society of Civil Engineers. 1972. *A Biographical Dictionary of American Civil Engineers*. Vol. I. New York: American Society of Civil Engineers.

———. 1991. *A Biographical Dictionary of American Civil Engineers*. Vol. II. New York: American Society of Civil Engineers.

———. 1993. *Official Register*. New York: American Society of Civil Engineers.

———. [n.d.] *ASCE Guide to History and Heritage Programs*. New York: American Society of Civil Engineers.

American Society of Mechanical Engineers. 1980. *Mechanical Engineers in America Born Prior to 1861: A Biographical Dictionary*. New York: American Society of Mechanical Engineers.

———. [1991.] *ASME History and Heritage Guide*. New York: American Society of Mechanical Engineers.

Ammann, O. H. 1918. "The Hell Gate Arch Bridge and Approaches of the New York Connecting Railroad over the East River in New York City," with discussion. *Transactions of the American Society of Civil Engineers* 82: 852–1004, 1005–39.

———. 1922. "For Students and Practicing Engineers" [review of *A Practical Treatise on Suspension Bridges*, by D. B. Steinman]. *Engineering News-Record*, December 21, pp. 1080–81.

———. 1923. "Possibilities of the Modern Suspension Bridge for Moderate Spans." *Engineering News-Record*, June 21, pp. 1072–78.

———. 1924. "The Problem of Bridging the Hudson River at New York with Particular Reference to the Proposed Bridge Between Fort Washington Point and Fort Lee." *Papers and Transactions for . . . and Proceedings of the . . . Annual Meeting* (Connecticut Society of Civil Engineers).

[———]. 1926. "Tentative Report of Bridge Engineer on Hudson River Bridge at New York Between Fort Washington and Fort Lee." New York: Port of New York Authority. [Reprinted in Miller and Saidla, *Engineers as Writers*, pp. 237–51.]

———. 1929a. "Suspension Bridges Revisited" [review of *A Practical Treatise on Suspension Bridges*, 2nd ed., by D. B. Steinman]. *Engineering News-Record*, May 26, p. 800.

———. 1929b. "Specifications for Design of Bridges Carrying Highway and Electric Rail Passenger Traffic." The Port of New York Authority Bridge Department, July 1.

———. 1931a. "Brobdingnagian Bridges." *Technology Review*, July, pp. 441–44, 464.

———. 1931b. "Speech" [plus introduction and discussion at a general meeting held November 4, 1931, in New York City]. *Proceedings of the Annual Meeting* (American Institute of Consulting Engineers), pp. 3–27.

———. 1933a. "Advances in Bridge Construction." *Civil Engineering*, August, pp. 428–32.

———. 1933b. "George Washington Bridge: General Conception and Development of Design." *Transactions of the American Society of Civil Engineers* 97: 1–65.

———. 1939. "Planning and Design of Bronx-Whitestone Bridge." *Civil Engineering*, April, pp. 217–20.

———. 1940. "Present Trends in Structural Design." *Civil Engineering*, January, pp. 21–24.

———. 1941. "Stay Systems for Suspension Bridges" [response to letter from D. B. Steinman]. *Engineering News-Record*, February 27, pp. 36–37.

———. 1946. "Additional Stiffening of Bronx-Whitestone Bridge." *Civil Engineering*, March, pp. 101–3.

———. 1953. "Present Status of Design of Suspension Bridges with Respect to Dynamic Wind Action." *Journal of the Boston Society of Civil Engineers* 40: 231–53.

———. 1963. "Planning and Design of the Verazano-Narrows [*sic*] Bridge." *Transactions of the New York Academy of Sciences* ser. II, vol. 25, pp. 598–620.

———. 1966. "Verrazano-Narrows Bridge: Conception of Design and Construction Procedure." *Journal of the Construction Division: Proceedings of the American Society of Civil Engineers*, CO2: 5–21.

Ammann, O. H., Theodore von Kármán, and Glenn B. Woodruff. 1941. *The Failure of the Tacoma Narrows Bridge*. Washington, D.C.: Federal Works Agency.

Anderson, Graham, and Ben Roskrow. 1994. *The Channel Tunnel Story*. London: E. & F. N. Spon.

Anderson, Norman D. 1992. *Ferris Wheels: An Illustrated History*. Bowling Green, Ohio: Bowling Green State University Popular Press.

Andrade, Jorge Carrera. 1941. *To the Bay Bridge: Canto al Puente de Oakland*. Translated by Eleanor L. Turnbill. Stanford, Calif.: Hoover Library on War, Revolution and Peace.

Austin, Teresa. 1994. "The Art of the Infrastructure." *Civil Engineering*, September, pp. 40–43.

Ausubel, Jesse H., and Robert Herman, eds. 1988. *Cities and Their Vital Systems: Infrastructure, Past, Present, and Future*. Washington, D.C.: National Academy Press.

Bacow, Adele Fleet, and Kenneth E. Kruckemeyer. 1986. *Bridge Design: Aesthetics and Developing Technologies*. Boston: Massachusetts Department of Public Works and Massachusetts Council on the Arts and Humanities.

Baird, Howard C. 1924. "Bear Mountain Suspension Bridge over Hudson River." *Engineering News-Record*, December 4, pp. 914–18.

Baker, B. 1873. *Long-Span Railway Bridges*. Rev. ed. London: E. & F. N. Spon.

———. 1882. "The Forth Bridge." *Report of Meeting*, British Association for the Advancement of Science, pp. 419–33.

———. 1887. "Bridging the Firth of Forth." In several installments, *Engineering*, July 29, p. 116; August 5, p. 148; August 12, pp. 170–71; August 19, p. 210; August 26, p. 238. See also *Proceedings of the Royal Institution, London* (1887), pp. 142–49.

Baker, Charles Whiting. 1917. "The Story of 'Engineering News.'" *Engineering News-Record*, April 5, pp. 6–10.

Ball, Norman R. 1988. *"Mind, Heart, and Vision": Professional Engineering in Canada 1887 to 1987*. Ottawa: National Museum of Science and Technology, National Museums of Canada.

Barker, Harry. 1917. "Quebec Suspended Span Successfully Hung from Cantilevers." *Engineering News-Record*, September 27, pp. 580–84.

Baxter, James Phinney, 3rd. 1933. *The Introduction of the Ironclad Warship*. Cambridge, Mass.: Harvard University Press.

Beckett, Derrick. 1984. *Stephensons' Britain*. Newton Abbot, Devonshire: David & Charles.

Biggart, A. S. 1898. "Sir William Arrol." *Cassier's Magazine* 15: 2–19.

Billings, Henry. 1966. *Bridges*. New York: Viking.

Billington, David P. 1977. "History and Esthetics in Suspension Bridges." *Journal of the Structural Division: Proceedings of the American Society of Civil Engineers* 103:1655–72. Discussion, 104(1978): 246–49, 378–80, 619, 732–33, 1027–35, 1174–76. Closure, 105 (1979): 671–87.

———. 1979. *Robert Maillart's Bridges: The Art of Engineering*. Princeton, N.J.: Princeton University Press.

———. 1980. "Wilhelm Ritter: Teacher of Maillart and Ammann." *Journal of the Structural Division: Proceedings of the American Society of Civil Engineers* ST5: 1103–16.

———. 1983. *The Tower and the Bridge: The New Art of Structural Engineering*. New York: Basic Books.

———. 1990. *Robert Maillart and the Art of Reinforced Concrete*. New York/Cambridge, Mass.: Architectural History Foundation/MIT Press.

Billington, David P., and Aly Nazmy. 1990. "History and Aesthetics of Cable-Stayed Bridges." *Journal of Structural Engineering* 117: 3103–34.

Black, Archibald. 1936. *The Story of Bridges*. New York: Whittlesey House.

Blum, Howard. 1980. "Trump: The Development of a Manhattan Developer." *The New York Times*, August 26, pp. B1–B4.

Boller, Alfred P. 1876. *Practical Treatise on the Construction of Iron Highway Bridges for the Use of Town Committees, Together with a Short Essay upon the Application of the Principles of the Lever to a Ready Analysis of the Strains upon the More Customary Forms of Beams and Trusses*. New York: John Wiley & Sons.

———. 1883. *Report upon the Construction of a High Level Bridge over the Harlem River, North of High Bridge, New York, Made to the Commissioners of the Department of Public Parks*. New York: Evening Post Job Printing Office.

Bonanos, Christopher. 1992. "The Father of Modern Bridges." *American Heritage of Invention and Technology*, Summer, pp. 8–20.

Bowden, E. Warren. 1936. "The Triborough Bridge Project." *Civil Engineering*, August, pp. 515–19.

Brooklyn Museum, The. 1983. *The Great East River Bridge, 1883–1983*. New York: Harry N. Abrams.

Brown, David J. 1993. *Bridges*. New York: Macmillan.

Bruschi, Maria Grazia, and Terry L. Koglin. 1994. "Main Cable Preservation for the Williamsburg Bridge." *Structural Engineering International* 4: 122–24.

Buchanan, R. A. 1989. *The Engineers: A History of the Engineering Profession in Britain, 1750–1914*. London: Kingsley.

———. 1992. *The Power of the Machine: The Impact of Technology from 1700 to the Present*. London: Viking.

Buck, L. L. 1881. "The Re-enforcement of the Anchorage and Renewal of Suspended Superstructure of Niagara Railroad Suspension Bridge." *Transactions of the American Society of Civil Engineers* 10: 195–224.

———. 1894. "A Few Remarks About the Niagara Gorge," with discussion. *Transactions of the American Society of Civil Engineers* 32: 205–8, 208–13.

Buck, R. S. 1898. "The Niagara Railway Arch," with discussion. *Transactions of the American Society of Civil Engineers* 40: 125–50, 151–77.

Buckley, Tom. 1991. "The Eighth Bridge." *The New Yorker*, January 14, pp. 37–59.

Building Arts Forum/New York. 1991. *Bridging the Gap: Rethinking the Relationship of Architect and Engineer*. New York: Van Nostrand Reinhold.

Burke, Martin P., Jr. 1989. "Bridge Design and the 'Bridge Aesthetics Bibliography.' " *Journal of Structural Engineering* 115: 883–99.

Burr, William H., and Boller & Hodge, Consulting Engineers. 1908. *Queensboro Bridge: Report on Design and Construction*. New York: [Board of Estimate and Apportionment].

Calder, Bill. 1986. "Conde Balcom McCullough: Oregon's Master Bridge Builder." *Oregon Coast*, June/July, pp. 49–53.

Calhoun, Daniel Hovey. 1960. *The American Civil Engineer: Origins and Conflict*. Cambridge, Mass.: Technology Press.

California Department of Transportation. 1990. *Historic Highway Bridges of California*. Sacramento, Calif.: California Department of Transportation.

Carnegie, Andrew. 1924. *Autobiography*. Popular ed. Boston: Houghton Mifflin.

Caro, Robert A. 1974. *The Power Broker: Robert Moses and the Fall of New York*. New York: Alfred A. Knopf.

Carswell, Charles. 1926. *The Building of the Delaware River Bridge, Connecting Philadelphia, Pa., and Camden, N.J.* Burlington, N.J.: Enterprise Publishing Co.

Chanute, O., and George Morison. 1870. *The Kansas City Bridge, with an Account of the Regimen of the Missouri River, and a Description of Methods Used for Founding in That River*. New York: D. Van Nostrand.

[Chisholm Bro.] 1891. *Niagara: Chisholm's Complete Guide to the Grand Cataract*. Portland, Maine: Chisholm Brothers.

Chrimes, M. M. 1993. "Sir John Fowler—Engineer or Manager?" *Proceedings of the Institution of Civil Engineers* 97 (Civil Engineering): 135–43.

Cilley, Frank H. 1904. "General Methods for the Calculation of Statically Indeterminate Bridges, As Used in the Check Calculations of Designs for the Manhattan Bridge and the Blackwell's Island Bridge, New York." *Transactions of the American Society of Civil Engineers* 53: 413–51.

City of Bayonne, New Jersey. 1931. *The Bayonne Bridge: Dedication Ceremonies*, November 14.

Clark, Edwin. 1850. *The Britannia and Conway Tubular Bridges: With General Inquiries on Beams and on the Properties of Materials Used in Construction*. 2 vols. London: Day and Son.

Clarke, Thomas Curtis. 1896. "Science and Engineering." *Transactions of the American Society of Civil Engineers* 35: 508–19.

Cohen, Edward. 1967. "Notes on the Professional Career of Othmar Hermann Ammann." *Annals of the New York Academy of Sciences* 136: 717–40.

Condit, Carl W. 1960. *American Building Art: The Nineteenth Century*. New York: Oxford University Press.

———. 1961. *American Building Art: The Twentieth Century*. New York: Oxford University Press.

Consortium of Civil Engineering Departments of New York City Colleges and Universities. 1990. *Preventive Maintenance Management System for New York City Bridges*. New York: Columbia University, Department of Civil Engineering and Mechanics, Center for Infrastructure Studies.

Cooper, Theodore. 1874. "The Erection of the Illinois and St. Louis Bridge." *Transactions of the American Society of Civil Engineers* 3: 239–54.

———. 1878. "Observations on the Stresses Developed in Metallic Bars by Applied Forces." *Transactions of the American Society of Civil Engineers* 7: 174–82.

———. 1880. "The Use of Steel for Railway Bridges." *Transactions of the American Society of Civil Engineers* 8: 261–77.

———. 1888. *General Specifications for Iron and Steel Railroad Bridges and Viaducts*. New and rev. ed. New York: Engineering News Publishing Co.

———. 1889. "American Railroad Bridges," with discussion. *Transactions of the American Society of Civil Engineers* 21: 1–58, 566–607.

———. 1894. "Train Loads for Railroad Bridges." *Transactions of the American Society of Civil Engineers* 31: 174–84.

———. 1906. "New Facts about Eye-bars," with discussion. *Transactions of the American Society of Civil Engineers* 56: 411–28, 429–50.

Court of Inquiry. 1880. *Report of the Court of Inquiry, and Report of Mr. Rothery, upon the Circumstances Attending the Fall of a Portion of the Tay Bridge on the 28th December 1879*. London: Her Majesty's Stationery Office.

Crane, Hart. 1946. *The Collected Poems of Hart Crane*. Edited by Waldo Frank. New York: Liveright.

Daley, Robert. 1958. "Dreamer in Concrete and Steel." *The New York Times Magazine*, January 5, pp. 33, 36.

Dana, Allston, Aksel Anderson, and George M. Rapp. 1933. "George Washington Bridge: Design of Superstructure." *Transactions of the American Society of Civil Engineers* 97: 97–163.

Danko, George Michael. 1979. "The Evolution of the Simple Truss Bridge, 1790 to 1850: From Empiricism to Scientific Construction." Ph.D. Thesis, University of Pennsylvania.

Darnell, Victor C. 1984. *A Directory of American Bridge-Building Companies, 1840–1900*. Washington D.C.: Society for Industrial Archeology.

———. 1989. "The Other Literature of Bridge Building." *IA: The Journal of the Society for Industrial Archeology* 15 (no. 2): 40–56.

Davis, Kenneth S. 1989. "Crisis Behind the TVA." *American Heritage of Invention and Technology*, Spring–Summer, pp. 8–16.

Davis, Mac. 1956. *Jews at a Glance*. New York: Hebrew Publishing Company.

The Delaware River Bridge: Twenty-fifth Anniversary. 1951.

DeLony, Eric. 1993. *Landmark American Bridges*. New York: American Society of Civil Engineers.

———. 1994. "The Golden Age of the Iron Bridge." *American Heritage of Invention and Technology*, Fall, pp. 8–13, 16–22.

Derleth, C., Jr. 1925. "Cantilever Highway Bridge Across Carquinez Strait." *Engineering News-Record*, September 24, pp. 504–7.

Doig, Jameson W. 1990. "Politics and the Engineering Mind: O. H. Ammann and the Hidden Story of the George Washington Bridge." *Yearbook of German-American Studies* 25: 151–99.

———. [1992.] "Empire on the Hudson: Idealism and Political Power at the Port of New York Authority." Manuscript, courtesy of Jameson W. Doig.

Doig, Jameson W., and David P. Billington. 1994. "Ammann's First Bridge: A Study in Engineering, Politics, and Entrepreneurial Behavior." *Technology and Culture* 35: 537–570.

Dorsey, Florence. 1947. *Road to the Sea: The Story of James B. Eads and the Mississippi River*. New York: Reinhart & Company.

Dunham, Clarence Whiting. 1979. *Saga of an Engineer*. Privately printed.

Durand, W. F. 1945. "Biographical Memoir of Ralph Modjeski, 1861–1940." *Biographical Memoirs* (National Academy of Sciences) 23, pp. 243–261. Washington, D. C.: National Academy Press.

Durrer, Margot Ammann. 1979. "Memories of My Father." *Swiss-American Historical Society Newsletter* 15: 26–33.

Eads, James B. 1868. *Report of the Engineer-in-Chief of the Illinois and St. Louis Bridge Company*. St. Louis: Missouri Democrat Book and Job Printing House.

———. 1881. *Review of Captain Phelps' Pamphlet Entitled, "Transportation of Ships on Railways."* [Washington, D.C.: National Republican Print.]

———. 1884. *Addresses and Papers of James B. Eads, Together with a Biographical Sketch*. Edited by Estill McHenry. St. Louis: Slawson & Co.

The Eads Bridge. 1974. Princeton, N.J.: The Art Museum and the Department of Civil Engineering, Princeton University.

Edwards, Llewellyn Nathaniel. 1959. *A Record of History and Evolution of Early American Bridges*. Orono, Maine: Printed at the University Press.

Ellet, Charles, Jr. 1843. *A Popular Notice of Suspension Bridges, with a Brief Description of the Wire Bridge Across the Schuylkill, at Fairmount*. Philadelphia: John C. Clark.

Embury, Aymar, II. 1938a. "The Architect and the Engineer." *Civil Engineering*, January, pp. 3–4.

———. 1938b. "Esthetics of Bridge Anchorages." *Civil Engineering*, February, pp. 85–89.

———. 1938c. "Esthetic Design of Steel Structures." *Civil Engineering*, April, pp. 261–65.

Emmerson, George S. 1973. *Engineering Education: A Social History*. Newton Abbot, Devonshire: David & Charles.

Engle, Harry J. 1934. "Art in Bridge Building." *Civil Engineering*, December, pp. 627–31.

Farquharson, F. B. 1949. *Aerodynamic Stability of Suspension Bridges: With Special Reference to the Tacoma Narrows Bridge*. Report, University of Washington, Structural Research Laboratory.

Fenves, Steven J. 1989. "A History of Pittsburgh's Bridges." *Pittsburgh Engineer*, May–June, pp. 14–19, 32, 34–36.

Ferguson, Eugene S. 1992. *Engineering and the Mind's Eye*. Cambridge, Mass.: MIT Press.

Finch, James Kip. 1936. "Engineering and Architecture." *Civil Engineering*, June, pp. 377–81.

———. 1941. "Wind Failures of Suspension Bridges, or, Evolution and Decay of the Stiffening Truss." *Engineering News-Record*, March 13, pp. 74–79.

———. 1951. *Engineering and Western Civilization*. New York: McGraw-Hill.

———. 1954. *A History of the School of Engineering, Columbia University*. New York: Columbia University Press.

———. 1978. *Engineering Classics*. Edited by Neal FitzSimons. Kensington, Md.: Cedar Press.

Fisher, John W. 1984. *Fatigue and Fracture in Steel Bridges: Case Studies*. New York: John Wiley & Sons.

Florman, Samuel C. 1987. *The Civilized Engineer*. New York: St. Martin's Press.

Ford, Peter A. 1993. "Charles S. Storrow, Civil Engineer: A Case Study of European Training and Technological Transfer in the Antebellum Period." *Technology and Culture* 34: 271–99.

Fowler, Charles Evan. 1919. "Revisions of the Niagara Railway Arch Bridge," with discussion. *Transactions of the American Society of Civil Engineers* 83: 1919–99, 2000–2025.

Frampton, Kenneth, Anthony C. Webster, and Anthony Tischhauser. 1993. *Calatrava: Bridges*. Zurich: Artemis.

Frankland, F. H. 1932. "Tunnels or Bridges—Which?" *Civil Engineering*, February, pp. 85–90.

Fraser, Chelsea. 1928. *The Story of Engineering in America*. New York: Thomas Y. Crowell.

Fredrich, Augustine J., ed. 1989. *Sons of Martha: Civil Engineering Readings in Modern Literature*. New York: American Society of Civil Engineers.

[———, ed.] 1993. " 'Great Engineering Is Great Art': Joseph Pennell's Images of American Civil Engineering." Catalog of an Exhibit for the 1993 National Convention of the American Society of Civil Engineers, Dallas, October 24–29. [Evansville: University of Southern Indiana.]

Freeman, Ralph. 1932. "Sydney Harbor Bridge Completed." *Civil Engineering*, April, pp. 217–22.

French, Thomas L. 1989. "Horace King, Bridge Builder." *Alabama Heritage*, no. 11 (Winter), pp. 35–47.

Galileo. 1638. *Dialogues Concerning Two New Sciences*. Translated by Henry Crew and Alfonso de Salvio. New York: Dover Publications.

Gangewere, R. Jay. 1993. "Bridges to the City." *Carnegie Magazine*. July–August, pp. 24–35.

Gilroy, Harry. 1953. "To Bridge a Dilemma." *The New York Times Magazine*, October 11, pp. 62–63.

Goldberger, Paul. 1985. *On the Rise: Architecture and Design in a Postmodern Age*. New York: Penguin Books.

Golden, Dixie W. 1987. "A Man and His Bridge." In *A Golden Gate Jubilee, 1937–1987*. Cincinnati: College of Engineering, University of Cincinnati.

Golden Gate Bridge, Highway and Transportation District. [c. 1987.] *Golden Gate Bridge*. San Francisco: Golden Gate Bridge, Highway and Transportation District.

———. 1994. *Highlights, Facts & Figures of the Golden Gate Bridge, Highway and Transportation District*. San Francisco: Golden Gate Bridge, Highway and Transportation District.

Gordon, J. E. 1978. *Structures: Or Why Things Don't Fall Down*. New York: Da Capo Press.

Government Board of Engineers, The. 1918. *The Quebec Bridge over the St. Lawrence River near the City of Quebec on the Line of the Canadian National Railways*. Ottawa: Department of Railways and Canals.

Graff, Frederic. 1885. "[Presidential] Address." *Transactions of the American Society of Civil Engineers* 14: 227–37.

Grayson, Lawrence P. 1993. *The Making of an Engineer: An Illustrated History of Engineering Education in the United States and Canada*. New York: John Wiley & Sons.

Harbison, Robert. 1992. *Creatures from the Mind of the Engineer: The Architecture of Santiago Calatrava*. Zurich: Artemis.

Hardesty, Shortridge, and Harold E. Wessman. 1939. "Preliminary Design of Suspension Bridges," with discussion. *Transactions of the American Society of Civil Engineers* 104: 579–608, 609–34.

Hardesty & Hanover. [c. 1986.] *Hardesty & Hanover, Consulting Engineers*. [New York: Hardesty & Hanover.]

———. 1987a. *Centennial, 1887–1987*. Nos. 1–3.

———. 1987b. *One Hundred Years of Bridge Engineering*. New York: [Hardesty & Hanover.]

Hartenberg, R. S., ed. 1979. *National Historic Mechanical Engineering Landmarks*. New York: American Society of Mechanical Engineers.

Hawkes, Nigel. 1990. *Structures: The Way Things Are Built*. New York: Macmillan.

Henry, Dorothy, and J. A. Jerome. 1965. *Modern British Bridges*. London: C. R. Books.

Herbertson, Elizabeth Taylor. 1970. *Pittsburgh Bridges*. New York: Exposition Press.

[Homer, Truman J.] 1865. *Report of the City Engineer to the Board of Common Council of the City of St. Louis, in Relation to a Bridge Across the Mississippi River at St. Louis*. St. Louis: M'Kee, Fishback and Co., printers.

Hood, Clifton. 1993. *722 Miles: The Building of the Subways and How They Transformed New York*. New York: Simon & Schuster.

Hoover, Herbert. 1952. *Memoirs: Years of Adventure, 1874–1920*. New York: Macmillan.

Hopkins, H. J. 1970. *A Span of Bridges: An Illustrated History*. New York: Praeger.

Howard, Needles, Tammen & Bergendoff, Consulting Engineers. 1953. *The Delaware Memorial Bridge over the Delaware River: Final Engineering Report*. Wilmington: Delaware State Highway Department.

Hughes, Thomas P. 1989. *American Genesis: A Century of Invention and Technological Enthusiasm, 1870–1970*. New York: Viking.

[Hungerford, Edward.] 1903. "The Building of the Bridge." In *The Williamsburg Bridge: An Account of the Ceremonies Attending the Formal Opening of the Structure, December the Nineteenth, MDCCCCIII, Together with an Illustrated Historical and Descriptive Sketch of the Enterprise . . . and Certain Statistical Tables*. Brooklyn–New York: Eagle Press.

Hunt, Charles Warren. 1897. *Historical Sketch of the American Society of Civil Engineers*. New York: American Society of Civil Engineers.

Hunt, Donald. 1994. *The Tunnel: The Story of the Channel Tunnel, 1802–1994*. Upton-upon-Severn, Worcestershire: Images.

Ito, Manabu, et al., eds. 1991. *Cable-Stayed Bridges: Recent Developments and Their Future*. (Proceedings of the Seminar, Yokohama, Japan, December 10–11, 1991.) Amsterdam: Elsevier.

Jackson, Donald C. 1988. *Great American Bridges and Dams*. Washington, D.C.: Preservation Press.

Jacobs, David, and Anthony E. Neville. 1968. *Bridges, Canals & Tunnels*. New York: American Heritage.

Jakkula, A. A. 1941. "A History of Suspension Bridges in Bibliographical Form." *Bulletin of the Agricultural and Mechanical College of Texas*, 4th ser., vol. 12, no. 7.

Jones, Jonathan. 1928. "Design of Great International Suspension Bridge over Detroit River." *Engineering News-Record*, September 27, pp. 460–66.

Jurecka, Charlotte. 1979. *Brücken: Historische Entwicklung—Faszination der Technik*. Vienna: Verlag Anton Schroll.

von Kármán, Theodore, with Lee Edson. 1967. *The Wind and Beyond: Theodore von Kármán, Pioneer in Aviation and Pathfinder in Space*. Boston: Little, Brown.

Katz, Leon. 1988. "A Poet in Steel." *Portfolio: A Quarterly Review of Trade and Transportation* 1: 31–39.

Kemp, Emory L. 1977–78. "Thomas Paine and His 'Pontifical Matters.'" *Transactions of the Newcomen Society* 49: 21–40.

———. 1979. "Links in a Chain: The Development of Suspension Bridges, 1801–70." *The Structural Engineer* 57A: 255–63.

———. 1989. "The Fabric of Historic Bridges." *IA: The Journal of the Society for Industrial Archaeology* 15 (no. 2): 3–22.

Ketchum, Mark A., and Al Heldermon. 1991. "Probing the Golden Gate." *Civil Engineering*, June, pp. 43–45.

Keystone Bridge Company, The. [1874.] *Descriptive Catalogue of Wrought-Iron Bridges* . . . Pittsburgh: Keystone Bridge Co.

Kipling, Rudyard. 1987. *The Day's Work*. Edited by Thomas Pinney. Oxford: Oxford University Press.

Kirby, Richard Shelton, and Philip Gustave Laurson. 1932. *The Early Years of Modern Civil Engineering*. New Haven, Conn.: Yale University Press.

Kirby, Richard S., Sidney Withington, Arthur B. Darling, and Frederick G. Kilgour. 1956. *Engineering in History*. Reprint ed. New York: Dover Publications, 1990.

Koerte, Arnold. 1992. *Two Railway Bridges of an Era: Firth of Forth and Firth of Tay*. Basel: Birkhäuser.

Kouwenhoven, John A. 1974. "Eads Bridge: The Celebration." *Bulletin of the Missouri Historical Society* 30: 159–80.

———. 1982. "The Designing of the Eads Bridge." *Technology and Culture* 23: 535–68.

Kuesel, Thomas R. 1990. "Whatever Happened to Long-Term Bridge Design?" *Civil Engineering*, February, pp. 57–60.

Kunz, F. C. 1915. *Design of Steel Bridges: Theory and Practice for the Use of Civil Engineers and Students*. New York: McGraw-Hill.

Kutler, Stanley I. 1990. *Privilege and Creative Destruction: The Charles River Bridge Case*. Baltimore: Johns Hopkins University Press.

Kyle, John M. 1966. "A Tribute to Othmar H. Ammann," *The Port Authority Review* 4: 3–8.

———. 1967. "Tribute to Othmar Hermann Ammann." *Annals of the New York Academy of Sciences* 136: 741–46.

Labrum, E. A., ed. 1994. *Civil Engineering Heritage: Eastern and Central England*. London: Thomas Telford.

Landes, Cheryl. 1991. "McCullough Bridge: Engineering as Art." *Oregon Coast*, July–August, pp. 22–23.

Layton, Edwin T. 1986. *The Revolt of the Engineers*. Baltimore: Johns Hopkins University Press.

Leonhardt, Fritz. 1984a. *Baumeister in einer unwälzenden Zeit*. Stuttgart: Deutsche Verlags-Anstalt.

———. 1984b. *Bridges: Aesthetics and Design*. Cambridge, Mass.: MIT Press.

Levy, Matthys, and Mario Salvadori. 1992. *Why Buildings Fall Down: How Structures Fail*. New York: W. W. Norton.

Lewis, Gene D. 1968. *Charles Ellet, Jr., 1810–1862: The Engineer as Individualist*. Urbana: University of Illinois Press.

Lindenthal, Gustav. 1883. "Rebuilding of the Monongahela Bridge, at Pittsburgh, Pa.," with discussion. *Transactions of the American Society of Civil Engineers* 12: 353–85, 386–92.

———. 1887. *The Proposed New York City Terminal Railroad, Including North River Bridge and Grand Terminal Station, in New York City*. [New York:] privately printed.

———. 1888. "A Discussion of Long Span Bridges." In three parts, *Engineering News*, March 3, pp. 153–155; March 10, pp. 174–75; March 24, p. 226.

———. 1889a. "A Reply to Certain Criticisms on the Proposed North River Suspension Bridge at New York City." *Engineering News*, July 20, pp. 58–59.

———. 1889b. "The Economic Conditions of Long-Span Bridges with Special Reference to the Proposed North River Bridge at New York City." In three parts, *Engineering News*, November 9, pp. 435–37; November 16, pp. 464–65; November 23, pp. 486–87.

———. 1905. "A Rational Form of Stiffened Suspension Bridge," with discussion. *Transactions of the American Society of Civil Engineers* 55: 1–15, 16–93.

———. 1911a. "Notes on the Quebec Bridge Competition.—I." *Engineering News*, November 16, pp. 581–86.

———. 1911b. "Notes on the Quebec Bridge Competition.—II. The Competitive Suspension Design." *Engineering News*, November 23, pp. 613–19.

———. 1912. "An Opportunity for a Bridge Across the North River at New York City." *Engineering News*, December 19, pp. 1142–43.

———. 1916. "Dr. Waddell Makes Notable Addition to Bridge Books." *Engineering News*, December 21, pp. 1174–78.

———. 1918a. *The North River Bridge: How to Finance It and Who Should Build It*. New York: privately printed.

———. 1918b. "Plan for North River Span with Federal Backing." *The New York Times Magazine*, February 24, pp. 6–7.

———. 1919. *Railroad Terminal Plans as a Part of the Port Problem of New York*. [New York:] privately printed.

———. 1921. "Some Thoughts on Long-Span Bridge Design." *Engineering News-Record*, November 24, pp. 861–64.

———. 1922. "The Continuous Truss Bridge over the Ohio River at Sciotoville, Ohio, of the Chesapeake and Ohio Northern Railway," with discussion. *Transactions of the American Society of Civil Engineers* 85: 910–53, 954–75.

———. 1924a. "Lindenthal Outlines Hudson Bridge Plan." *The New York Times*, January 13, p. 6.

———. 1924b. "Bridge Engineering." *Engineering News-Record*, April 17, pp. 652–57.

———. 1932. "Bridges with Continuous Girders." *Civil Engineering*, July, pp. 421–24.

———. 1933. *A Sound Scientific Money System: As a Cure for Unemployment*. Rev. ed. Boston: Stratford Company.

Macdonald, Charles. 1877. "The Failure of the Ashtabula Bridge," with discussion. *Transactions of the American Society of Civil Engineers* 6: 74–87, 195–222, 231–34.

Mackay, Sheila. 1990a. *Bridge Across the Century: The Story of the Forth Bridge*. Edinburgh: Moubray House.

———. 1990b. *The Forth Bridge: A Picture History*. Edinburgh: Moubray House.

MacKaye, Milton. 1934. "Poet in Steel." *The New Yorker*, June 2, pp. 23–27.

Maryland Department of Transportation. 1993. *Aesthetic Bridges: Users Guide*. Baltimore: State Highway Administration.

Matteo, John, George Deodatis, and David P. Billington. 1994. "Safety Analysis of Suspension-Bridge Cables: Williamsburg Bridge." *Journal of Structural Engineering* 120: 3197–3211.

McBeth, D. G. 1990. "The Forth Bridge." *The Structural Engineer* 68: 93–100.

McClintic-Marshall Company. [c. 1918.] *Sciotoville Bridge over the Ohio River at Sciotoville, Ohio, for the Chesapeake & Ohio Northern Railway Company*. Pittsburgh: McClintic-Marshall Co.

McCullough, C. B. 1922. "Old Suspension Bridge Used in Erecting New Arch." *Engineering News-Record*, November 2, pp. 730–33.

———. 1932. "Timber Highway Bridges in Oregon." *Engineering News-Record*, August 25, pp. 213–14.

McCullough, Conde B., and John R. McCullough. 1946. *The Engineer at Law: A Resumé of Modern Engineering Jurisprudence*. 2 vols. Ames: Iowa State College Press.

McCullough, David. 1972. *The Great Bridge*. New York: Simon & Schuster.

———. 1977. *The Path Between the Seas: The Creation of the Panama Canal, 1870–1914*. New York: Simon & Schuster.

———. 1992. *Brave Companions: Portraits in History*. New York: Prentice-Hall.

McGraw, James H. 1917. "The Technical Journal as an Institution." *Engineering News-Record*, April 5, pp. 1–2.

McKibben, Frank P. 1932. "Bridges and Poetry." *Civil Engineering*, February, pp. 71–75.

Mehren, E. J. 1917. "Development of the 'Engineering Record.'" *Engineering News-Record*, April 5, pp. 2–5.

Melan, J. 1913. *Theory of Arches and Suspension Bridges*. Translated by D. B. Steinman. Chicago: Myron C. Clark.

———. 1917. *Plain and Reinforced Concrete Arches*. Translated by D. B. Steinman. New York: John Wiley & Sons.

"Memoirs of Deceased Members." *Transactions of the American Society of Civil Engineers*:

Othmar Hermann Ammann (abstract), vol. 133 (1968), pp. 734–35.

Leffert Lefferts Buck, vol. 73 (1911), pp. 493–97.

Octave Chanute, vol. 74 (1911), pp. 483–89.

Thomas Curtis Clarke, vol. 50 (1903), pp. 495–99.

Theodore Cooper, vol. 84 (1921), pp. 828–30.

Elmer Lawrence Corthell, vol. 81 (1917), pp. 1658–64.

Milton Harvey Freeman, vol. 89 (1926), pp. 1596–98.

George Washington Goethals, vol. 93 (1929), pp. 1813–17.

Henry Wilson Hodge, vol. 83 (1919–20), pp. 2224–28.

Clifford Milburn Holland, vol. 89 (1926), pp. 1625–29.

Gustav Lindenthal, vol. 105 (1940), pp. 1790–94.

Ralph Modjeski, vol. 106 (1941), pp. 1624–28.

Leon Solomon Moisseiff, vol. 111 (1946), pp. 1509–12.

Charles Walker Raymond, vol. 77 (1914), pp. 1894–1901.

Holton Duncan Robinson, vol. 111 (1946), pp. 1530–36.

John Augustus Roebling, vol. 98 (1933), pp. 1614–18.

Charles Conrad Schneider, vol. 81 (1917), pp. 1665–70.

Arthur Mellen Wellington, *Proceedings*, vol. 21 (1895), pp. 199–202.

Squire Whipple, vol. 36 (1896), pp. 527–30.

Metz, Tracy. 1990. "Structural Dynamics." *Architectural Record*, October, pp. 54–60.

Michigan Department of Transportation. 1987. "The Zilwaukee Bridge: From the Beginning." Report. Lansing: Michigan Department of Transportation.

Miller, Walter J., and Leo E. A. Saidla. 1953. *Engineers as Writers: Growth of a Literature*. New York: Van Nostrand.

Miller, William J., Jr. 1990. *Crossing the Delaware: The Story of the Delaware Memorial Bridge, The Longest Twin-Suspension Bridge in the World*. Wilmington, Del.: Gauge Corporation.

Modjeska, Helena. 1910. *Memories and Impressions: An Autobiography*. New York: Macmillan.

Modjeski, Ralph. 1909. *Report on Manhattan Bridge*. New York: City of New York, Department of Bridges.

———. 1919. *The McKinley Bridge Across the Mississippi River at St. Louis, Mo.* Chicago: H. C. Sherman & Co.

Möhringer, Karl. 1931. *The Bridges of the Rhine: Roman, Medieval, and Modern. An Account of Notable Bridges over the Rhine River and Its Tributaries, with a Description of Those Recently Built*. Messkirch, Baden: Möhringer.

Moisseiff, Leon S. 1905. "Theory and Formulas for the Analytical Computation of a Three-Span Suspension Bridge with Braced Cable," with discussion. *Transactions of the American Society of Civil Engineers* 55: 94–113, 114–28.

———. 1925. "The Towers, Cables and Stiffening Trusses of the Bridge over the Delaware River Between Philadelphia and Camden." *Journal of the Franklin Institute*, October, pp. 436–66.

———. 1928a. "Designing the Towers of the Hudson River Bridge." *Engineering News-Record*, May 24, pp. 819–822.

———. 1928b. "Esthetics of Bridges." *Engineering News-Record*, November 15, p. 741.

———. 1933. "George Washington Bridge: Design of the Towers." *Transactions of the American Society of Civil Engineers* 97: 164–205.

Moisseiff, Leon S., and Frederick Lienhard. 1933. "Suspension Bridges Under the Action of Lateral Forces," with discussion. *Transactions of the American Society of Civil Engineers* 98: 1080–95, 1096–1141.

[Moore, John Hammond.] 1982. *Wiley: One Hundred and Seventy Five Years of Publishing*. New York: John Wiley & Sons.

Morgan, Arthur E. 1971. *Dams and Other Disasters: A Century of the Army Corps of Engineers in Civil Works*. Boston: Porter Sargent.

Morison, Elting E. 1986. "The Master Builder." *American Heritage of Invention and Technology*, Fall, pp. 34–40.

Morison, George S. 1896. "Suspension Bridges—A Study," with discussion. *Transactions of the American Society of Civil Engineers* 36: 359–416, 417–82.

Munich Reinsurance Company. 1992. *Bridges—Technology and Insurance*. Munich: Munich Reinsurance Co.

Museum of Modern Art, The. 1964. *Twentieth Century Engineering*. New York. Museum of Modern Art.

Nelson, Lee H. 1990. *The Colossus of 1812: An American Engineering Superlative*. New York: American Society of Civil Engineers.

Noble, David F. 1977. *America by Design: Science, Technology, and the Rise of Corporate Capitalism*. New York: Alfred A. Knopf.

Normile, Dennis. 1994. "A Short Course in Modern Bridges." *Technology Review*, November/December, pp. 52–59.

O'Neill, Bill. 1991. "Bridge Design Stretched to the Limits." *New Scientist*, October 26, pp. 36–43.

[O'Shaughnessy, M. M., and J. B. Strauss. 1921.] *Bridging "The Golden Gate."* [San Francisco]: privately printed.

Otter, R. A., ed. 1994. *Civil Engineering Heritage: Southern England*. London: Thomas Telford.

Outerbridge, Graeme, and David Outerbridge. 1989. *Bridges*. New York: Harry N. Abrams.

Pagon, W. Watters. 1934a. "What Aerodynamics Can Teach the Civil Engineer." *Engineering News-Record*, March 15, pp. 348–53.

———. 1934b. "Aerodynamics and the Civil Engineer—II. Vibration Problems in Tall Stacks Solved by Aerodynamics." *Engineering News-Record*, July 12, pp. 41–43.

———. 1934c. "Aerodynamics and the Civil Engineer—III. Drag Coefficients for Structures Studied in Wind Tunnel Model Tests." *Engineering News-Record*, October 11, pp. 456–58.

———. 1934d. "Aerodynamics and the Civil Engineer—IV. Wind Tunnel Studies Reveal Pressure Distribution on Buildings." *Engineering News-Record*, December 27, pp. 814–18.

———. 1935a. "Aerodynamics and the Civil Engineer—V. Vortices, Eddies and Turbulence as Experienced in Air Movements." *Engineering News-Record*, April 25, pp. 582–86.

———. 1935b. "Aerodynamics and the Civil Engineer—VI. Engineering Meteorology." *Engineering News-Record*, May 9, pp. 665–68.

———. 1935c. "Aerodynamics and the Civil Engineer—VII. "Wind Velocity in Relation to Height Above Ground." *Engineering News-Record*, May 23, pp. 742–45.

———. 1935d. "Aerodynamics and the Civil Engineer—VIII. Using Aerodynamic Research Results in Civil Engineering Practice." *Engineering News-Record*, October 31, pp. 601–7.

Palladio, Andrea. 1738. *The Four Books on Architecture*. New York: Dover Publications, 1965.

Paxton, Roland, ed. 1990. *100 Years of the Forth Bridge*. London: Thomas Telford.

Paxton, Roland, John Kerr, and Douglas McBeth. 1981. *Our Engineering Heritage: Three Notable Examples in the Edinburgh Area. Dean Bridge, Leith Docks, Forth Rail Bridge*. Edinburgh: Institution of Civil Engineers.

Paxton, Roland, and Ted Ruddock. [1980.] *A Heritage of Bridges Between Edinburgh, Kelso and Berwick*. Edinburgh: Institution of Civil Engineers.

Penfold, Alastair, ed. 1980. *Thomas Telford: Engineer*. London: Thomas Telford.

Pennsylvania Historical and Museum Commission and Pennsylvania Department of Transportation. 1986. *Historic Highway Bridges in Pennsylvania*. [Harrisburg:] Commonwealth of Pennsylvania.

Peters, Tom F. 1987. *Transitions in Engineering: Guillaume Henri Dufour and the Early 19th Century Cable Suspension Bridges*. Basel: Birkhäuser.

Petroski, Henry. 1985. *To Engineer Is Human: The Role of Failure in Successful Design*. New York: St. Martin's Press.

———. 1994. *Design Paradigms: Case Histories of Error and Judgment in Engineering*. New York: Cambridge University Press.

Plowden, David. 1974. *Bridges: The Spans of North America*. New York: Viking.

Pope, Gregory T. 1994. "Alaska-Siberian Bridge." *Popular Mechanics*, April, pp. 56–58.

Pope, Thomas. 1811. *A Treatise on Bridge Architecture*. New York: A. Niven.

Popular Science Monthly, The. 1886. "Sketch of James B. Eads." February, pp. 544–53.

Port of New York Authority, The. [c. 1929.] *The Kill van Kull Bridge Between Bayonne, New Jersey, and Port Richmond, Staten Island, New York*. New York: Port of New York Authority.

———. [c. 1930.] *The Hudson River Bridge Between Fort Lee, New Jersey, and Manhattan, New York* [brochure]. New York: Port of New York Authority.

———. [c. 1960.] *George Washington Bridge: Lower Level*. New York: Port of New York Authority.

Port of New York Authority, The, Bridge Department. 1929. *Specifications for Design of Bridges Carrying Highway and Electric Rail Passenger Traffic*. New York: Port of New York Authority.

Port of New York Authority, The, and Triborough Bridge and Tunnel Authority. 1955. *Arterial Facilities*. New York: Port of New York Authority and Triborough Bridge and Tunnel Authority.

Porter, Peter A. 1891. *Niagara Falls, River, Frontier Guide: Scenic, Botanic, Electric, Historic, Geologic, Hydraulic*. Buffalo, N.Y.: Matthews Northrup Works.

Prebble, John. 1975. *The High Girders*. London: Book Club Associates.

Pugsley, Sir Alfred. 1968. *The Theory of Suspension Bridges*. 2nd ed. London: Edward Arnold.

Purcell, C. H. 1934. "San Francisco–Oakland Bay Bridge." *Civil Engineering*, April, pp. 183–87.

Purcell, C. H., Chas. E. Andrew, and Glenn B. Woodruff. 1934. "San Francisco–Oakland Bay Bridge: A Review of Preliminaries." *Engineering News-Record*, March 22, pp. 371–77.

Quebec Bridge and Railway Company, The. 1907. *The Quebec Bridge: The Longest Single Span Bridge in the World Crossing the St. Lawrence River Seven Miles Above Quebec, Canada*. New Liverpool, Quebec.: E. R. Kinloch and N. R. McClure.

Ratigan, William. 1959. *Highways over Broad Waters: Life and Times of David B. Steinman, Bridgebuilder*. Grand Rapids, Mich.: Wm. B. Eerdmans.

Reavis, L. U. 1874. *A History of the Illinois and St. Louis Bridge*. St. Louis: Tribune Publishing Company.

Reier, Sharon. 1977. *The Bridges of New York*. New York: Quadrant Press.

Rensselaer Polytechnic Institute. 1925. *The Centennial Celebration of Rensselaer Polytechnic Institute, Troy, New York, October 3rd and 4th, 1924*. Troy, N.Y.: Rensselaer Polytechnic Institute Board of Trustees.

Rensselaer Polytechnic Institute Library. 1955. *The Writings of David Barnard Steinman Relating to the Design, Construction and History of Bridges, Including Contributions to the Development of the Engineering Profession, 1909–1954*. Troy, N.Y.: Rensselaer Polytechnic Institute.

Réthi, Lili. 1965. *The Great Bridge: The Verrazano-Narrows Bridge*. New York: Ariel/Farrar, Straus & Giroux.

Reynolds, Terry S. 1992. "The Education of Engineers in America Before the Morrill Act of 1862." *History of Education Quarterly* 32: 459–82.

Reynolds, Terry S., ed. 1991. *The Engineer in America: A Historical Anthology from "Technology and Culture."* Chicago: University of Chicago Press.

Rezneck, Samuel. 1968. *Education for a Technological Society: A Sesquicentennial History of Rensselaer Polytechnic Institute*. Troy, N.Y.: Rensselaer Polytechnic Institute.

Ritter, W. 1894. *Der Brückenbau in den Vereinigten Staaten Amerikas*. Bern: Haller'sche Buchdruckerei.

Robbins, Paul H. [1984.] *Building for Professional Growth: A History of the National Society of Professional Engineers, 1934–1984*. Washington, D.C.: National Society of Professional Engineers.

Robinson & Steinman, Engineers. [c. 1932.] *Bridges Lasting and Beautiful*. New York: Robinson & Steinman.

Roebling, John A. 1841. "Remarks on Suspension Bridges, and on the Comparative Merits of Cable and Chain Bridges." *American Railroad Journal, and Mechanics' Magazine*, n.s., vol. 6, pp. 193–96.

———. 1852. *Report to the Directors of the Niagara Falls International and Suspension Bridge Companies*. Buffalo, N.Y.: Steam Press of Jewett, Thomas & Co.

———. 1869. *Long and Short Span Railway Bridges*. New York: Van Nostrand.

Roebling, W. A. 1873. *Pneumatic Tower Foundations of the East River Suspension Bridge*. New York: Averell & Peckett.

———. 1877. *Report of the Chief Engineer of the New York & Brooklyn Bridge*. Brooklyn: Eagle Print.

Royal Commission. 1908. "Quebec Bridge Inquiry Report." Sessional Paper No. 154, 7–8 Edward VII. Ottawa.

Rubin, Lawrence A. 1958. *Mighty Mac: The Official Picture History of the Mackinac Bridge*. Detroit: Wayne State University Press.

Ruddock, Ted. 1979. *Arch Bridges and Their Builders, 1735–1835*. Cambridge: Cambridge University Press.

Russell, J. S. 1839. "On the Vibration of Suspension Bridges and Other Structures and the Means of Preventing Injury from this Cause." *Transactions, Royal Scottish Society of Arts* 1: 304–14.

Sandstrom, Gösta E. 1963. *Tunnels*. New York: Holt, Reinhart and Winston.

Schlager, Neil, ed. 1993. *When Technology Fails: Significant Technological Disasters, Accidents, and Failures of the Twentieth Century*. Detroit: Gale Research.

Schneider, Charles C. 1885. "The Cantilever Bridge at Niagara Falls," with discussion. *Transactions of the American Society of Civil Engineers* 14: 499–541, 543–606.

———. 1908. "Report on Design of Quebec Bridge." Sessional Paper No. 154, 7–8 Edward VII, pp. 153–206. Ottawa.

Schneider & Co., Cheusot Iron Works, and H. Hersent. 1889. "On the Proposed Channel Bridge," with discussion. *The Journal of the Iron and Steel Institute*, pt. 2, pp. 40–127, 128–38.

Schodek, Daniel L. 1987. *Landmarks in American Civil Engineering*. Cambridge, Mass.: MIT Press.

Schwarz, Jordan A. 1993. *The New Dealers: Power Politics in the Age of Roosevelt*. New York: Alfred A. Knopf.

Scott, Quinta, and Howard S. Miller. 1979. *The Eads Bridge*. Columbia: University of Missouri Press.

Secretary of Transportation. 1993. *The Status of the Nation's Highways, Bridges, and Transit: Conditions and Performance*. Washington, D.C.: U.S. Government Printing Office.

Senior, Michael. 1991. *The Crossing of the Conwy: And Its Role in the Story of North Wales*. Llanrwst, Gwynedd: Gwasg Carreg Gwalch.

Shallat, Todd. 1994. *Structures in the Stream: Water, Science, and the Rise of the U.S. Army Corps of Engineers*. Austin: University of Texas Press.

Shank, William H. 1973. *Vanderbilt's Folly: A History of the Pennsylvania Turnpike*. York, Pa.: American Canal and Transportation Center.

———. 1990. *Historic Bridges of Pennsylvania*. Rev. ed. York, Pa.: American Canal and Transportation Center.

Shanor, Rebecca Read. 1988. *The City That Never Was: Two Hundred Years of Fantastic and Fascinating Plans That Might Have Changed the Face of New York City*. New York: Viking.

Shapiro, Mary J. 1983. *A Picture History of the Brooklyn Bridge*. New York: Dover Publications.

Shipway, J. S. 1987. *The Tay Railway Bridge, Dundee, 1887–1987: A Review of Its Origin*. Edinburgh: Institution of Civil Engineers.

———. 1989. "Tay Rail Bridge Centenary—Some Notes on Its Construction, 1882–87." *Proceedings of the Institution of Civil Engineers* 86 (pt. 1): 1089–09.

———. 1990. "The Forth Railway Bridge Centenary, 1890–1990: Some Notes on Its Design." *Proceedings of the Institution of Civil Engineers* 88 (pt. 1): 1079–1107.

Sibly, P. G. 1977. "The Prediction of Structural Failure." Ph.D. Thesis, University of London.

Sibly, P. G., and A. C. Walker. 1977. "Structural Accidents and Their Causes." *Proceedings of the Institution of Civil Engineers* 62 (pt. 1): 191–208.

Silverberg, Robert. 1966. *Bridges*. Philadelphia: Macrae Smith Company.

Sinclair, Bruce. 1980. *A Centennial History of the American Society of Mechanical Engineers, 1880–1980*. Toronto: University of Toronto Press.

Sivewright, W. J. 1986. *Civil Engineering Heritage: Wales and Western England*. London: Thomas Telford.

Smiles, Samuel. 1904. *Lives of the Engineers*. Popular ed., in 5 vols. London: John Murray.

———. 1966. *Selections from Lives of the Engineers, with an Account of Their Principal Works*. Edited by Thomas Parke Hughes. Cambridge, Mass.: MIT Press.

Smith, Dwight A., James B. Norman, and Pieter T. Dykman. 1989. *Historic Highway Bridges of Oregon*. Portland: Oregon Historical Society Press.

Smith, H. Shirley. 1965. *The World's Great Bridges*. Rev. ed. New York: Harper & Row.

Smith, William Sooy. 1882. "The Hudson River Tunnel," with discussion. *Transactions of the American Society of Civil Engineers* 11: 314–23, 323–24.

Society for Industrial Archaeology. 1989. "Theme Issue: Bridges." *IA: The Journal of the Society for Industrial Archeology* 15 (no. 2).

Spanning Niagara: The International Bridges, 1848–1962. 1984. Niagara Falls, N.Y.: Buscaglia-Castellani Art Gallery of Niagara University.

Spielmann, Arthur, and Charles B. Brush. 1880. "The Hudson River Tunnel," with discussion. *Transactions of the American Society of Civil Engineers* 9: 259–72, 273–77.

Stackpole, Peter. 1984. *The Bridge Builders: Photographs and Documents of the Raising of the San Francisco Bay Bridge, 1934–1936*. Corte Madera, Calif.: Pomegranate Art Books.

Steinman, David B. 1911. *Suspension Bridges and Cantilevers: Their Economic Proportions and Limiting Spans*. New York: Van Nostrand.

———. 1918. "Stress Measurements on the Hell Gate Arch Bridge," with discussion. *Transactions of the American Society of Civil Engineers* 82: 1040–76, 1077–1137.

———. 1922. *A Practical Treatise on Suspension Bridges: Their Design, Construction and Erection*. New York: John Wiley & Sons.

———. 1924. "Design of the Florianópolis Suspension Bridge." *Engineering News-Record*, November 13, pp. 778–82.

———. 1925. "Outstanding Practice Problems of the Profession." *Engineering News-Record*, May 21, pp. 851–54.

———. 1927. "Designing the Carquinez Cantilever Bridge." *Engineering News-Record*, May 12, pp. 777–81.

———. [c. 1929.] *Fifty Years of Progress in Bridge Engineering*. New York: American Institute of Steel Construction.

———. 1930. "Rope Strand Cables Used in New Bridge at Portland, Oregon." *Engineering News-Record*, February 13, pp. 272–77.

———. 1932. *The Wichert Truss*. New York: Van Nostrand.

———. 1935a. "What the National Society of Professional Engineers Can Accomplish." *The American Engineer*, January, pp. 6, 14, 19.

———. 1935b. "Registration of Engineers." *Electrical Engineering*, August, pp. 876–81.

———. 1939a. "Do You Know . . . Pertinent Facts Concerning the Professions. 4. Accounting." *The American Engineer*, March, p. 15.

———. [1939b.] *The Place of the Engineer in Civilization*. Raleigh: School of Enginering, North Carolina State College of Agriculture and Engineering.

———. 1941a. "Stay Systems for Suspension Bridges." *Engineering News-Record*, February 27, p. 36.

———. 1941b. "Why Some Applicants Were Rejected." *Engineering News-Record*, October 23, pp. 95–96.

———. 1943. "Rigidity and Aerodynamic Stability of Suspension Bridges," with discussion. *Transactions of the American Society of Civil Engineers* 110: 439–75, 476–580.

———. 1945a. "Design of Bridges Against Wind. I. General Considerations—Aerostatic Stability." *Civil Engineering*, October, pp. 469–72.

———. 1945b. "Design of Bridges Against Wind. II. Aerodynamic Instability—Historical Background." *Civil Engineering*, November, pp. 501–4.

———. 1945c. "Design of Bridges Against Wind. III. Elementary Explanation of Aerodynamic Instability." *Civil Engineering*, December, pp. 558–60.

———. 1946a. "Design of Bridges Against Wind. IV. Aerodynamic Instability—Prevention and Cure." *Civil Engineering*, January, pp. 20–23.

———. 1946b. "Design of Bridges Against Wind. V. Criteria for Assuring Aerodynamic Stability." *Civil Engineering*, February, pp. 66–68.

[———. c. 1947]. *Bridges*. [New York: D. B. Steinman, Engineer.]

———. 1948. "The World's Most Notable Bridges." *Engineering News-Record*, December 9, pp. 92–94.

———. 1950. *The Builders of the Bridge: The Story of John Roebling and His Son*. 2nd ed. New York: Harcourt, Brace.

———. 1953. "Bridges and Man's Increased Mobility." *Transactions of the American Society of Civil Engineers*, CT [Centennial Transactions]: 767–81.

———. 1954a. "Bridges and Aerodynamics." *The Virginia Engineer*, Spring, pp. 4–7, 23–24.

———. 1954b. "The Woman Who Helped Build the Brooklyn Bridge." *The Transit*, Spring–Fall, pp. 24–29.

———. 1954c. "Suspension Bridges: The Aerodynamic Problem and Its Solution." *American Scientist*, July, pp. 397–438, 460.

———. 1954d. "World's Longest Suspension Span." *Roads and Engineering Construction*, July, pp. 84–87, 100–103.

———. 1954e. "Bridges." *Scientific American*, November, pp. 61–71.

———. 1955. *I Built a Bridge, and Other Poems*. New York: Davidson Press.

———, in collaboration with John T. Nevill. 1957. *Miracle Bridge at Mackinac*. Grand Rapids, Mich.: Wm. B. Eerdmans.

———. 1959. *Songs of a Bridgebuilder*. Grand Rapids, Mich.: Wm. B. Eerdmans.

Steinman, Boynton, Gronquist & London, Consulting Engineers. [1960.] *Bridges*. [New York.: Steinman, Boynton, Gronquist & London.]

Steinman, D. B., and C. H. Gronquist. 1932. "Building First Long-Span Bridge in Maine." *Engineering News-Record*, March 17, pp. 386–89.

Steinman, David B., and Sara Ruth Watson. 1941. *Bridges and Their Builders*. New York: G. P. Putnam's Sons. [Rev. and expanded ed. New York: Dover Publications, 1957.]

Straub, Hans. 1964. *A History of Civil Engineering: An Outline from Ancient to Modern Times*. Translated by Erwin Rockwell. Cambridge, Mass.: MIT Press.

Strauss, Joseph B. 1937. *The Golden Gate Bridge: Report of the Chief Engineer to the Board of Directors of the Golden Gate Bridge and Highway District, California*. San Francisco: Golden Gate Bridge and Highway District.

Strauss Bascule Bridge Company. [c. 1920.] *Bascule and Direct Lift Bridges*. [Chicago: Strauss Bascule Bridge Company.]

Stüssi, Fritz. 1974. *Othmar H. Ammann: Sein Beitrag zur Entwicklung des Brückenbaus*. Basel: Birkhäuser.

Talese, Gay. 1964. *The Bridge*. New York: Harper & Row.

Timoshenko, S. 1930. "The Stiffness of Suspension Bridges," with discussion. *Transactions of the American Society of Civil Engineers* 94: 377–91, 392–405.

Triborough Bridge and Tunnel Authority. 1964. *Spanning the Narrows*. New York: Triborough Bridge and Tunnel Authority.

[Tugby, Thomas.] 1890. *Tugby's Guide to Niagara Falls, Being a Complete Guide to All the Points of Interest Around and in the Immediate Neighbourhood*. Niagara Falls, N.Y.: Thomas Tugby.

Turner, Roland, and Steven L. Goulden, eds. 1981. *Great Engineers and Pioneers in Technology. Vol. I: From Antiquity through the Industrial Revolution*. New York: St. Martin's Press.

Tyrrell, Henry Grattan. 1911. *History of Bridge Engineering*. Chicago: published by the author.

———. 1912. *Artistic Bridge Design: A Systematic Treatise on the Design of Modern Bridges According to Aesthetic Principles*. Chicago: Myron C. Clark.

United States Steel. 1936. *San Francisco–Oakland Bay Bridge*. Pittsburgh, Pa.: American Bridge Company.

———. 1937. *Suspension Bridges*. Pittsburgh, Pa.: American Bridge Company.

Van der Zee, John. 1986. *The Gate: The True Story of the Design and Construction of the Golden Gate Bridge*. New York: Simon & Schuster.

Van der Zee, John, and Russ Cone. 1993. "The Case of the Missing Engineer." *Civil Engineering Transitions* (Purdue University newsletter), Summer. [Reprinted from San Francisco *Examiner Image*, May 31, 1992.]

Vincent, George S. 1958. "Golden Gate Bridge Vibration Studies." *Journal of the Structural Division: Proceedings of the American Society of Civil Engineers* 84: 1817-1–1817-40.

Vincenti, Walter G. 1990. *What Engineers Know and How They Know It: Analytical Studies from Aeronautical History*. Baltimore: Johns Hopkins University Press.

[Vollmar, Joseph E., Jr.] 1974. *James B. Eads and the Great St. Louis Bridge*. St. Louis: Engineers Club of St. Louis.

Waddell, J. A. L. 1895. "The Halsted Street Lift-Bridge," with discussion. *Transactions of the American Society of Civil Engineers* 33: 1–36, 37–60.

———. 1905. *The Principal Professional Papers*. Edited by J. L. Harrington. New York: Virgil H. Hewes.

———. 1916. *Bridge Engineering*. New York: John Wiley & Sons.

———. 1921. "Bridge Versus Tunnel for the Proposed Hudson River Crossing at New York City," with discussion. *Transactions of the American Society of Civil Engineers* 84: 570–74, 575–79.

———. 1927. "Quantities and Materials and Costs per Square Foot of Floor for Highway and Electric-Railway Long-Span Suspension Bridges," with discussion. *Transactions of the American Society of Civil Engineers* 91: 884–910, 911–45.

———. 1928. *Memoirs and Addresses of Two Decades*. Edited by F. W. Skinner. Easton, Pa.: Mack Printing Company.

———. 1931. "Engineering." *The North American Review*, June, pp. 560–62.

Waddell & Harrington, Consulting Engineers, ed. 1911. *Addresses to Engineering Students*. Kansas City, Mo.: Waddell & Harrington.

Waddell & Son. [c. 1917.] *Catalog*. [Kansas City, Mo.: Waddell & Son.]

Wade, Herbert T. 1921. "The World's Greatest Bridge." *The American Review of Reviews*. August, pp. 187–93.

Walker, Derek. 1987. *The Great Engineers: The Art of British Engineers, 1837–1987*. New York: St. Martin's Press.

Walther, René. 1988. *Cable Stayed Bridges*. London: Thomas Telford.

Waters, Tony. 1989. *Bridge by Bridge Through London: The Thames from Tower Bridge to Teddington*. Whitstable, Kent: Pryor Publications.

Watson, J. G. 1982. *A Short History*. London: Institution of Civil Engineers.

Watson, Philip P. 1987. *The Ambassador Bridge: A Monument to Progress*. Detroit: Wayne State University Press.

Watson, Sara Ruth, and Emily Watson. 1950. *Famous Engineers*. New York: Dodd, Mead.

Watson, Wilbur J., and Sara Ruth Watson. 1937. *Bridges in History and Legend*. Cleveland: J. H. Jansen.

Webster, Anthony C. 1992. "Utility, Technology and Expression." *Architectural Review*, November, pp. 68–74.

Wellington, A. M. 1889. *The Economic Theory of the Location of Railways: An Analysis of the Conditions Controlling the Layout of Railways to Effect the Most Judicious Expenditure of Capital*. 4th ed. New York: John Wiley & Sons.

Westhofen, W. 1890. "The Forth Bridge." *Engineering*, February 28, pp. 213–83.

White, Joseph, and M. W. von Bernewitz. 1928. *The Bridges of Pittsburgh*. Pittsburgh, Pa.: Cramer Printing & Publishing.

Widmer, Urs C. 1979. "Othmar Hermann Ammann, 1879–1965: His Way to Great Bridges." *Swiss-American Historical Society Newsletter* 15: 4–25.

Wischart, M. K. 1928. "The Greatest Bridge in the World and the Man Who Is Building It." *The American Magazine*, June, pp. 34–35, 183–86, 189.

Wisely, William H. 1974. *The American Civil Engineer, 1852–1974: The History, Traditions and Development of the American Society of Civil Engineers*. New York: American Society of Civil Engineers.

Wolfert, Ira. 1956. "A Boy's Dream: The World's Longest Bridge." *Popular Science*, October, pp. 104–7, 252, 254, 256, 258.

Wood, Sharon. 1989. *The Portland Bridge Book*. Portland: Oregon Historical Society Press.

Woodruff, Glenn B. 1935. "From the Viewpoint of the Bridge Designer." *Civil Engineering*, April, pp. 214–15.

Woodward, C. M. 1881. *A History of the St. Louis Bridge; Containing a Full Account of Every Step of Its Construction and Erection, and Including the Theory of the Ribbed Arch and the Tests of Materials*. St. Louis: G. I. Jones.

Woolf, S. J. 1934. "A Master Bridge Builder Looks Ahead." *The New York Times Magazine*, April 15, pp. 7, 19.

Yager, Rosemary. 1968. *James Buchanan Eads: Master of the Great River*. Princeton, N.J.: Van Nostrand.

ILLUSTRATIONS

(with sources and acknowledgments)

INDEX

Italicized page numbers refer to illustrations and their captions.

ALSO BY HENRY PETROSKI

Available from Vintage Books

THE EVOLUTION OF USEFUL THINGS

How did the table fork acquire a fourth tine? What advantage does the Phillips-head screw have over its single-grooved predecessor? Why does the paper clip look the way it does? What makes Scotch tape Scotch? In this delightful book, Henry Petroski takes a microscopic look at artifacts that most of us count on but rarely contemplate, including such icons of the everyday as pins, Post-its, and fast-food "clamshell" containers.

History/Science/0-679-74039-2

TO ENGINEER IS HUMAN
The Role of Failure in Successful Design

How did a simple design error cause one of the great disasters of the 1980s—the collapse of the walkways at the Kansas City Hyatt Hotel? What made the innovative Tacoma Narrows Bridge twist apart in a mild wind in 1940? How did an oversized waterlily inspire the magnificent Crystal Palace, the crowning achievement of Victorian architecture and engineering? In *To Engineer Is Human*, Henry Petroski examines these cases and produces a work that looks at our deepest notions of progress and perfection.

Science/Technology/0-679-73416-3

Also available from Alfred A. Knopf

THE PENCIL
A History of Design and Circumstance

"Very engaging and wonderfully informative. . . . *The Pencil* unfolds a history of invention, craftsmanship, engineering, manufacture and business that is also at times a history of cultural life on both sides of the Atlantic. . . . No reader of this book will ever be able to pick up a pencil again without marveling."
—*Newsday*

Science/Engineering/0-679-73415-5

Available at your local bookstore, or call toll-free to order:
1-800-793-2665 (credit cards only).